Green Organic Chemistry in Lecture and Laboratory

Sustainability: Contributions through Science and Technology

Series Editor: Michael C. Cann, Ph.D.
Professor of Chemistry and Co-Director of Environmental Science
University of Scranton, Pennsylvania

Preface to the Series

Sustainability is rapidly moving from the wings to center stage. Overconsumption of non-renewable and renewable resources, as well as the concomitant production of waste has brought the world to a crossroads. Green chemistry, along with other green sciences technologies, must play a leading role in bringing about a sustainable society. The **Sustainability: Contributions through Science and Technology** series focuses on the role science can play in developing technologies that lessen our environmental impact. This highly interdisciplinary series discusses significant and timely topics ranging from energy research to the implementation of sustainable technologies. Our intention is for scientists from a variety of disciplines to provide contributions that recognize how the development of green technologies affects the triple bottom line (society, economic, and environment). The series will be of interest to academics, researchers, professionals, business leaders, policy makers, and students, as well as individuals who want to know the basics of the science and technology of sustainability.

Michael C. Cann

Published Titles

Microwave Heating as a Tool for Sustainable Chemistry
Edited by Nicholas E. Leadbeater, 2010

Green Chemistry for Environmental Sustainability
Edited by Sanjay Kumar Sharma, Ackmez Mudhoo, 2010

Forthcoming Title

A Novel Green Treatment for Textiles:
Plasma Treatment as a Sustainable Technology
C. W. Kan, 2012

Sustainability: Contributions through Science and Technology

Series Editor: Michael Cann

Green Organic Chemistry in Lecture and Laboratory

Edited by Andrew P. Dicks

CRC Press
Taylor & Francis Group
Boca Raton London New York

CRC Press is an imprint of the
Taylor & Francis Group, an **informa** business

CRC Press
Taylor & Francis Group
6000 Broken Sound Parkway NW, Suite 300
Boca Raton, FL 33487-2742

First issued in paperback 2016

Version Date: 20110708

ISBN 13: 978-1-138-19928-6 (pbk)
ISBN 13: 978-1-4398-4076-4 (hbk)

Visit the Taylor & Francis Web site at
http://www.taylorandfrancis.com

and the CRC Press Web site at
http://www.crcpress.com

Contents

Foreword

If you do not change direction, you may end up where you are heading.

—Lao Tzu, the founder of Taoism

Although this quote is more than 2,000 years old, it is more appropriate now than ever before. The direction that humankind is now moving is not sustainable. We are rapidly being engulfed by a growing environmental, social, and economic storm. The combination of our expanding world population, rising affluence, and technological advances has brought the world to the brink of environmental bankruptcy. Our ecological footprint now significantly exceeds the carrying capacity of the earth. Without serious mid-course corrections of our unsustainable lifestyles, humankind will face some very serious threats to world order. Challenges include the following: How do we feed, clothe, shelter, and provide potable water to the current 7 billion people on the planet, and the 9 billion that will inhabit the earth by mid-century? How do we curb the threat of climate chaos, while meeting the demands of an increasingly affluent population whose energy demands are projected to increase by more than 30% through 2050? Just like information technology has swept the world by storm, sustainable technology will *as of necessity* be the next thunderbolt that encompasses all of humanity. Novel scientific applications, along with conservation, offer a pathway to sustainability.

Education is, of course, the key to launching and maintaining a wholesale shift in the way we develop and enact technologies. New approaches must efficiently and effectively utilize our natural resources in a cyclical manner, reduce our energy demands, and eliminate the use and production of toxic materials, all while utilizing renewable resources and energy. "Green chemistry" or "sustainable chemistry" is at the heart of a revolution in the discipline that has the potential to do all of these. This book helps to bring the world of green chemistry to not only the scientists and engineers of the future, but also to our prospective political leaders, economists, business leaders, teachers, and world citizens. The development and implementation of sustainable technology offers a mighty challenge to humankind, but it also provides a wonderful opportunity to those with the proper skills and knowledge. The term *green chemistry* was coined in 1991, and significant educational endeavors have taken place since then, particularly in the venue of organic chemistry. It is our understanding that these enterprises have never been reviewed, compiled, and presented in a single volume as in this work. The editor and the chapter authors sincerely hope this book will excite and provoke the minds of those individuals who explore its pages, will sow the seeds for tomorrow's sustainable applications, and will stimulate further pedagogical efforts in green chemistry.

Michael C. Cann
University of Scranton

Preface

> We know of no published green experiments designed for use in the organic teaching laboratory.
>
> —**Scott Reed and Jim Hutchison,**
> *Journal of Chemical Education,*
> Volume 77, December 2000, 1627–1629

How times have changed! Since these words were written ten years ago as part of an article describing the environmentally benign preparation of adipic acid, the volume of pedagogical green chemistry literature has grown to impressive proportions. Much of the credit for this goes to Ken Doxsee and Jim Hutchison, who published their excellent, motivating textbook (*Green Organic Chemistry: Strategies, Tools, and Laboratory Experiments*) in 2004. The majority of the hands-on activities and lecture case studies have been designed for undergraduates taking organic courses at college or university. There are, of course, many students worldwide who are enrolled in such offerings. As we rapidly approach the 2011 International Year of Chemistry, the opportunity to teach future generations about green and sustainable principles has never been more important.

As part of the CRC Press book series "Sustainability: Contributions through Science and Technology," this publication is unlike others in the realm of green chemical education. It is primarily written for organic chemistry educators who are instructing at either introductory or advanced levels. Faculty teaching first-year general chemistry will also find it useful, particularly if their course has even a small organic component to it. As the title implies, the book is comprehensive in its coverage of teaching green organic chemistry from both practical and theoretical standpoints. Previous titles have tended to focus on one or the other of these perspectives. An instructor may wish to develop an upper-level stand-alone course in the subject, or to simply "green up" aspects of an existing syllabus. Both approaches will be made much easier upon consultation of the experiments and case studies outlined within these pages. Adding green components to a current program is often the route of choice, and incorporating even one element that showcases sustainability into a chemistry curriculum is a positive step forward.

Chapter 1, "Introduction to Teaching Green Organic Chemistry," appropriately focuses on the twelve principles, and how they can be used to focus classroom and laboratory discussions. Time is also taken to outline the plethora of resources available in the field. The second chapter, "Designing a Green Organic Chemistry Lecture Course," provides a fascinating firsthand account of the challenges and rewards an instructor may experience. The remaining contributions are based upon areas where much didactic research has taken place during the last decade. These are solventless and aqueous reactivity, greener reagents, greener nonaqueous solvents, waste management/recycling, and energy efficiency (microwave heating). Each of these chapters contains a discussion of undergraduate laboratory experiments and, in most cases, exemplary lecture case studies taken from real-world industrial and research laboratories. Related material presented in different sections is closely linked to provide as much cohesiveness as possible.

The reader will immediately notice that practical details are not included for any highlighted experiments. Primary references are included for all peer-reviewed articles, so it is straightforward to consult the literature and adapt laboratory work according to local glassware and equipment availability. A comprehensive appendix (the "Greener Organic Chemistry Reaction Index") catalogs almost 180 reactions according to mechanistic type, with required techniques and greener features noted. This index includes many experiments developed since 2000, along with some older examples (as an illustration, reactions were being performed in water long before the green movement began!). These processes typically indicate that green chemistry is not "everything or nothing," and that it is crucial our students are taught to critically appraise any and all new reactions they encounter.

I sincerely acknowledge the efforts of all the chapter authors, who have unfailingly been a pleasure to collaborate with during the preparation of this work. I am also grateful to Mike Cann for his inspirational foreword and Hilary Rowe at Taylor and Francis for planting the seed for this book in my mind. Finally, I thank my family for their continuous love and support.

Andrew P. Dicks
Toronto, Canada

About the Editor

Andrew P. Dicks (Andy) joined the University of Toronto Chemistry Department in 1997. After undergraduate and graduate study in the United Kingdom, he became an organic chemistry sessional lecturer in 1999, and was hired as part of the university teaching stream faculty two years later. He has research interests in undergraduate laboratory instruction that involve designing novel and stimulating experiments, particularly those that showcase green chemistry principles. This work has led to over twenty peer-reviewed publications in the chemical education literature. Following promotion in 2006, he became associate chair for undergraduate studies for two years and developed an ongoing desire to improve the student experience in his department. He has won several pedagogical awards, including the University of Toronto President's Teaching Award, the Canadian Institute of Chemistry National Award for Chemical Education, and most recently, a 2011 American Chemical Society—Committee on Environmental Improvement Award for Incorporating Sustainability into Chemistry Education.

Contributors

Sudhir B. Abhyankar
Department of Chemistry
Memorial University of Newfoundland
Corner Brook, Newfoundland, Canada

John Andraos
Department of Chemistry
York University
Toronto, Ontario, Canada

Marsha R. Baar
Chemistry Department
Muhlenberg College
Allentown, Pennsylvania

Loyd D. Bastin
Department of Chemistry
Widener University
Chester, Pennsylvania

Andrew P. Dicks
Department of Chemistry
University of Toronto
Toronto, Ontario, Canada

Amanda R. Edward
Heathfield School
Pinner, Middlesex, United Kingdom

Leo Mui
Department of Chemistry
University of Toronto
Toronto, Ontario, Canada

Effiette L. O. Sauer
Department of Physical and
Environmental Sciences
University of Toronto Scarborough
Scarborough, Ontario, Canada

1 Introduction to Teaching Green Organic Chemistry

Dr. Sudhir B. Abhyankar

CONTENTS

1.1 INTRODUCTION

A sustainable future cannot be attained without sustainable chemistry, and progress in this area is critically dependent upon advances in green chemistry. Green chemistry is the utilization of a set of fundamental principles that relate to all chemical subdisciplines. These principles seek to reduce or eliminate the role of hazardous substances in the design, manufacture, and application of chemical products.[1]

Green chemistry has been described in a number of ways. A few of these include the following:

- Pollution prevention at the molecular level[2]
- Chemistry that is "benign by design"[3]
- Chemistry for a sustainable future[4]
- Stopping pollution before it starts[5]
- Preventive medicine for the environment, and the right prescription for chemical education[2]

The main goal of green chemistry is to minimize or eradicate the use of hazardous materials in chemical processes, thereby reducing their impact on human health and the environment. In this regard, the term *hazardous* applies not only to toxic substances, but also to those that might be described as flammable, explosive, or environmentally persistent. This provides clarity to the definition of the word *hazardous*, which is not restricted to toxic materials but is much broader in its meaning.

It is now generally accepted that mainstream education in green chemistry is essential to prepare the next generation of students to take on the environmental challenges that will arise in the future. In a recent editorial in the *Journal of Chemical Education*, Mary Kirchhoff stated that "green chemistry is not a field solely under the purview of green chemists: it is an approach that is applicable to all areas of chemistry, and it is the responsibility of all practicing chemists," and "we must do a better job of educating our students with respect to green chemistry, sustainability and environmental issues."[6] Similarly, the value of green chemistry education was expressed by Walter Leitner in a *Green Chemistry* journal editorial where he wrote: "As the principles of green chemistry and the concepts of sustainability in chemical manufacturing are becoming part of the explicit corporate policy and aims in the chemical industry, there is a rapidly growing need for the education of chemists in the field."[7]

It is critical that both undergraduate and graduate students be much more than simply informed about and familiar with principles and practices of green chemistry. They should also be able to apply these principles in the design and manufacture of chemical compounds using innovative methodologies. In doing so, students will appreciate that green chemistry is not a stand-alone discipline operating in isolation from other chemical fields. Rather, the approaches of green chemistry can (and must) be integrated into every aspect of the chemical and engineering worlds and become a "way of life" among practicing scientists. This book is intended to fulfill this mandate by integrating advances in green chemistry research into the teaching of green organic chemistry in both lecture and laboratory environments. Some of the chapters that follow outline design and implementation of an upper-level green organic chemistry lecture course, discussion of solvent elimination in organic laboratories, and chemical waste management and recycling approaches. All the expounded principles can be interwoven into a four-year undergraduate program, or alternatively, select examples from each chapter chosen to promote a curricular "step in the green direction."[8]

1.2 EARLY DEVELOPMENTS IN GREEN CHEMISTRY

Even though he did not conceive of or use the term *green chemistry*, Italian chemist Giacomo Luigi Ciamician (1857–1912) has been described as a founder of the field.[9] In his address to the Eighth International Congress of Applied Chemistry in 1912, Ciamician stated: "And if in the distant future the supply of coal becomes completely exhausted, civilization will not be checked by that, for life and civilization will continue as long as the sun shines"[10] (at that time coal was the most widely used fossil fuel). In comparison, modern-day green chemistry has its origins in the environmentally friendly approaches of the early 1970s. It was during this decade that the Environmental Protection Agency (EPA) became established in the United States. A number of significant regulatory laws, such as the Clean Air Act (CAA), the Clean

Water Act (CWA), and the Toxic Substances Control Act (TSCA), were passed to protect the environment from release of hazardous substances into the air, soil, and water. It is important to note that the emphasis during these times was much more on pollution regulation rather than pollution prevention. In 1990 the U.S. government passed the Pollution Prevention Act (PPA), and during the early 1990s the term *green chemistry* was coined at the U.S. EPA. In one of the earliest publications about introducing green principles in teaching and research, Collins described a lecture course entitled "Introduction to Green Chemistry" in 1995.[11] The course was delivered to upper-level undergraduates and graduate students in 1992 and 1993. Lecture topics included the role of catalysis in green chemistry, traditional energy sources, pollution and green energy, atmospheric pollution, biocatalysis, and bioremediation. Students designed and delivered presentations in areas such as vulcanized rubber recycling, toxic chemical degradation using catalytic antibodies, and biological degradation of effluent. This offering was part of a broader initiative called "Environment across the Curriculum," and environmental modules were prepared for inclusion in courses at Carnegie Mellon University in the United States.[11]

In 1997, the Green Chemistry Institute was founded, and it became part of the American Chemical Society in 2001.[12] The Royal Society of Chemistry launched the internationally renowned journal *Green Chemistry* in 1999, which is acknowledged as the leading publication in the area. The last ten years have seen considerable progress in green chemistry research and education through national and international conferences, meetings, and symposia, and information dissemination that includes public awareness. For example, the Thirty-Fourth International Chemistry Olympiad for high school students (held in Groningen, the Netherlands) promoted green chemistry as a central theme in 2002. The International Union of Pure and Applied Chemistry (IUPAC) currently arranges a biennial International Conference on Green and Sustainable Chemistry (ICGC). This was organized for the first time in 2006 in Dresden, Germany. Russia hosted the event in 2008, with the third conference taking place in Ottawa, Canada, during August 2010. In addition, a number of focused initiatives have been taken up by many countries around the world. Some of these include establishment of the Presidential Green Chemistry Awards in the United States, the European Green and Sustainable Chemistry Award, and the formation of the Canadian Green Chemistry Network (www.greenchemistry.ca). Many international organizations, including the Organization for Economic Cooperation and Development (OECD), as well as IUPAC, have adopted green and sustainable chemistry as a part of their ongoing missions.[13,14] The United Nations Educational, Scientific, and Cultural Organization (UNESCO) and IUPAC have successfully collaborated in designating 2011 as the International Year of Chemistry (IYC). IYC 2011 events will emphasize that chemistry is a creative science essential for sustainability and improvements to the lifestyles of humans.[15] The chemical industry has also taken an active role in using principles and practice of green chemistry in the manufacturing sector. This is evidenced, for example, by the recent publication of a book and an article highlighting green chemistry in the pharmaceutical industry.[16,17] There are many industrial case studies that lend themselves to classroom discussion of green principles, a selection of which are described in the following chapters of this book.

1.3 THE TWELVE PRINCIPLES OF GREEN CHEMISTRY

Green chemistry is based upon a fundamental set of twelve principles to achieve its ambition of reduction or elimination of hazardous chemicals. These principles were first formulated almost twenty years ago.[1] An introduction to each of the principles is followed by a brief discussion in Section 1.4 of how each one can be expanded upon in the teaching of green organic chemistry at the undergraduate level, with some appropriate examples.*

1. Prevention. The first and most important principle is the prevention of waste. It is better to prevent waste rather than treat it or clean it up after it is formed.
2. Atom economy. Synthetic methods should be designed to maximize the incorporation of all materials used in the process into the final product.
3. Less hazardous chemical synthesis. Wherever practicable, synthetic methods should be designed to use and generate substances that possess little or no toxicity to people or the environment.
4. Designing safer chemicals. Chemical products should be designed to effect their desired function while minimizing their toxicity.
5. Safer solvents and auxiliaries. The use of auxiliary substances (e.g., solvents or separation agents) should be made unnecessary whenever possible and innocuous when used.
6. Design for energy efficiency. Energy requirements of chemical processes should be recognized for their environmental and economic impacts and should be minimized. If possible, synthetic methods should be conducted at ambient temperature and pressure.
7. Use of renewable feedstocks. A raw material or feedstock should be renewable rather than depleting whenever technically and economically practicable.
8. Reduce derivatives. Unnecessary derivatization (use of blocking groups, protection/deprotection, and temporary modification of physical/chemical processes) should be minimized or avoided if possible, because such steps require additional reagents and can generate waste.
9. Catalysis. Catalytic reagents (as selective as possible) are superior to stoichiometric reagents.
10. Design for degradation. Chemical products should be designed so that at the end of their function they break down into innocuous degradation products and do not persist in the environment.
11. Real-time analysis for pollution prevention. Analytical methodologies need to be further developed to allow for real-time, in-process monitoring and control prior to the formation of hazardous substances.
12. Inherently safer chemistry for accident prevention. Substances and the form of a substance used in a chemical process should be chosen to minimize the potential for chemical accidents, including releases, explosions, and fires.

In 2001, an additional twelve principles were proposed by Winterton to "aid laboratory and research chemists, interested in applying green chemistry, to plan

* The principles were reprinted with permission from Oxford University Press. (Anastas, P. T., Warner J. C. *Green Chemistry: Theory and Practice*. Oxford University Press, New York, 1998.)

and carry out their work to include the collection of data that are of particular use to those wishing to assess the potential for waste minimization."[18] These principles are especially aimed at process chemists, chemical engineers, and chemical technologists and are focused on the following areas: identification and quantification of reaction by-products; reporting of reaction conversions, selectivities, and productivities; establishment of full mass balances for reactions; quantification of catalyst/solvent losses; research into thermochemistry concerns; prediction of heat/mass transfer issues; deliberation with engineers; consideration of overall process on type of chemistry; planning and practice of sustainability features; quantification and minimization of utility usage; recognition where safety and waste minimization are contradictory; and recording and diminishment of laboratory waste. The Winterton twelve principles are less well known than those devised by Anastas and Warner, yet there is much to learn regarding green chemistry from "cross-pollination" between chemists and chemical engineers. This observation is underscored in Section 2.5. A number of reports and review articles on the status of green chemistry have been written in the last few years,[19–22] with a recent and critical review of green chemistry principles and practice published by the Royal Society of Chemistry.[23]

1.4 THE TWELVE PRINCIPLES IN TEACHING GREEN ORGANIC CHEMISTRY

1. Prevention. Waste prevention is the first principle of green chemistry, and pollution prevention often (but not always) means avoiding waste. Waste can take many forms and can impact the environment depending upon its nature, toxicity, quantity, or the manner in which it is released. A simple measure of the mass of waste produced per kilogram of a desired reaction product, called the environmental impact factor (E-factor), was introduced by Sheldon in 1992.[24] A well-established case study that illustrates how E-factors can vary for the same overall transformation is the synthesis of oxirane (ethylene oxide). Oxirane is industrially important in the synthesis of many chemicals, including ethylene glycol (antifreeze), other glycols, polyglycol ethers, and ethanolamines. The E-factor for an early two-step synthesis of oxirane via a chlorohydrin intermediate was 5 (Scheme 1.1).[25] This means that for each and every kilogram of the oxirane product formed, 5 kg of waste was produced (including water, hydrochloric acid,

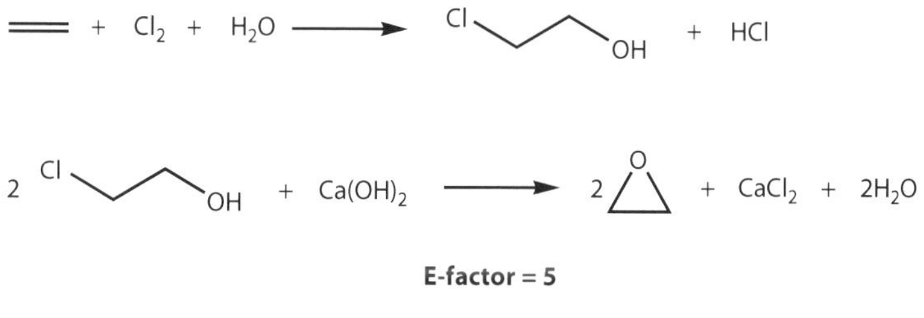

SCHEME 1.1 Two-step oxirane synthesis.

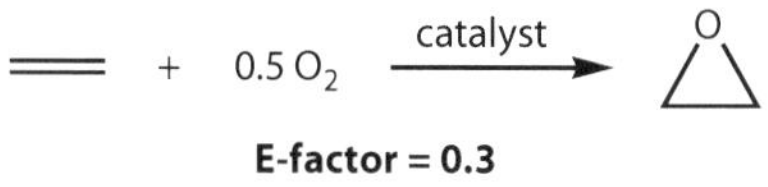

SCHEME 1.2 Single-step oxirane synthesis.

and calcium (II) chloride), which clearly must be disposed of. When the synthesis of oxirane was modified to use molecular oxygen and a catalytic surface (Scheme 1.2), thus removing the requirement for chlorine gas and calcium hydroxide, the value of the E-factor was reduced to 0.3. This represents a significant reduction in the amount of waste produced and hence released into the environment.

The E-factor is a straightforward and useful metric for students to understand and has been widely applied in the chemical industry.[26] It gives students an appreciation of the amount of waste that is produced in a chemical process and encourages them to seek alternative methods to minimize it. However, the E-factor metric does not discriminate between different *types* of waste. Indeed, it treats all waste equally, be it 1 kg of water or 1 kg of mercury produced by a chemical transformation. When waste cannot be avoided, innovative ways need to be considered to utilize the waste from one reaction to be used as raw materials for another process. This concept is discussed in more detail from an undergraduate laboratory perspective in Chapter 7.

2. Atom economy. One of the most important and fundamental principles of green chemistry is that of atom economy (AE). This term, also known as atom efficiency, was introduced by Trost in 1991.[27] It is intrinsically a measure of how many atoms from the starting materials are incorporated into the desired product(s) during a chemical transformation, by consideration of reactant and product molecular weights (Figure 1.1). The ideal reaction would integrate all of the reactant atoms into the product of interest, and the percentage atom economy would have a value of 100%. If, however, only

$$\%\ \text{Yield} = \frac{\text{experimental quantity of desired product}}{\text{theoretical maximum quantity of desired product}} \times 100$$

$$\underset{\text{(intrinsic)}}{\%\ \text{Atom Economy}} = \frac{\text{molar mass of desired product}}{\text{molar mass of all reactants}} \times 100$$

$$\underset{\text{(experimental)}}{\%\ \text{Atom Economy}} = \frac{\text{theoretical maximum quantity of desired product}}{\text{actual quantity of all reactants used}} \times 100$$

Overall Reaction Efficiency = % Yield × % Atom Economy (experimental)

FIGURE 1.1 Some measures of reaction efficiency.

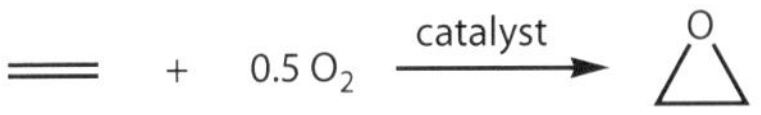

Molar mass (g/mol): 28.05   0.5 × 32.00   44.05

$$\textbf{\% Atom Economy} = \frac{44.05}{44.05} \times 100 = 100\%$$

SCHEME 1.3 Intrinsic atom economy calculation for single-step oxirane synthesis.

half of the reactant atoms are included in the desired product, the percentage atom economy will be 50%. This essentially means that half of the reactants end up in the formation of by-products, and if those by-products are not utilized in some way, they must be considered as waste. The intrinsic atom economy in the oxirane synthesis via direct oxidation using molecular oxygen is 100% (Scheme 1.3). To determine the intrinsic atom economy for the chlorohydrin intermediate pathway, one must first write a fully balanced equation for the overall transformation involving both reaction steps and all observed products in Scheme 1.1 (Scheme 1.4).

It is now possible to calculate an atom economy of nearly 35% for the two-step oxirane synthesis. When teaching incoming organic chemistry students, it is common to neglect any undesired by-products formed in a transformation (particularly inorganic ones). This approach must be addressed for a proper green analysis of any reaction to be meaningful. A real benefit of the atom economy concept lies in the fact that it can be calculated from a fully balanced reaction equation in the planning phase of any reaction. A percentage yield, on the other hand, can only be calculated after a reaction

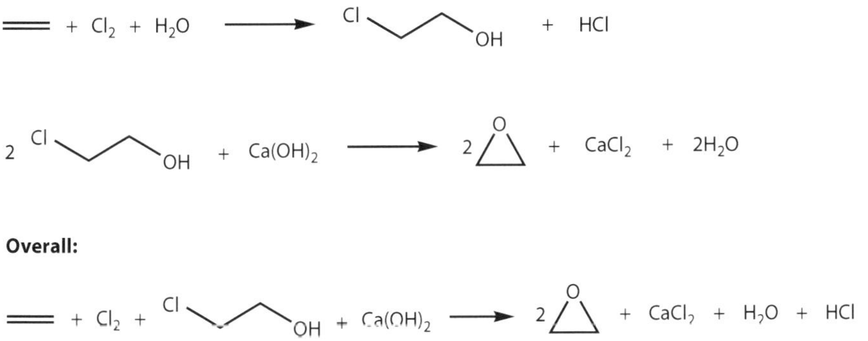

28.05   70.91   80.51   74.09   2 × 44.05   110.98   18.02   36.46

$$\textbf{\% Atom Economy} = \frac{88.1}{253.56} \times 100 = 34.7\%$$

SCHEME 1.4 Intrinsic atom economy calculation for two-step oxirane synthesis.

has been performed in the laboratory. One of the most important aspects of green chemistry is the concept of reaction design, and atom economy plays a crucial role in the preparatory stages of chemical syntheses.

All undergraduate chemistry students are very familiar with percentage yield calculations, and organic chemistry students unfailingly calculate the percentage yield for each reaction they personally undertake. Calculations involving percentage atom economy are simple, yet add an essential extra dimension to the study of chemical reactions. They require students to write a complete balanced equation for every chemical reaction performed and to identify the desired product and all by-products generated, if any. Figure 1.1 illustrates some important equations to be used in this regard. Students can now calculate the atom economy, along with the percentage yield, for every reaction they perform in the organic laboratory and quickly learn to distinguish between reactions that have favorable atom economies and those that do not. They come to realize that certain types of reactions (e.g., general additions) proceed with high atom economies. Similarly, reactions involving rearrangements also take place with high atom economies, as a simple reorganization of reactant atoms takes place to form a new product. In comparison, substitution reactions normally proceed with lower atom economies. Scheme 1.5 illustrates the intrinsic atom economy calculations for two of the common types of reactions encountered in introductory

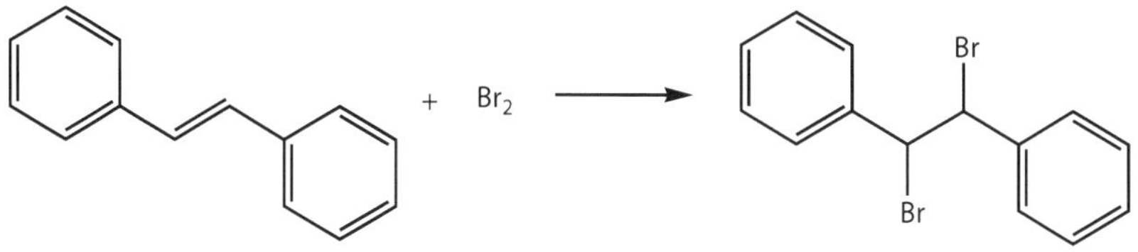

180.25 **159.8** **340.05**

$$\text{\% Atom Economy} = \frac{340.05}{180.25 + 159.8} \times 100 = 100\%$$

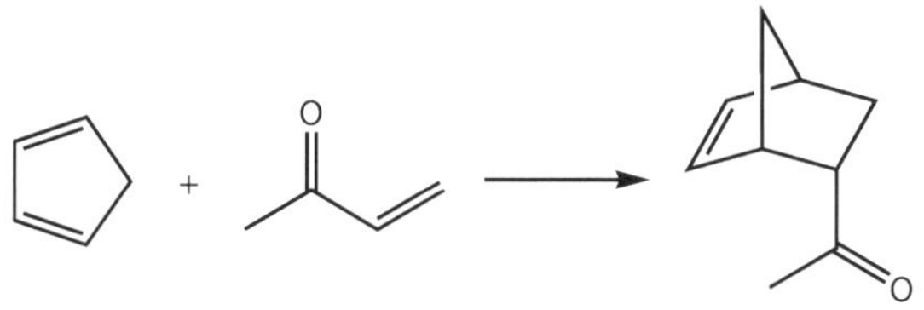

66.11 **70.1** **136.21**

$$\text{\% Atom Economy} = \frac{136.21}{66.11 + 70.1} \times 100 = 100\%$$

SCHEME 1.5 Intrinsic atom economy calculations for an electrophilic addition reaction and a Diels-Alder reaction.

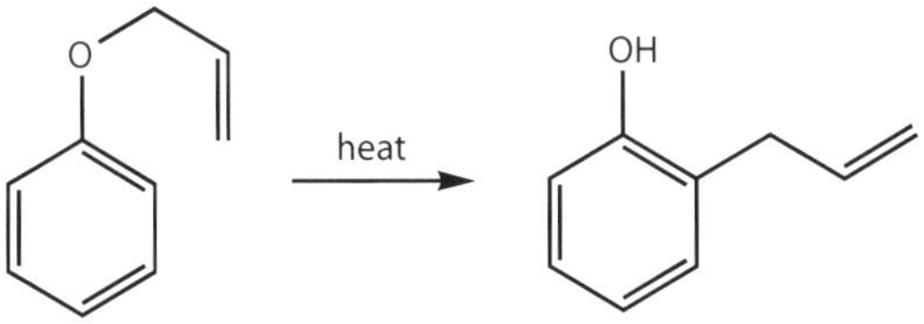

134.19 **134.19**

$$\% \textbf{ Atom Economy} = \frac{134.19}{134.19} \times 100 = 100\%$$

SCHEME 1.6 Intrinsic atom economy calculation for a Claisen allyl ether rearrangement.

organic chemistry. These are an electrophilic addition to an alkene and a Diels-Alder reaction. An example of a rearrangement reaction with excellent atom economy is shown in Scheme 1.6.

In comparison, substitution and elimination reactions often proceed with relatively lower atom economies. Both types of reactions are ubiquitous in introductory organic chemistry lectures and also performed in the laboratory. Two examples of intrinsic atom economy calculations involving a nucleophilic substitution reaction and an elimination reaction are shown in Scheme 1.7. Of particular note is the E2 reaction between chlorocyclohexane and potassium hydroxide to form cyclohexene as the elimination product. In this example, water and potassium chloride are formed as byproducts, meaning that the atom economy is only 47%.

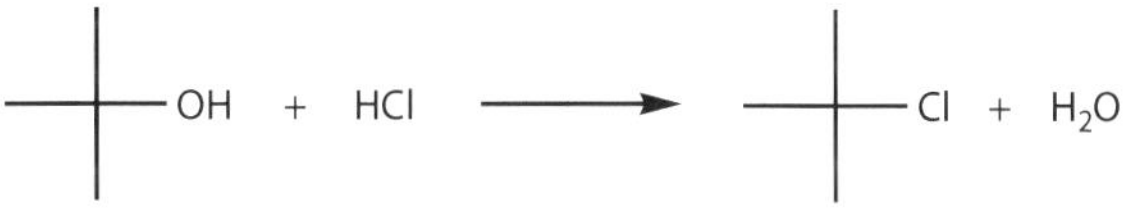

74.12 **36.46** **92.57** **18.02**

$$\% \textbf{ Atom Economy} = \frac{92.57}{74.12 + 36.46} \times 100 = 83.7\%$$

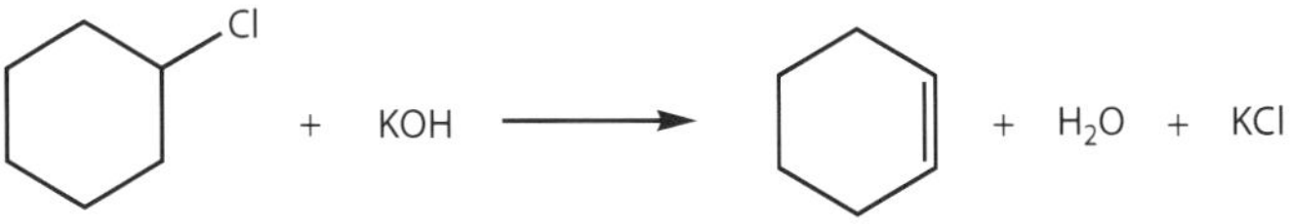

118.6 **56.11** **82.14** **18.02** **74.55**

$$\% \textbf{ Atom Economy} = \frac{82.14}{118.6 + 56.11} \times 100 = 47.0\%$$

SCHEME 1.7 Intrinsic atom economy calculations for a nucleophilic substitution reaction and an elimination reaction.

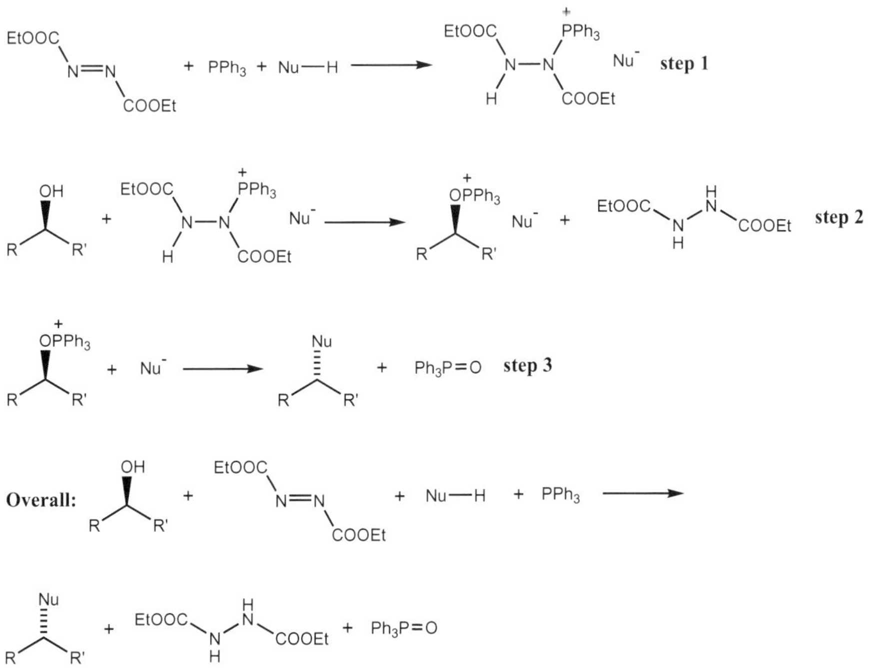

SCHEME 1.8 A generalized Mitsunobu reaction.

The Mitsunobu reaction is another widely used transformation that is typically covered in an upper-level organic chemistry lecture course.[28] A number of comprehensive reviews of this reaction have been published since it was first reported in 1967.[29–32] It is formally a condensation reaction of an alcohol with a compound having an active hydrogen atom (NuH) that is mediated by triphenylphosphine and a dialkyl azodicarboxylate (Scheme 1.8). The reaction has some important features that make it of great interest to synthetic chemists. Among these is the observation that chiral secondary alcohols are substituted with inversion of configuration and high stereospecificity. In addition, a variety of nucleophiles derived from nitrogen, oxygen, sulfur, and carbon can be employed, with the reaction being generally compatible with a broad range of functionalities. Finally, the necessary operations can be undertaken with ease in the laboratory, with only simple addition of reagents to a reaction vessel near room temperature.

The major disadvantage of the Mitsunobu reaction from a green chemistry perspective is the very poor intrinsic atom economy. Indeed, the conversion of (*S*)-2-butanol into (*R*)-(1-methylpropyl) benzoate proceeds with only 28% atom economy (Scheme 1.9). The reason why the reaction exhibits such a low atom economy is the use of stoichiometric quantities of diethyl azodicarboxylate and triphenylphosphine (combined molecular weight of 436), which overall function to eliminate a molecule of water (molecular weight of 18) from this condensation reaction. During the process, diethyl

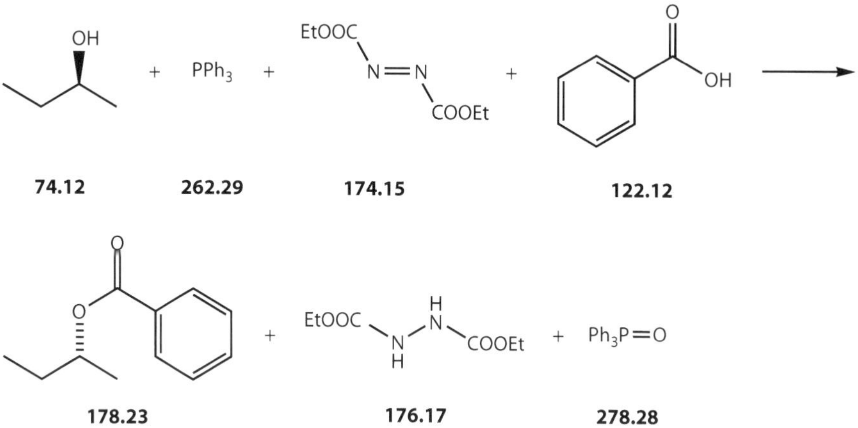

$$\% \text{ Atom Economy} = \frac{178.23}{74.12 + 262.29 + 174.15 + 122.12} \times 100 = 28.2\%$$

SCHEME 1.9 Intrinsic atom economy calculation for a Mitsunobu reaction.

azodicarboxylate is converted to diethyl hydrazodicarboxylate and triphenylphosphine reacts to form triphenylphosphine oxide. In the latter case, the Mitsunobu reaction is similar to the more familiar Wittig olefination of aldehydes and ketones. Calculations of the intrinsic atom economics for Wittig reactions have been discussed in many pedagogical articles[33] and are treated in more detail in Section 3.5.4.

In 2007, a group of pharmaceutical manufacturers (AstraZeneca, Eli Lilly & Company, GlaxoSmithKline, Johnson & Johnson, Merck & Co., Inc., Pfizer, Inc., and Schering-Plough Corporation) contributed toward an article regarding important green research areas.[34] The goal was to summarize how green chemists and green engineers could collaborate and solve the big issues currently facing the pharmaceutical industry. Two years previously, these seven global corporations partnered with the American Chemical Society Green Chemistry Institute (ACS GCI) to form the ACS GCI Pharmaceutical Roundtable. Key research areas were identified within the following three general categories: (1) reactions currently used but better reagents preferred, (2) more aspirational reactions, and (3) solvent themes. In the first category, great emphasis was placed on finding a safer and more environmentally friendly Mitsunobu reaction. Aside from the low atom economy issue, chromatography is generally necessary to separate the unwanted by-products from the desired nucleophilic substitution product. Diethyl azodicarboxylate is also thermally unstable, toxic, and shock sensitive. Use of polymer-bound triphenylphosphine has found favor, as this permits recycling and reduces required solvent volumes, along with simplifying purification procedures.[35–37] Ideally, novel Mitsunobu reactions would be catalytic with the production of benign by products. A step in this direction is to use iodosobenzene diacetate as a stoichiometric oxidant, which leads to the preferable iodobenzene and acetic acid as by-products instead of a dialkyl hydrazodicarboxylate.[38] There is clearly some way to go to make the Mitsunobu reaction routinely viable in a commercial setting.

It is important to appreciate that even though one can classify processes as atom economic or non-atom economic, each reaction should be considered individually and evaluated for its efficiency. This is particularly useful when a chemical compound can be prepared using two different reaction pathways. The atom economy for each pathway can be calculated and a direct comparison can be made. Significantly, students must also be taught that calculations involving atom economy assume molar equivalents of reactants. If the actual reaction utilizes an excess of one reactant, the excess will not generally end up in the desired product. It is therefore essential to calculate an "experimental atom economy" using the actual mass of all reactants and the theoretical maximum mass of the desired product[39] (Figure 1.1). The experimental atom economy is therefore based upon the actual quantities of reagents used in a synthetic experiment.

A typical example in an introductory organic laboratory experiment involves the conversion of 2-naphthol to butyl 2-naphthyl ether using sodium hydroxide and 1-iodobutane, via a Williamson ether synthesis. An undergraduate experimental procedure requires a student to combine 0.56 g of sodium hydroxide and 1.0 g of 2-naphthol in 20 mL of ethanol. Following a short reflux period, 1.62 g (1.0 mL) of 1-iodobutane is added, and after further heating, the ether product is precipitated in ice water.[40] The maximum reported student mass of the isolated desired product, butyl 2-naphthyl ether, is measured to be 1.29 g. Once the balanced equation for the reaction is written, the percentage yield, the percentage intrinsic atom economy, the percentage experimental atom economy, the overall reaction efficiency, and the reaction E-factor can be calculated (Scheme 1.10). Here, the reaction efficiency is defined as the percentage yield multiplied by the percentage experimental atom economy (Figure 1.1). A value of 41% indicates that this percentage of the starting material atoms make their way into the butyl 2-naphthyl ether product, providing a very different efficiency perspective than the maximum student reaction yield (95%). If one uses the median student yield (reported as 31%) rather than the maximum yield, the typical reaction efficiency drops to only 13%! The reaction E-factor is calculated as 1.5 if the ethanol solvent is recycled and any workup is ignored, meaning that for every 1.0 g of butyl 2-naphthyl ether synthesized, 1.5 g of waste is generated. In this instance, waste is composed of unreacted starting materials and reaction by-products (sodium iodide and water). If the reaction workup is included in the calculation, and no solvent recycling takes place, the E-factor is significantly greater than 1.5.

These simple calculations provide some useful insights in using the principles of green chemistry and are valuable tools when comparing multiple synthetic pathways to prepare the same product. For example, if a product

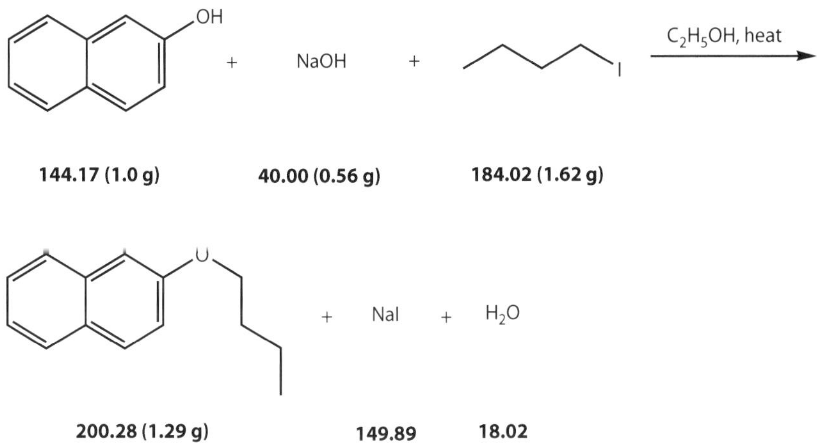

$$\text{Percentage Yield} = \frac{1.29\ \text{g}}{1.36\ \text{g}} \times 100 = \mathbf{95\%}$$

$$\underset{\text{(intrinsic)}}{\%\ \text{Atom Economy}} = \frac{200.28}{144.17 + 40.00 + 184.02} = \mathbf{54\%}$$

$$\underset{\text{(experimental)}}{\%\ \text{Atom Economy}} = \frac{1.36\ \text{g}}{1.0\ \text{g} + 0.56\ \text{g} + 1.62\ \text{g}} = \mathbf{43\%}$$

Overall Reaction Efficiency = 95% × 43% = 41%

$$\textbf{E-Factor} = \frac{\mathbf{(1.0\ g + 0.56\ g + 1.62\ g) - 1.29\ g}}{\mathbf{1.29\ g}} = \mathbf{1.5}$$

SCHEME 1.10 Reaction efficiency and waste calculations for conversion of 2-naphthol to butyl 2-naphthyl ether.

can be obtained by a reaction that proceeds with 90% yield and 50% experimental atom economy, while the same product can be obtained by a different reaction that proceeds with a 70% yield and has 85% experimental atom economy, which reaction is more desirable to be performed in the laboratory? A more detailed analysis of other sustainable chemistry metrics has been published recently.[41]

3. Less hazardous chemical synthesis. Interest in green chemistry has resulted in a number of creative and innovative ways to synthesize organic molecules. Many new reactions, reagents, catalysts, and experimental conditions have been developed with the ideal aim of eliminating noxious substances. A noteworthy example that illustrates this principle is modification of the

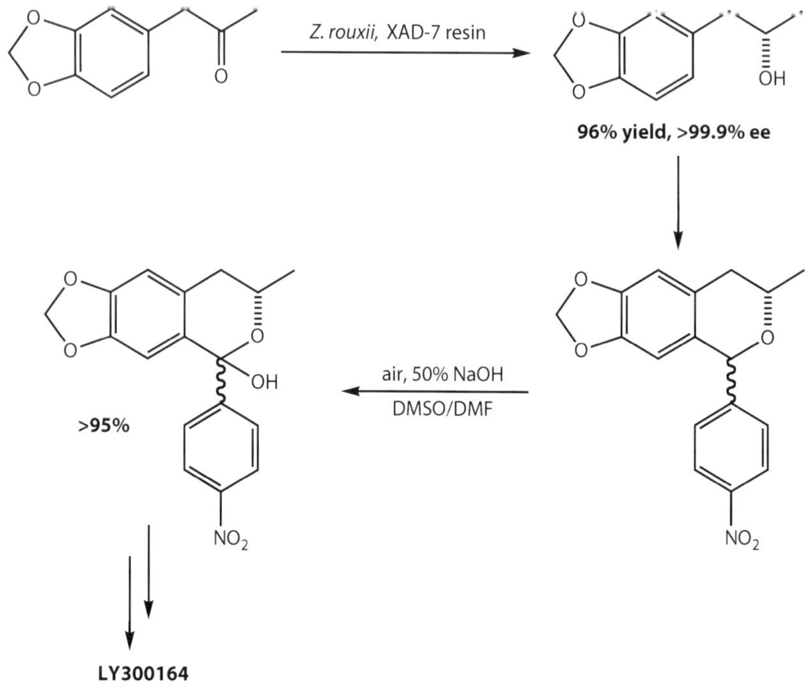

SCHEME 1.11 Synthesis of LY300164 highlighting two greener approaches.

synthetic pathway toward an anticonvulsant drug candidate, LY300164, by Lilly Research Laboratories.[42] The first step in the redesigned synthesis uses a type of yeast (*Zygosaccharomyces rouxii*) to perform a biocatalytic ketone reduction within a novel three-phase reaction system with a 96% yield and greater than 99.9% enantiomeric excess (ee) (Scheme 1.11). This allows for removal of organic reaction components from the aqueous waste system by employing a slurry containing glucose, a polymethyacrylate ester resin, and a buffer. A second key step is selective oxidation using compressed air, dimethylsulfoxide, and sodium hydroxide, which negates the use of chromium trioxide (a known carcinogen) and prevents chromium waste. The revised chemical synthesis eliminates use of 340 L of solvent and 3 kg of chromium waste for each kilogram of LY300164 synthesized. The new methodology also exhibits improved efficiency, with the percentage yield climbing from 16% to 55%. Use of greener reagents in the undergraduate organic laboratory is the major focus of Chapter 6.

4. Designing safer chemicals. An understanding of the fundamental relationships between chemical properties and toxicity[43] has better enabled researchers to design safer substances.[44] The primary intention of designing safer chemicals is to strike the right balance between maximizing the desired performance and the function of the chemical product while minimizing its impact on the environment, human health, and wildlife health. One example of designing safer chemicals is the compound 4,5-dichloro-2-*n*-octyl-4-isothiazoline-3-one

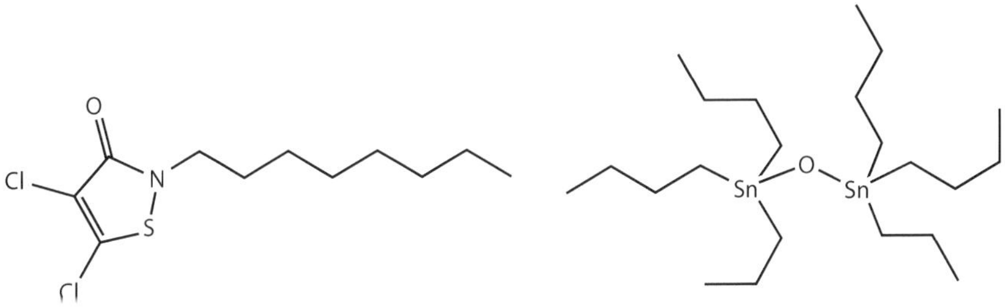

DCOI TBTO

FIGURE 1.2 4,5-Dichloro-2-*n*-octyl-4-isothiazoline-3-one (DCOI) and tributyltin oxide (TBTO).

(DCOI), which finds use as an antifouling agent to reduce the deposition of various microorganisms and marine precipitation on the hulls of cargo ships[45] (Figure 1.2). This compound was developed by Rohm and Haas to replace the antifouling agent tributyltin oxide (TBTO), which has a tendency to bioaccumulate in the marine environment and is toxic to many organisms. In comparison, DCOI is found to exhibit less bioaccumulation and toxicity. A metrics analysis of two synthesis plans to determine the "greenest" preparation of DCOI is outlined in Section 2.6.4.

5. Safer solvents and auxiliaries. Solvents invariably constitute an important, integral part of chemical reactions. Two of the common groups of solvents historically used in organic synthesis include particular halogenated hydrocarbons and aromatic compounds (Figure 1.3). Their adverse effects on human health and the environment are well documented. It is therefore unsurprising to note that solvents are perhaps one of the most active areas of research in green chemistry. Since the best solvent for a chemical reaction is no solvent at all,[46] solvent-free systems[47,48] have been a major focus of green research. A number of reactions can now be easily carried out in the solid state, thereby eliminating the need for any solvents. Similarly, research using water as a solvent for organic reactivity is well established.[49–51] Utilization of supercritical fluids[52] and ionic liquids[53–55] in place of traditional solvents has additionally proved beneficial in designing less hazardous chemical syntheses. More efficient modifications of well-known reactions have also contributed significantly in this area. C-H bond activation using new and innovative methods in catalysis is at the forefront of research.[56] From a pedagogical perspective, Chapters 3 to 5 focus on how several of these approaches can be integrated into a teaching laboratory environment.

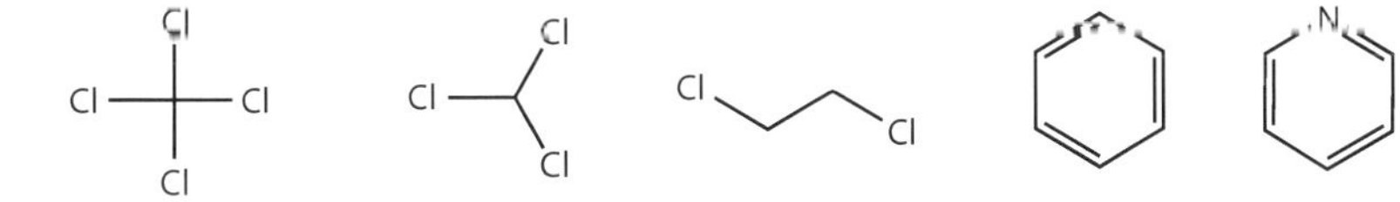

carbon tetrachloride chloroform 1,2-dichloroethane benzene pyridine

FIGURE 1.3 Some solvents historically used in organic synthesis.

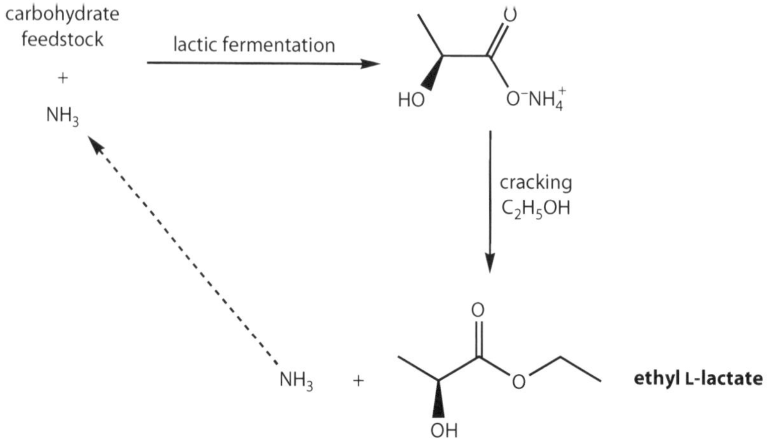

SCHEME 1.12 Ethyl L-lactate synthesis from carbohydrate feedstocks.

Argonne National Laboratory was awarded the U.S. Presidential Green Chemistry Award in 1998 in the Alternative Solvents/Reaction category.[57] Chemists developed a novel process to synthesize a biodegradable organic solvent such as ethyl L-lactate via sugar fermentation. This synthetic strategy is carbohydrate based rather than petrochemical based, and the innovative technology requires little energy, is highly efficient, eliminates large volumes of salt waste, and reduces pollution and emissions. The process cracks ammonium L-lactate under catalytic and thermal conditions in the presence of ethanol to generate ethyl L-lactate and ammonia, which is recycled for use in fermentation (Scheme 1.12). Ethyl L-lactate is miscible with both organic and aqueous solvents and has been approved for incorporation into foodstuffs by the U.S. Food and Drug Administration. It can replace halogenated solvents (e.g., dichloromethane and chloroform) and others (including chlorofluorocarbons) in paints, cleaners, and additional industrial applications. Further discussion on this topic can be found in Section 5.8.

6. Design for energy efficiency. An important aspect of chemical reactivity is analysis of energy requirements. Energy input for a chemical reaction is typically achieved by three classical methods: thermal, photochemical, and electrochemical. Design of a reaction that does not require a great amount of energy is highly desirable, and other energy input methods (e.g., sonification, mechanical stirring, and microwave irradiation) should also be considered. In this venue, green student organic reactions under microwave conditions are discussed in detail in Chapter 8. Research in green chemistry has opened the door for a number of reactions to be carried out at ambient temperature and pressure, thereby reducing energy needs and increasing energy efficiencies. Further work in exploring alternative energy sources (such as solar power) could lead to more energy-efficient chemical reactions and processes. Bristol-Myers Squibb earned the Greener Synthetic Pathways Award as part of the Presidential Green Chemistry Challenge in

2004 for its redesigned synthesis of the anticancer drug Taxol®.[58] Paclitaxel, the active ingredient in Taxol, was historically semisynthesized from a naturally occurring compound (10-deacetylbaccatin III) in an eleven-step process requiring thirteen solvents and thirteen reagents. A considerable amount of energy is conserved by the modern strategy that extracts paclitaxel directly from plant cell cultures, which avoids any chemical transformations. As such, ten solvents and six drying steps have been eliminated from commercial operations.

7. Use of renewable feedstocks. At present, it is estimated that the majority of the world's manufacturing products are derived from nonrenewable fossil fuels. As the name implies, these will become depleted over a period of time. It is therefore prudent to focus attention on renewable feedstocks, which are predominantly associated with biological and plant-based starting materials. Cellulose, lignin, and other wood compounds, as well as starch, chitin, and L-lactic acid, are some characteristic renewable feedstocks. A recent example of using a renewable feedstock (instead of a petroleum-based chemical) is during the preparation of 1,6-hexanedioic acid (adipic acid).[59] Adipic acid is used in large quantities in the commercial production of nylon and was originally prepared from benzene, a known carcinogen. This approach also utilized a final oxidation reaction with nitric acid, generating nitrous oxide (a greenhouse gas) as a by-product. Methods have been developed to produce adipic acid from D-glucose, which can be obtained from starch, a renewable feedstock. D-Glucose is initially converted into (*Z*,*Z*)-muconic acid by biocatalysis using a genetically engineered microbe, with further hydrogenation leading to adipic acid (Scheme 1.13). Water can be used as the reaction solvent under mild conditions of temperature and pressure.

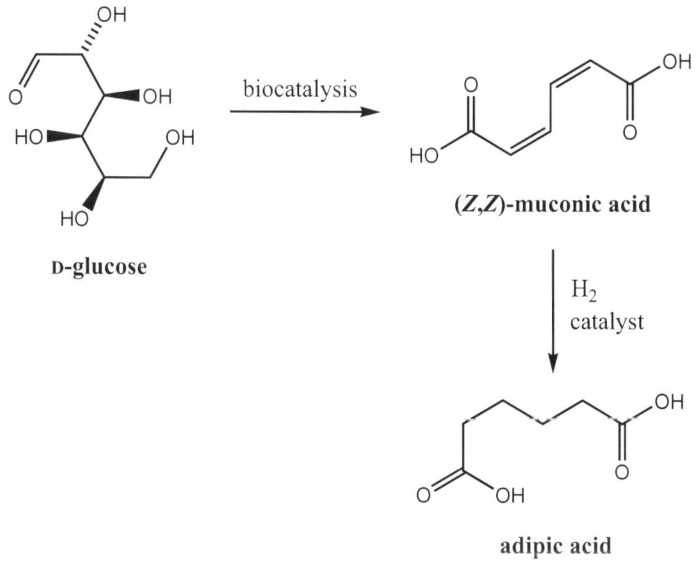

SCHEME 1.13 Adipic acid synthesis from D-glucose.

8. Reduce derivatives. Use of protecting groups is quite common in organic synthesis when effecting a reaction in the presence of a labile functional group. The common practice is to protect a functional group, carry out the required reaction at a desired site, and then deprotect to generate the original functional group. This would normally require extra synthetic steps. An innovative concept, known as noncovalent derivatization, has been developed by Cannon and Warner.[60] This method does not employ covalent bonding to form derivatives but uses intermolecular forces to achieve required molecular transformations. An early example of noncovalent derivatization is illustrated by the controlled diffusion and solubility of hydroquinones, which are used as developers in photographic systems.[60] At sufficiently elevated pH, hydroquinones are fully deprotonated, forming anionic species that are both soluble and mobile in aqueous media, including thin-film photographic systems. At neutral pH, by comparison, hydroquinones are nonionic, and therefore insoluble and immobile in such systems. Quinones, the oxidation products of hydroquinones, cannot undergo deprotonation in alkaline solution. These phenomena have been utilized in diffusion-controlled silver halide photographic imaging systems such as Polaroid instant photography. Some photographically useful hydroquinones, however, have sufficient aqueous solubility in their protonated state to present a particular problem. In these cases, the marginal solubility leads to migration of reagents in the multilayer film structure prior to pH elevation. This compromises the performance of the imaging system and must be corrected. Researchers at Polaroid have directed their efforts toward more effectively controlling hydroquinone immobilization at neutral pH while not interfering with the solubility (and therefore reactivity) of the deprotonated species. Instead of relying on base-labile covalent protective groups, which would be the traditional approach, they have developed a noncovalent protecting group in the form of a cocrystal between hydroquinones and bis-(*N*,*N*-dialkyl)terephthalamides (Figure 1.4). This approach has solved the problem without chemical modification of the original hydroquinone structures and has minimized waste material and energy usage. Undergraduate synthesis of cocrystals in a solvent-free environment is outlined in Section 3.5.2.3.
9. Catalysis. Catalysts are used in small amounts and can carry out a single reaction many times. They typically increase reaction rates by lowering the energy of activation and providing an alternative path for a reaction to proceed. This is a much better approach to chemical reactions that solely depend upon stoichiometric amounts of reactants (and sometimes require an excess of one reactant). Use of catalysis is prevalent in both industrial manufacturing and in academic laboratories since the advantages afforded are numerous. Enzymes are commonly used as catalysts to effect a desired chemical reaction in chemical syntheses. Catalysts have the ability to drive selective synthetic pathways, can often be recycled and can reduce energy requirements of chemical reactions as well as the amount of materials used. Practical undergraduate examples of organocatalysis, biocatalysis, and

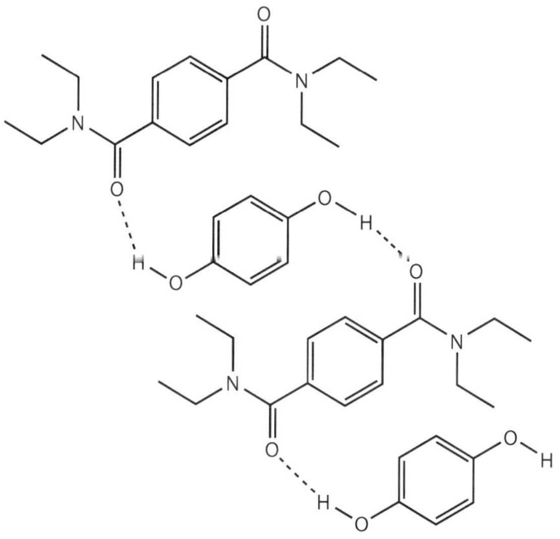

FIGURE 1.4 A hydroquinone protected by noncovalent interactions with a bis-(*N*,*N*-dialkyl)terephthalamide.

metal catalysis are highlighted throughout this book, with several commercial examples previously detailed in this current chapter section.

10. Design for degradation. Chemical substances that persist in the environment remain available to exert their adverse effects on human and wildlife health and may bioaccumulate. Designing biodegradable chemicals should therefore be a primary consideration in the planning stages of chemical synthesis. Certain classes of chemical compounds, such as those containing halogens, are known to possess enhanced environmental persistence. One successful example of design for degradation consists of the development of laundry detergents. Synthetic laundry detergents traditionally consisted of branched-chain alkylbenzene sulfonates (ABSs). These compounds were initially synthesized from the hydrocarbons propene and benzene. Their typical structure is shown in Figure 1.5.

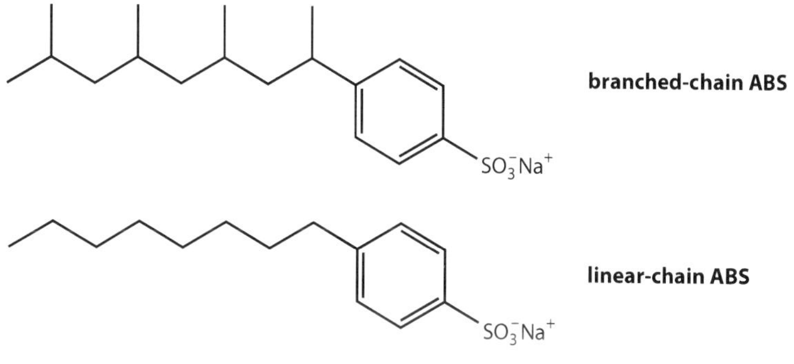

FIGURE 1.5 Branched-chain and linear-chain alkylbenzene sulfonates.

Branched-chain ABSs are incompletely biodegraded in municipal sewage treatment systems. As a result, excessive foaming has been observed in activated sludge aeration tanks as well as receiving rivers. People have even witnessed a head of foam on their drinking water from the tap in some areas.[61] In comparison, linear ABSs are almost completely biodegradable in sewage treatment plants (Figure 1.5). This example illustrates it is indeed possible to design chemical compounds for effective degradation.

11. Real-time analysis for pollution prevention. Real-time, in-process analysis is vitally important in the chemical manufacturing sector. Using appropriate techniques, generation of hazardous by-products and side reactions can be monitored and controlled. The analytical techniques used should be consistent with the principles of green chemistry in that they should avoid generation of waste and hazardous substances and minimize use of solvents or use greener solvents during analysis. An example of green analytical chemistry in undergraduate laboratories employs a flow-injection spectrophotometric method for measurement of creatinine in urine, where consumption of reagents is reduced by 60% compared to the traditional batch method.[62] A creatinine-picrate complex is formed under basic conditions that is easily quantified by UV-visible spectrophotometry (Scheme 1.14). Solution waste is consequently irradiated with UV light and photodegraded as a greener waste management strategy.
12. Inherently safer chemistry for accident prevention. The potential for a chemical disaster can never be understated. There have been some notable chemical accidents that have resulted in the loss of human lives. Thousands of lives

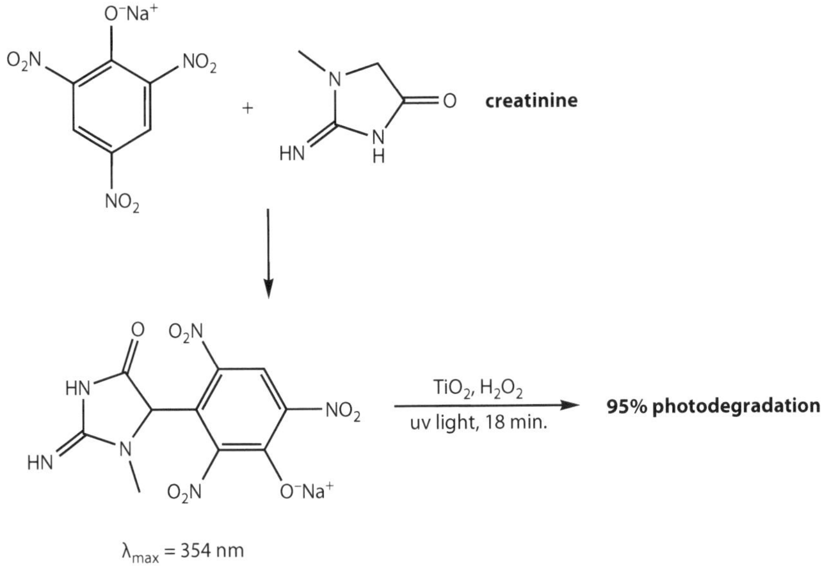

SCHEME 1.14 Analysis of creatinine in urine samples with subsequent photodegradation.

were lost in Bhopal, India, in 1984, due to the accidental release of the poisonous gas methyl isocyanate.[63] The hazards posed by toxicity, flammability, and potential for explosions should be carefully evaluated at the design stage.

The twelve principles of green chemistry, originally formulated almost twenty years ago, have been instrumental for most advances in the discipline. These principles are meant to operate as an integrated cohesive system working toward the broader goal of sustainable development. Teaching introductory organic chemistry classes provides an excellent opportunity for instructors to acquaint their students with the field of green chemistry. This leads to discussions of many principles in both the classroom and the laboratory. Students are trained to write balanced chemical equations to include *all* products, not just the organic product of interest. They learn the intention of doing a chemical reaction (to obtain a desired product) and to identify others as by-products. They can routinely calculate the experimental atom economy for every reaction they do in the laboratory, in addition to the traditional percentage yield of the desired product. The introductory organic laboratory also affords occasion to integrate some of the more recent research findings in green chemistry into teaching. Here students get multiple chances to perform certain experiments under solid-state conditions, as well as in aqueous media. Other reactions can be carried out in greener organic solvents, such as polyethylene glycol or ionic liquids. Students can also perform experiments utilizing biomaterials and learn about microwave-assisted organic synthesis (MAOS). The laboratory component of the organic chemistry course truly offers a wide variety of various newer approaches that undergraduates can gain valuable experience in.

Advanced organic chemistry offerings can further reinforce both theory and practice of green chemistry. Senior courses in organic synthesis can highlight the role catalysis plays and include a discussion of the evaluation of entire synthesis "greenness." When students plan on doing their own independent multistep synthesis, they can apply the principles of green chemistry under three distinct categories:

1. Starting materials
   a. Renewable
   b. Simple structure
   c. Nonhazardous
2. Reaction conditions
   a. Minimize number of reaction steps
   b. Few or no by-products (atom-economic reactions)
   c. Solid-state, aqueous-medium, greener organic solvents
   d. Low energy input
   e. Catalytic reactions where possible
   f. Avoid protecting groups
3. Products
   a. Replace polluting materials
   b. Replace petroleum-based materials
   c. Aim to synthesize biodegradable substances

This leads to the development of several sustainable synthesis optimization rules, as the outlined by Diehlmann et al.[64] Students undertaking upper-level research in organic chemistry have even greater opportunities to apply their knowledge and understanding of the theory and practices of green chemistry that could and should be incorporated in their research projects.

1.5 GREEN ORGANIC CHEMISTRY TEACHING RESOURCES

A decade ago there were very few green organic chemistry teaching resources available for use in the classroom or the laboratory. The landmark paper by Collins[11] in 1995 was the first pedagogical article on green chemistry to appear in the *Journal of Chemical Education*. Indeed, on publishing an environmentally benign synthesis of adipic acid in 2000, Reed and Hutchison stated that "while many chemistry courses now cover environmental issues as a part of their curriculum, few integrate such concepts into their laboratory sections, owing in part to a lack of published material in this field."[65] Since that time there has been an explosion in the number of resources available in teaching green organic chemistry in lecture and practical venues, with more being developed on a regular basis.

1.5.1 Textbooks

Green chemistry textbooks tend to fall into one of two categories: those focusing on theoretical principles and real-world case studies, and those presenting green laboratory experiments. In the former category, books written by Matlack[66] (now in its second edition) and Lancaster[67] provide excellent, thorough coverage of introductory concepts. Cann and Connelly have described opportunities to weave green examples into lecture courses across the undergraduate curriculum.[68] A recent publication from the American Chemical Society, edited by Anastas, Levy, and Parent, focuses on many topics of current interest. These include student-motivated endeavors enhancing green literacy, K–12 outreach and science literacy through green chemistry, and linking hazard reduction to molecular design.[69] The seminal textbook written from a laboratory perspective is *Green Organic Chemistry: Strategies, Tools, and Laboratory Experiments*, published in 2004.[70] Doxsee and Hutchison detail nineteen experiments that have been successfully implemented at the University of Oregon, along with background information about green chemistry metrics and greener reaction conditions, among other topics. A similar approach has been taken by Kirchhoff and Ryan.[71] Latterly, *Experiments in Green and Sustainable Chemistry* came onto the market in 2009, and includes practical details at a level such that no other literature resources are required to perform experimental work.[72] The appendix of this current textbook is a comprehensive repository of peer-reviewed undergraduate green organic chemistry reactions. Indexing is arranged by mechanistic type, with further inclusion of experimental techniques and green principles exemplified.

1.5.2 Journals

Several scholarly journals currently feature green chemistry articles from a pedagogical perspective. The *Journal of Chemical Education* has published a green chemistry feature column since the December 2001 issue.[73] Edited by Mary Kirchhoff, submissions are accepted in areas that include laboratory experimentation, new course implementation, demonstrations, teaching case studies, and designing modules for existing curricula. Some of the published articles are suitable for inclusion at the high school level.[74] More recently, the journal *Green Chemistry Letters and Reviews* has introduced a specific education section subdivided into "Educational Materials" and "Perspectives on Implementation."[75] Examples of "Educational Materials" suitable for this journal are classroom demonstrations and outreach activities, along with laboratory exercises. Examples of "Perspectives on Implementation" are green impact assessments and case studies that describe how green chemistry can be incorporated into undergraduate and graduate courses. The *Chemical Educator* is a third journal that has published articles with a green element, from both practical and theoretical angles.

The Royal Society of Chemistry journal *Green Chemistry* has an important role to play in education, both directly and indirectly. Some laboratory experiments have been profiled in this venue.[76] Just as importantly, cutting-edge discoveries are disclosed that can sometimes be adapted for a teaching environment. The industrially oriented *Organic Process Research and Development* (published by the American Chemical Society) is replete with thorough and varied case studies about how companies have "greened" their operations, from both scientific and engineering perspectives. These can be integrated into stand-alone green courses, or simply included in a more traditional delivery, such as an upper-level modern organic synthesis offering. A final journal of note is *ChemSusChem*, published by Wiley-VCH, which features research at the boundary of sustainability with engineering, chemistry, biotechnology, and materials science.

1.5.3 Online Resources

Instructors can additionally connect with several online resources to enrich their green teaching repertoire. Several well-established examples are highlighted here. A standout, rich contribution from the University of Oregon is the Greener Education Materials for Chemists (GEMs) web site.[77] This fully searchable database allows the user to find teaching resources according to academic level, and to fine-tune searches in terms of green principles, experimental techniques, authors, and other filters. At the time of writing, over one hundred laboratory protocols, case studies, and additional items are accessible. In a similar vein, NOP ("Nachhaltigkeit im Organisch-chemischen Praktikum," or "Sustainability in the Organic Chemistry Laboratory Course") is an open-access compendium of lab experiments from Germany used to integrate sustainability principles into existing curricula.[78] From a high school perspective, Sally Henrie at Union University in Tennessee has compiled twenty-four greener experiments that can each be performed in less than forty-five minutes (e.g., "Ideal Gas Law: Finding % H_2O_2 with Carrot Juice"). Sample experiments can be downloaded for free with teacher and student manuals available for purchase on CD.[79] In comparison, faculty members at St. Olaf College, Minnesota, have designed the Green Chemistry Assistant web site.[80] This online application is primarily available for students to perform green calculations on a reaction of interest (atom economy, theoretical yield, percentage yield).

An Internet conference with the title "Educating the Next Generation: Green and Sustainable Chemistry" ran from April to June 2010, and was organized by the ACS Committee on Computers in Chemical Education.[81] Seven presented papers focused on instruction, with titles including "Education Resources from the American Chemical Society Green Chemistry Institute" and "Development of an Undergraduate Catalytic Chemistry Course." These articles will remain online and available for perusal. The ACS has also set up a Green Chemistry Resource Exchange to profile and share new developments and examples.[82] In addition, several green chemistry networks are currently active across the world.[83]

1.5.4 Green Chemistry Summer Schools

A number of institutions operate workshops where faculty and graduate students can learn the principles of green chemistry in a hands-on environment. These include the University of Oregon and the University of Scranton (Pennsylvania). The ACS has further arranged an annual summer school since 2003 at locations across North and South America. Over 425 graduate students and postdoctoral fellows have attended these schools in total, where they have participated in laboratory work and interacted with pioneers and practitioners in the field. It is truly gratifying to see that there are so many ongoing, varied, and concerted efforts being made to further comprehension of how to incorporate principles and practices of green chemistry in the classroom and the laboratory.

1.6 CONCLUSION

Green chemistry is based upon a cohesive set of principles to achieve its goal of reducing or eliminating the role of hazardous materials in the design and manufacture of chemical products. Applications of these principles and the utilization of ensuing greener technologies form a solid foundation for sustainable development. It is important that the students of today are prepared to accept the challenges of tomorrow so that a sustainable future can become reality. They need to be equipped with the necessary knowledge, skills, and expertise in green chemistry to become critically thinking scientists and engineers. This chapter provides an introduction to many important concepts, such as the environmental impact factor, atom economy, catalysis, energy-efficient reactions, eco-friendly solvents, and renewable feedstocks. The chapters that follow are written to reinforce these fundamental notions with practical examples and case studies.

REFERENCES

1. Anastas, P. T., Warner J. C. *Green Chemistry: Theory and Practice*. Oxford University Press, New York, 1998.
2. Parent, K., Kirchhoff, M., Godby, S., Eds. *Going Green: Integrating Green Chemistry into the Curriculum*. American Chemical Society, Washington, DC, 2004, 1–2.

3. Anastas, P., Williamson, T., Eds. *Green Chemistry: Designing Chemistry for the Environment*. American Chemical Society Symposium Series 626. American Chemical Society, Washington, DC, 1994, 1.
4. Sheldon, R. A. *Green Chem.* 2007, 9, 1273–1283.
5. La Merrill, M., Parent, K., Kirchhoff, M. *ChemMatters* April 2003, 7–10.
6. Kirchhoff, M. M. *J. Chem. Educ.* 2010, 87, 121.
7. Leitner, W. *Green Chem.* 2004, 6, 351.
8. Bennett, G. D. *J. Chem. Educ.* 2006, 83, 1871–1872.
9. Nebbia, G., Kauffman, G. *Chem. Educator* 2007, 12, 362–369.
10. Ciamician, G. *Science* 1912, 36, 385–394.
11. Collins, T. *J. Chem. Educ.* 1995, 72, 965–966.
12. American Chemical Society Green Chemistry Institute. www.epa.gov/gcc/pubs/gcinstitute.html (accessed December 23, 2010).
13. Sustainable Chemistry: Organization for Economic Cooperation and Development. www.oecd.org/dataoecd/16/25/29361016.pdf (accessed December 23, 2010).
14. Isobe, M. Report of the IUPAC Organic and Biomolecular Chemistry Division (III). August 2007. http://old.iupac.org/news/archives/2007/44th_council/Item_11-DivIII_2007.pdf (accessed December 23, 2010).
15. (a) International Year of Chemistry 2011. www.chemistry2011.org/about-iyc/introduction (accessed December 23, 2010). (b) Kirchhoff, M. M. *J. Chem. Educ.* 2011, 88, 1–2.
16. Dunn, P., Wells, A., Williams, M., Eds. *Green Chemistry in the Pharmaceutical Industry*. Wiley-VCH, Weinheim, Germany, 2010.
17. Andrews, I., Cui, J., DaSilva, *J. Org. Process Res. Dev.* 2010, 14, 19–29.
18. Winterton, N. *Green Chem.* 2001, 3, G73–G75.
19. Anastas, P. T., Kirchhoff, M. M. *Acc. Chem. Res.* 2002, 35, 686–694.
20. Warner, J. C., Cannon A. S., Dye, K. M. *Environ. Impact Assess. Rev.* 2004, 24, 775–799.
21. Clark, J. H. *Green Chem.* 2006, 8, 17–21.
22. Horvath, I. T., Anastas, P. T. *Chem. Rev.* 2007, 107, 2169–2173.
23. Anastas, P. T., Eghbali, N. *Chem. Soc. Rev.* 2010, 39, 301–312.
24. Sheldon, R. A. *Chem. Ind.* 1992, 903–906.
25. Sheldon, R. A. *Green Chem.* 2007, 9, 1273–1283.
26. Sheldon, R. A. *Chem. Tech.* 1994, 24, 38–47.
27. Trost, B. M. *Science* 1991, 254, 1471–1477.
28. Mitsunobu, O., Yamada, Y. *Bull. Chem. Soc. Jpn.* 1967, 40, 2380–2382.
29. Mitsunobu, O. *Synthesis* 1981, 1–28.
30. Hughes, D. *Org. React.* 1992, 42, 335–656.
31. Hughes, D. *Org. Prep. Proced. Int.* 1996, 28, 127–164.
32. Kumara Swamy, K., Bhuvan Kumar, N., Pavan Kumar, K. *Chem. Rev.* 2009, 109, 2551–2651.
33. For an example, see Cann, M., Dickneider, T. *J. Chem. Educ.* 2004, 81, 977–980.
34. Constable, D. J. C., Dunn, P. J., Hayler, J. D., Humphrey, G. R., Leazer, Jr., J. L., Linderman, R. J., Lorenz, K., Manley, J., Pearlman, B. A., Wells, A., Zaks, A., Zhang, T. Y. *Green Chem.* 2007, 9, 411–420.
35. Amos, R. A., Emblidge, R. W., Havens, N. J. *Org. Chem.* 1983, 48, 3598–3600.
36. Tunoori, A. R., Dutta, D., Georg, G. I. *Tetrahedron Lett.* 1998, 39, 8751–8754.
37. Janda, K. D., Wentworth, P., Vandersteen, A. M. *Chem. Commun.* 1997, 759–760.
38. But, T. Y. S., Toy, P. H. *J. Am. Chem. Soc.* 2006, 128, 9636–9637.
39. Atom Economy: A Measure of the Efficiency of a Reaction. http://academic.scranton.edu/faculty/cannm1/organicmodule.html (accessed December 23, 2010).

40. Esteb, J. J., Magers, J. R., McNulty, L., Morgan, P., Wilson, A. M. *J. Chem. Educ.* 2009, 86, 850–852.
41. Calvo-Flores, F. *ChemSusChem* 2009, 2, 905–919.
42. The Presidential Green Chemistry Challenge Award Recipients 1996–2009, 1999 Greener Synthetic Pathways Award. U.S. Environmental Protection Agency, Office of Pollution Prevention and Toxics, Washington, DC, 2009, 108–109.
43. Voutchkova, A., Ferris, L., Zimmerman, J., Anastas, P. *Tetrahedron* 2010, 66, 1031–1039.
44. DeVito, S., Garrett, R. *Designing Safer Chemicals: Green Chemistry for Pollution Prevention.* American Chemical Society Symposium Series 640. American Chemical Society, Washington, DC, 1996.
45. The Presidential Green Chemistry Challenge Award Recipients 1996–2009, 1996 Designing Greener Chemicals Award. U.S. Environmental Protection Agency, Office of Pollution Prevention and Toxics, Washington, DC, 2009, 144–145.
46. Sheldon, R. A. *Green Chem.* 2005, 7, 267–278.
47. Tanaka, K. *Solvent-free Organic Synthesis.* Wiley-VCH, Weinheim, Germany, 2003.
48. Tanaka, K., Toda, F. *Chem. Rev.* 2000, 100, 1025–1074.
49. Li, C.-J. *Chem. Rev.* 2005, 105, 3095–3165.
50. Lindström, U. M. *Chem. Rev.* 2002, 102, 2751–2772.
51. Lindström, U. M., Ed. *Organic Reactions in Water: Principles, Strategies and Applications.* Blackwell, Oxford, 2007.
52. Arai, Y., Sako, T., Takebayashi, Y., Eds. *Supercritical Fluids: Molecular Interactions, Physical Properties and New Applications.* Springer, New York, 2002.
53. Welton, T. *Chem. Rev.* 1999, 99, 2071–2083.
54. Wasserscheid, P., Welton, T., Eds. *Ionic Liquids in Synthesis*, 2nd ed. Wiley-VCH, Weinheim, Germany, 2007.
55. Plechkova, N., Seddon, K. *Chem. Soc. Rev.* 2008, 37, 123–150.
56. Crabtree, R. H. *Chem. Rev.* 2010, 110, 575.
57. The Presidential Green Chemistry Challenge Award Recipients 1996–2009, 1998 Greener Reaction Conditions Award. U.S. Environmental Protection Agency, Office of Pollution Prevention and Toxics, Washington, DC, 2009, 122–123.
58. The Presidential Green Chemistry Challenge Award Recipients 1996–2009, 2004 Greener Synthetic Pathways Award. U.S. Environmental Protection Agency, Office of Pollution Prevention and Toxics, Washington, DC, 2009, 58–59.
59. The Presidential Green Chemistry Challenge Award Recipients 1996–2009, 1998 Academic Award. U.S. Environmental Protection Agency, Office of Pollution Prevention and Toxics, Washington, DC, 2009, 116–117.
60. Cannon, A. S., Warner, J. C. *Cryst. Growth Des.* 2002, 2, 255–257.
61. Hill, J. W., Kolb, D. *Chemistry for Changing Times*, 10th ed. Pearson Prentice Hall, Upper Saddle River, NJ, 2004, 532–534.
62. Correia, P. R. M., Siloto, R. C., Cavicchioli, A., Oliveira, P. V., Rocha, F. R. P. *Chem. Educator* 2004, 9, 242–246.
63. Broughton, E. *Environ. Health* 2005, 4, 6–11.
64. Diehlmann, A., Kreeisel, G., Gorges, R. *Chem. Educator* 2003, 8, 102–106.
65. Reed, S., Hutchison, J. *J. Chem. Educ.* 2000, 77, 1627–1629.
66. Matlack, A. S. *Introduction to Green Chemistry*, 2nd ed. CRC Press, Boca Raton, FL, 2010.
67. Lancaster, M. *Green Chemistry: An Introductory Text.* The Royal Society of Chemistry, Cambridge, 2002.
68. (a) Cann, M. C., Connelly, M. E. *Real-World Cases in Green Chemistry*, American Chemical Society, Washington, DC, 2000. (b) Cann, M. C., Umile, T. P. *Real-World Cases in Green Chemistry*, Vol. 2. American Chemical Society, Washington, DC, 2008.

69. Anastas, P. T., Levy, I. J., Parent, K. E., Eds. *Green Chemistry Education: Changing the Course of Chemistry*. American Chemical Society Symposium Series 1011. American Chemical Society, Washington, DC, 2009.
70. Doxsee, K., Hutchison, J. *Green Organic Chemistry: Strategies, Tools, and Laboratory Experiments*. Brooks/Cole, Pacific Grove, CA, 2004.
71. Kirchhoff, M., Ryan, M. A., Eds. *Greener Approaches to Undergraduate Chemistry Experiments*. American Chemical Society, Washington, DC, 2002.
72. Roesky, H., Kennepohl, D., Eds. *Experiments in Green and Sustainable Chemistry*, Wiley-VCH, Weinheim, Germany, 2009.
73. Kirchhoff, M. M. *J. Chem. Educ.* 2001, 78, 1577.
74. For examples of high school green chemistry experiments: (a) Cacciatore, K. L., Amado, J., Evans, J. J., Sevian, H. *J. Chem. Educ.* 2008, 85, 251–253. (b) Cacciatore, K. L., Sevian, H. *J. Chem. Educ.* 2006, 83, 1039–1041.
75. Haack, J. A. *Green Chem. Lett. Rev.* 2007, 1, 7.
76. Examples of pedagogical experiments in *Green Chemistry*: (a) McKenzie, L. C., Thompson, J. E., Sullivan, R., Hutchison, J. E. *Green Chem.* 2004, 6, 355–358. (b) Warner, M. G., Succaw, G. L., Hutchison, J. E. *Green Chem.* 2001, 3, 267–270.
77. University of Oregon Greener Education Materials for Chemists. http://greenchem.uoregon.edu/gems.html (accessed December 23, 2010).
78. (a) Ranke, J., Bahadir, M., Eissen, M., König, B. *J. Chem. Educ.* 2008, 85, 1000–1005. (b) Sustainability in the Organic Chemistry Laboratory Course. Universität Regensburg, Germany. www.oc-praktikum.de/en-entry (accessed December 23, 2010).
79. Green Chemistry Labs, Union University, Tennessee. www.greenchemistrylabs.com (accessed December 23, 2010).
80. St. Olaf College Green Chemistry Assistant. http://fusion.stolaf.edu/gca/ (accessed December 23, 2010).
81. Spring 2010 ConfChem. Educating the Next Generation: Green and Sustainable Chemistry. www.ched-ccce.org/confchem/2010/Spring2010/Index.html (accessed December 23, 2010).
82. American Chemical Society Green Chemistry Resource Exchange. www.greenchemex.org (accessed December 23, 2010).
83. (a) University of Oregon Green Chemistry Education Network. http://cmetim.ning.com (accessed December 23, 2010). (b) Green Chemistry Network. www.greenchemistrynetwork.org (accessed December 23, 2010). (c) Canadian Green Chemistry Network. www.greenchemistry.ca (accessed December 23, 2010).

2 Designing a Green Organic Chemistry Lecture Course

Dr. John Andraos

CONTENTS

2.1 INTRODUCTION

This chapter introduces and discusses various aspects of launching and teaching a green chemistry lecture course in the undergraduate curriculum. First, several challenges are identified. This is followed by a detailed description of the course I taught at York University over an eight-year period, including feedback from students and library staff. Explanations are offered for both what went right and what could be improved in my experience. A number of exemplary case studies from the course highlighting pedagogical merits of discussing green chemistry principles in the classroom are discussed. The concluding section provides useful questions that instructors may think about and hopefully use in making up their own problem sets and assignments.

2.2 CHALLENGES IN LAUNCHING AND TEACHING A GREEN CHEMISTRY COURSE

The movement to incorporate green chemistry thinking in the chemistry curriculum has been ongoing since the publication of Anastas and Warner's seminal textbook.[1] Most of the examples cited in this reference originated from the chemical industry, and hence historically there has been a strong connection between green chemistry philosophy and new developments in industrial chemistry (particularly from the pharmaceutical industry) which has taken a significant leading role in promoting the field. Moreover, organic chemistry is the major subject that links green and industrial chemistry as evidenced by the majority of literature publications. In Canada, progress has been slow and scattered in the adoption of green chemistry ideas in the formal education of future chemists in colleges and universities. Table 2.1 summarizes courses offered at the upper undergraduate level whose titles or syllabi cover or mention either of these areas of study. Surprisingly, most courses are not required for a chemistry degree, and therefore are classified as electives or "educational" courses. All of them, however, require an introductory organic chemistry course as a prerequisite since it is obvious that students need a foundation in the basics of

TABLE 2.1
List of Courses at Canadian Universities in Green or Industrial Chemistry

	ENVIRONMENTAL / GREEN CHEMISTRY	
University	**Course Number**	**Course Title**
Calgary	CHEM 421	Environmental Chemistry
McGill	CHEM 462	Green Chemistry
Queen's	CHEM 326	Environmental and Green Chemistry
Toronto	CHM 343	Organic Experimental Techniques
York	CHEM 3070	Industrial and Green Chemistry
	INDUSTRIAL CHEMISTRY	
University	**Course Number**	**Course Title**
Calgary	CHEM 425	Industrial Chemistry
Carleton	CHEM 3700	Industrial Applications of Chemistry
Concordia	CHEM 445	Industrial Catalysis
Guelph	CHEM 4010	Chemistry and Industry
Laurentian	CHMI 3031	Industrial Inorganic Chemistry
McMaster	CHEM 3I03	Industrial Chemistry
Saskatchewan	CHEM 377	Industrial Chemistry
Ontario Institute of Technology	CHEM 4010	Industrial Chemistry
Prince Edward Island	CHEM 468	Advanced Inorganic Chemistry
Victoria	CHEM 478	Introduction to Chemical Process Industries
Western	CHEM 3330f	Industrial Chemistry
York	CHEM 3070	Industrial and Green Chemistry

organic chemistry before tackling advanced topics encountered in green chemistry. The Department of Chemical and Biochemical Engineering at the University of Western Ontario offers a green processing engineering program in which courses in green chemistry are given in the second year. Other green chemistry resources available are the Canadian Green Chemistry Network launched by McGill University and mentioned in Section 1.2, and Green Centre Canada (www.greencentrecanada.com) launched by Queen's University. In the United States the Green Chemistry Institute (Washington, D.C.), the Warner Babcock Institute for Green Chemistry (Wilmington Massachusetts), the Greener Education Materials for Chemists (GEMs) resource at the University of Oregon, the Berkeley Center for Green Chemistry (UC Berkeley), and green chemistry teaching modules launched by the University of Scranton (Pennsylvania) are important centers of activity in green chemistry teaching.

As this emerging field continues to evolve there have been a number of challenges and problems that have cropped up and continue to persist. A key question that arises is: Why have research faculty not caught on or are slow on the uptake? Green chemistry has not been sold strongly enough in terms of educating "smarter" young chemists who represent the feeder population of future high-end research graduate students that senior research faculty will draw upon to carry out their research. Despite the notion that green chemistry could be used as a good marketing tool to reverse sliding enrollments in chemistry programs, research faculty often perceive the subject as a "soft" science couched mainly in descriptive language. This view is partly a result of how the field has developed since the early 1990s in the scientific literature, and still continues by and large today. Green chemistry is portrayed as a curiosity, and is therefore not yet included in the set of ideas known as "the fundamentals" that form the mainstream way of thinking in the study of chemistry as, for example, bonding theory or structure determination and elucidation. Newer editions of introductory organic textbooks include green chemistry examples in margin notes, boxes, and short vignettes that are not part of the mainstream material that students are responsible for. An example is the seventh edition of John McMurry's textbook, which discusses the synthesis of pravadoline, a nonsteroidal anti-inflammatory pharmaceutical, in ionic liquids.[2]

The drive to legitimize green chemistry in more quantitative terms by introducing metrics has lead to another confusion in the literature, namely, the excessive number of metrics introduced, each with their own names by their respective authors. There are several instances where more than one name has been used to describe the very same metric, thus artificially expanding and needlessly complicating the field. As examples, the terms *balance yield* and *overall reaction mass efficiency* are identical, and *effluent load factor* is identical to the *E-factor*.[3,4] The task of choosing which metrics to use becomes ever more difficult for chemists who genuinely wish to discover and use green chemistry principles. This situation may have further fueled skeptics of green chemistry, and alienated others away from the field altogether. Since green chemistry has been dominated by organic chemistry, particularly synthetic organic chemistry, bias toward incorporating any numerical analysis in this area has also fueled skepticism and resistance to implement metrics concepts in synthetic organic chemistry beyond reaction yield and number of steps. Also, a lack of standardization of metrics usage in analyzing synthesis plans and processes has hampered getting green chemistry getting the status it deserves as a bona fide rigorous discipline.

Another challenge is the misconception that environmental chemistry and green chemistry are synonymous. Table 2.1 shows that a number of the listed course titles blend the two. Environmental chemistry tracks the fate of pollutants in the environment, namely, the atmosphere, bodies of water, soil sediments, and wildlife. In comparison, green chemistry deals with the invention of new chemical processes that prevent the production of pollutants in the first place. Problems exist in the way organic chemistry is traditionally taught that impact on the teaching of green chemistry. A first example of this is the noninclusion of balanced chemical equations when students learn about organic reactions for the first time in an introductory course. Second, oxidation number changes are often not routinely highlighted in chemical transformations. This is a valuable tool to parse bond-forming steps as interactions between electrophilic centers of high oxidation state and nucleophilic centers of low oxidation state, and to elucidate reaction mechanisms more effectively. Third, synthesis strategy and the problem of synthesis as a multivariable optimization problem with constraints where the variables may not all be known is not emphasized. Finally, toxicology is usually not taught as part of chemistry education.

Given these challenges, there are two main options in incorporating green chemistry subject matter in the chemistry undergraduate curriculum that have been attempted. One is to "green" existing courses by threading green chemistry principles or laboratory exercises among the topics and examples already covered.[5,6] The other choice is to launch brand new green chemistry courses with some combination of lecture and laboratory component, thereby expanding the existing curriculum.[7] Here the pros and cons of each approach are itemized in point form.

Option 1: "Greening" existing courses

Pros:

1. Less time-consuming for instructors to append existing material with green addenda.
2. Savings in terms of no extra funds required to be spent on hiring new faculty to teach existing courses. Labs may or may not incur new costs.
3. Savings on administrative costs.
4. Savings to students with respect to textbooks since the same book can be used as in the traditional course and supplemented where necessary.
5. No new course fees for students.

Cons:

1. Reduction in the overall content of types of reactions covered in the traditional course since time is devoted to comparing old and green ways of carrying out transformations to the same target molecules.
2. There is an obvious trade-off between content and depth covered.
3. Green add-ons tend to be of the "show and tell" variety with little accompanying quantitative reasoning developed. Associated metrics are often downplayed or kept to a bare minimum without

exceeding the sophistication of elementary atom economy calculations. However, these analyses still require balancing chemical equations.

Option 2: Launching a brand new green chemistry course

Pros:

1. For this option to be most effective it should be coupled with industrial chemistry. Typically the course title is "Industrial and Green Chemistry." Examples from industrial chemistry are rich with opportunities to talk about any of the twelve principles.
2. Course titles with the name *green* in them are appealing to modern students and can be an effective strategy to reverse declining enrollments in chemistry programs.
3. Examples are often guided by current events (*Chemical & Engineering News*, etc.), developments in the recent scientific literature, and green chemistry awards and citations given by various organizations.
4. A dedicated course provides more time to fully develop ideas in green chemistry and metrics analysis, and to show students the justification for this new thinking through the context of historical developments in the chemical industry.
5. Advanced problems and case studies may be presented to students in a real-world setting. Again, more time is advantageous in this regard.
6. In-depth written assignments are possible. Longer and deep-level critical thinking problem sets are possible where decision making is required, where no single "right" answer fits the problem discussed.
7. Greater student engagement, understanding, and appreciation of what chemists do for a living.

Cons:

1. Often needs to be launched from scratch since few departments offer any course on industrial chemistry, and these are often listed as electives or optional courses in course calendars.
2. Time-consuming for instructors, often with a significant learning curve to overcome.
3. Administrative costs associated with launching new courses, including the hiring of new teaching faculty.
4. Introduction of further scheduling complexities and tuition fees from students' perspective.
5. An overcluttered curriculum usually already exists with limited room for expansion. This option goes against the current trend of constantly cutting and trimming budgets, facilities, time, and personnel while attempting to maintain the same level of educational quality.
6. Accreditation and certification issues with chemical societies such as the Canadian Society of Chemistry and the American Chemical Society.
7. Convincing tenured senior faculty in particular to spend time mounting such a course.

One key skill that needs to be covered regardless of the option chosen is balancing chemical equations, since no credible and sensible green analysis with respect to waste management can begin without this. Unfortunately, this practice is not routinely undertaken in traditional organic chemistry courses. Development of such a skill is a big plus in gaining a deeper understanding of reaction mechanisms, and an important spin-off benefit of green chemistry education. Overall, green chemistry improves the basic chemistry literacy of students in addition to educating them about new technological approaches to mitigate waste management problems. It is through the exercise of reaction and synthesis plan optimization that students see the reason for making an effort to amass a thorough knowledge of basic organic chemistry and chemical reactivity principles and to develop a comprehensive personal library of organic transformations that they can draw upon. In summary, future success in applying green chemistry thinking to chemistry curricula will depend on the following factors:

1. Engaging tenured research faculty to accept and see benefits of green chemistry for themselves in their own research and for their students.
2. Greening existing courses (patching) or launching new stand-alone green chemistry courses.
3. Retooling all chemistry courses a department offers, especially lab components, with this underlying theme.
4. Moving away from a "show and tell" approach to a rigorous synthesis planning/optimization approach.
5. Increasing student engagement in learning chemistry.
6. Integrating chemistry courses with other scientific disciplines, business, law, and ethics.
7. Maintaining a vigorous upkeep of library resources.

From my experience the best course of action is to first make one simple change to the traditional second-year organic chemistry course: balance chemical equations so students get a head start on what is to come in future green chemistry courses of whatever description. This, if nothing else, will go a very long way in vastly improving students' conceptual understanding of reactivity and reaction mechanisms. After having done this, a combined industrial and green chemistry course in the third or fourth year should be offered, ideally with an accompanying lab component featuring one lab for each of the twelve principles. Such a course should be made mandatory for all chemistry degree students regardless of their particular chemistry discipline. The next section outlines in detail a lecture-based course that attempts to satisfy some of the objectives described above—some were met with resounding success, and others were not.

2.3 COURSE DESCRIPTION AND STRUCTURE

At York University I have taught a third-year undergraduate level semester course entitled "Industrial and Green Chemistry" (CHEM 3070) for eight years. Students who complete the course get three credits; however, there is no laboratory component.

It is an elective course for chemistry majors that is not a degree requirement, but is mandatory for the biotechnology degree program. (It should be noted that the biotechnology program makes it necessary to include any science course that has the keywords *industrial* or *industry* in its title. It is expected that students graduating from such a program will eventually end up launching their own start-up biotechnology companies.) Originally the course was called "Industrial and Environmental Chemistry," but its name was changed partly to incorporate the new growing field of research in green chemistry, and partly to increase enrollment in the chemical sciences with a title that would be appealing to students' interests. The only prerequisite course is a second-year introductory organic chemistry course worth six credits for lecture and lab components. The course is delivered weekly in a single three-hour lecture session in the evening. Class sizes have varied in number, from as low as seven students to as high as forty. As a consequence of positive feedback circulating among students, enrollments have steadily increased in recent years. The course syllabus and topics covered are summarized below. The course evaluation mechanism is summarized in Table 2.2.

Part I: Preliminaries—The Business of Doing Chemistry
- Timelines of discoveries of fundamental ideas
- Timelines of industrial developments
- Industrial accidents
- Markush structures and patents
- Confidentiality agreements

TABLE 2.2
Course Evaluation Breakdown for Industrial and Green Chemistry

Item	Mark(%)
2nd year organic chemistry prerequisite quiz	2.5
Steacie Science Library quiz (partnered with library staff to showcase full spectrum of library resources)	2.5
Problem sets (four are given; each assignment has ten questions to be completed in 14 days)	40
Term test (one hour)	25
Written assignment (given in four stages):	30
Stage 1: pick topic and reference list for synthesis schemes (Week 4)	
Stage 2: balanced chemical equations for schemes (Week 8)	
Stage 3: full metrics calculations and critique (Week 12)	
Stage 4: turning in of final paper and optional oral presentation (if class size permits) (Week 13)	
TOTAL	100

Part II: Survey of Modern Concerns

Thalidomide, phthalate esters, insecticides and herbicides, pharmaceuticals in the environment, acrylamide formation in fried and oven-cooked foods, insect repellent (DEET), Pb emissions, MTBE (methyl *t*-butyl ether), biodiesel, energy use for chemical processes

Part III: "Green" Chemistry

The twelve principles

New and emerging technologies

Quantification of "greenness": reaction metrics

Concept of minimum atom economy and maximum waste: optimization of parameters

Synthesis plan analysis and reaction networks

Part IV: Dyestuffs

Mauveine, indigo, anthraquinone dyes: alizarin, azo dyes, phenolphthaleine dyes, triphenylmethane dyes, phthalocyanines

Concept of color

Part V: Pharmaceuticals

Narcotics, therapeutics

Examples: Aspirin®, Ibuprofen (Advil®), Viagra®, Presidential Green Chemistry Challenge Awards, examples taken from *Organic Process Research and Development* journal

Part VI: Industrial Feedstocks

Industrial chemical trees

Group 1: Ethylene, ethylene oxide, propene, acrylonitrile, 1,3-butadiene, vinyl chloride

Group 2: Cyclohexane, nylon

Group 3: Benzene, toluene, ethylbenzene, styrene, cumene, phenol, aniline

Group 4: Acetone, ammonia, urea, ethyl acetate, acetonitrile, diethyl ether, tetrahydrofuran, ethanol, chloroform, carbon tetrachloride

The written assignment is a twenty-page paper in which students are required to evaluate and critique at least two synthesis plans to a chosen target molecule from a list of pharmaceuticals, cosmetics, food additives, agrichemicals, dyestuffs, natural products, fragrances, and other industrial commodity chemicals using green chemistry principles. Figure 2.1 shows the distribution of student interests in their selection of compound. To guarantee independent work, each student chooses a different molecule. The written assignment is evaluated according to the following criteria under content and presentation categories. Under content the criteria are (1) discussion with instructor on plans for chosen topic, including preliminary reference list, (2) satisfies chemistry audience, (3) satisfies nonchemistry audience, (4) extent of literature searches and documentation of references, (5) evidence of rough notes, and (6) thoroughness and relevance of analysis of chosen topic according to all

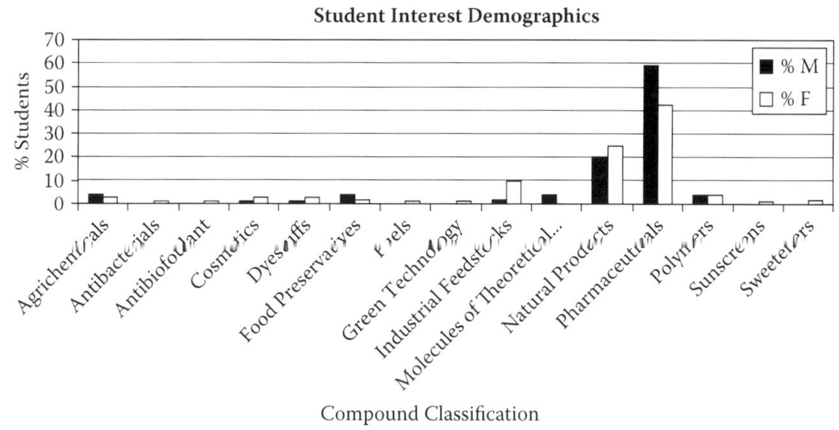

FIGURE 2.1 Distribution of student interests according to compound classification and gender.

parameters presented in class. Under presentation the criteria are (1) overall quality of presentation of document, (2) clarity and purposeful use of graphics, (3) clarity of writing, (4) brevity of writing, (5) ability to inform readers of ideas and arguments, and (6) ability to convince readers of ideas and arguments. Each category is worth 10 points. If students make an oral presentation they are given 10 bonus points. It should be noted that there were never any incidences of plagiarism or any other academic breaches encountered with the written assignment. The main reason for this is the very nature of the exercise, which was to evaluate in detail at least two synthesis plans from the literature for a given target molecule according to green metrics. First, no literature reports balance chemical equations, even to the present day. Second, the analysis of synthesis plans in a quantitative sense is completely new. This made it impossible for students to copy prior work because such analyses were never done before. Hence, students doing the assignment were the first in the world to carry out these tasks on their chosen molecule, and so their work was highly original. Some of the students produced excellent papers that were of publishable quality. Indeed, one of them was eventually published in a journal read by industrial chemists.[8] Overall, students, for the first time in their undergraduate educational experience, took pride and ownership of their work. A spin-off, yet disturbing, discovery that students often make when accessing and reading original journal articles and patents is the degree of missing information in reporting of experimental procedures. Students quickly make the connection that a fair assessment of greenness by any kind of metrics analysis necessitates the full disclosure of all masses of materials used in a chemical process, in addition to the mass of the final target product. Such gaps obviously weaken arguments that favor one plan or process over another. More broadly, this situation jeopardizes the legitimization and implementation of green chemistry principles if proper assessments cannot be made because of limited availability of essential data.

Within the first week of the course, students are required to attend a one-hour workshop given by science librarians on the use of library resources available to them. These include SciFinder®, World Patent Database EPO (European Patent Office), Merck Index, online journal references, chemical industry Web sites, chemical catalogs for commodity prices, *Chemical & Engineering News* (American Chemical Society magazine), and various dictionaries and encyclopedias (Ullmann's, Kirk-Othmer, Kleemann & Engel). Immediately following the workshop a quiz of ten questions is given so that students can demonstrate their newfound knowledge in searching various items using the resources they just learned about. These literature searching skills are essential for students in carrying out their independent written assignment successfully. Sample questions used in the course problem sets are given in Section 2.6.4. Topics for questions were selected from the literature, often in phase with current events, such as chemistry-related news items appearing in *Chemical & Engineering News* or follow-ups on subjects covered by a colloquium speaker who recently delivered a lecture to the department.

Following the philosophy that teaching and research inform each other and are therefore synergistically linked, the teaching of this course completely changed my own research direction from traditional mechanistic organic chemistry to synthesis optimization strategy analysis. As a result of this shift, original research work was published[9–15] that was a direct consequence of teaching green chemistry. The success of the course led to new research directions in the emerging fields of reaction metrics and synthesis plan analysis. Over a short period of time, the course became known outside of the York University community and a number of worldwide recognitions resulted. One of the most notable was an invitation by Paul Anastas to participate in the Green Chemistry Education Roundtable held at the United States National Academy of Sciences in Washington, D.C. in 2005,[16] which was the only contribution from Canada. Another milestone was an invitation by Scientific Update LLC to cohost a compressed three-day version of the course for industrial process chemists in 2007. Figure 2.2 summarizes the chain of events that led to opportunities that continue to the present day, including book chapter invitations (such as the present one), consulting opportunities with pharmaceutical companies, and conference invitations. Close inspection of this flowchart shows that the birth of the first idea based on unifying reaction metrics for reaction and synthesis plan analysis was originally met with resistance and rejection by academics. It was the industrial chemistry community, notably the late editor of *Organic Process Research and Development*, Dr. Chris R. Schmid of Eli Lilly, who first realized the potential and value of the concept in furthering understanding of synthesis optimization and green chemistry. I use this personal story to inform and motivate students at the outset of the course as I remind them that the most important characteristic of a good scientist is perseverance, not superior intelligence. Moreover, the most difficult task is not to come up with a good idea but to convince others that it is indeed a good idea.

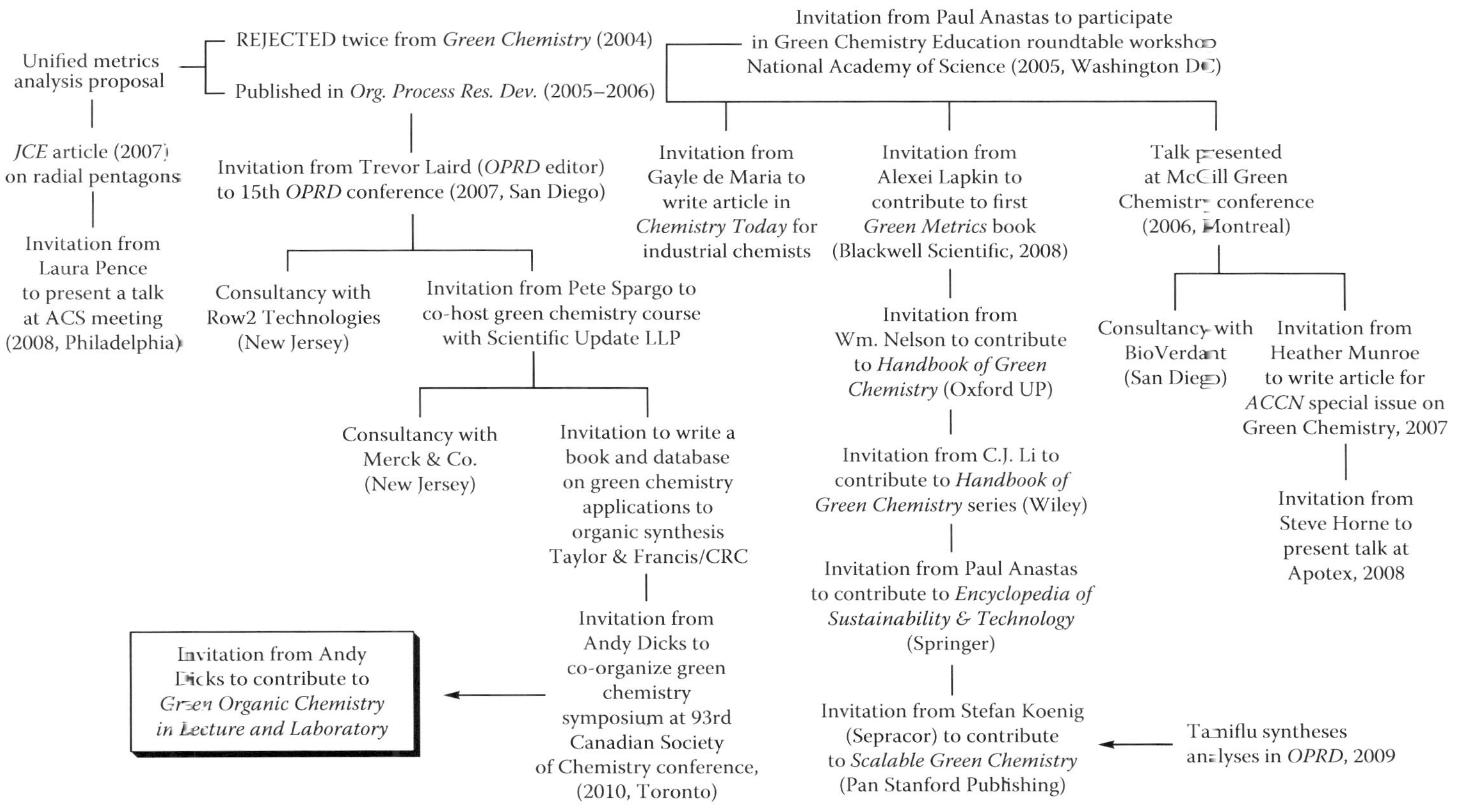

FIGURE 2.2 Flowchart indicating connections and successes spawned from teaching "Industrial and Green Chemistry" at York University.

2.4 FEEDBACK

In this section, feedback about the course from students and library staff is presented.

2.4.1 STUDENTS

Positive feedback

1. Acquisition of enhanced literature searching skills.
2. Historical anecdotes and context (explaining who, how, and why) of the subject made students more aware and appreciative of how we got to where we are today in the world of chemistry.
3. Students were turned on by the interdisciplinary nature of chemistry and for the first time in their education gained a big-picture view of their subject.
4. Students liked the real-life examples and current science events and news that were woven into the course.
5. Problem sets were judged to be a "great learning experience."
6. Students had a better appreciation of what chemistry is and what chemists do.
7. The course was a great motivation for self-learning (independence, uniqueness, ownership).
8. The written assignment was considered the "jewel in the crown" because it was a simulation of how real research is done.

Challenges

1. Students found it intimidating to question the literature. Critical thinking was viewed as "scary."
2. Students were still used to the classic model of having one textbook for a given course that was closely followed, where the "answers" would be found somewhere within.
3. Problem sets were considered difficult and time-consuming with respect to other assignments from other courses. Students were not used to multiple-solution or open-ended questions.
4. With respect to prerequisites, students discovered that no prior course prepared them for this one. In particular, they felt disadvantaged by their prior introductory organic chemistry course experience.
5. Students regretted that they were exposed to literature searching tools too late in their undergraduate education, as evidenced by their oft-mentioned comment: "If only I had known this when I was taking…."

2.4.2 LIBRARY STAFF

Positive feedback

1. Library workshop and quiz were considered a hit with staff and students.
2. Students greatly appreciated and benefited from knowing more about library resources available to them that they were not aware of previously.

3. Staff benefited from showcasing and promoting what resources they had, particularly underutilized material.
4. Increased usage of resources led to more computer database acquisitions and funding to purchase new resources (e.g., SciFinder site licensing to more computer terminals).

Challenges

1. Space and time restrictions to meet the demands of an increased enrollment in the green chemistry course.
2. Slow connections, nonstandard display of online journal holdings, licensing restrictions.

2.5 ADVICE ON LAUNCHING A GREEN CHEMISTRY COURSE AND EPILOGUE

Education is part of the "disruptive technology" that is required for substantive advances to be made in a field. The concepts in green chemistry are therefore disruptive to the traditional topics covered in organic chemistry. First and foremost, green chemistry is about decision making according to some specified criteria. This task can only be done properly by applying some form of metrics analysis to a given reaction or synthesis plan. The inevitable consequence of metrics analysis is ranking, and ranking in turn may not be so clear-cut in picking a "winning reaction" or plan. With this in mind, green chemistry decision making is definitely not done by a simple show-and-tell approach. When considering launching a green chemistry course, it should not be a token single course appearing in the curriculum as an elective catering to a general audience of students. Introducing a green chemistry course that stands independently and disconnected from other courses defeats its purpose. It is imperative that the feeder prerequisite courses be retooled as well. A key advantage in adopting a green chemistry course in a department is if the same instructor teaches both that course and its feeder prerequisites to ensure compatibility. The subject is highly sophisticated and demands a firm grasp of fundamentals. Students cannot hope to engage in decision making if they are still hesitant on identifying nucleophilic/electrophilic or acidic/basic sites in any given chemical structure, for example. Green chemistry will challenge the competency of a faculty member's basic knowledge of chemistry, particularly in organic chemistry. Depending on a department's culture and personalities, this could be a significant barrier to overcome in successfully introducing a green chemistry course. Instructors can go a very long way by teaching students how to balance chemical equations in introductory organic chemistry courses, as is customary in first-year general chemistry courses. This alone will be sufficiently adequate in preparing students for any courses in green chemistry, industrial chemistry, or other advanced courses in organic synthesis.

Green chemistry will definitely challenge students' comprehension of fundamental chemical concepts and will push their comfort zone toward unfamiliar territories where the "right answer" is not so clear-cut and not found in textbooks. This is a significant challenge for students with respect to getting appropriate credit in the form of marks. In order that the whole experience be fruitful for instructors and students, it is imperative that library resources be updated with the latest and best resources. This means adequate journal subscriptions and access to literature search engines and database compilations via print or online access. Since green chemistry is by its very nature interdisciplinary, instructors will quickly discover that many topics will be found in both the chemistry and chemical engineering literature. Often, the same topic will be called by different names in these two disciplines. Successful instructors will need to keep up with current literature, including reading news magazines such as *Chemical & Engineering News*, and broadening their readings to include literature from other fields. One needs to be aware that chemists and chemical engineers have traditionally been trained to view each other as adversaries rather than as having complementary skills in solving chemical problems.[17] If green chemistry is to have any widespread acceptance across disciplines, this way of thinking must change.

Before contemplating launching a green chemistry course, one first needs to investigate *how* faculty teach organic chemistry courses in their department. Characteristics that would both make green chemistry least and most likely to be accepted need to be identified. Key questions that should be asked are

1. Do instructors follow the course text *ad verbatim* without supplanting their own interpretation of a given concept?
2. Do instructors consistently revise their own knowledge of the subject as new discoveries are made?
3. Do instructors keep their problem set and test questions fresh from year to year?
4. Do instructors introduce real-world case scenarios to substantiate and illustrate the concepts introduced?
5. Do instructors pay careful attention to student questions and use them as springboards for new insights and possible new lines of inquiry for research?
6. Do instructors avoid topics because they themselves do not understand them well enough to teach them?

In considering a green chemistry course, a key individual to get on board is the undergraduate program director, since this person is charged with the duty of maintaining the highest level of curriculum quality for all courses offered by a department. Ideally, this person must be competent in his or her field of chemistry, must enjoy teaching and interacting with students, and have no personal insecurities that could hamper his or her job as a conveyor of knowledge. Along with the departmental chair, this person should be a visionary who sets goals and has the resources at his or her disposal to achieve them. Specific items in the introductory organic chemistry course that need revising or amending before embarking on launching a green chemistry course include

1. Balancing chemical equations
2. Making the complete classification of organic reaction types known to students from the outset: substitutions, rearrangements, redox reactions,

additions (including couplings, multicomponent reactions (MCRs), cyclizations, condensations), and eliminations
3. How to identify nucleophilic/electrophilic sites in any molecule
4. How to identify Brønsted type and Lewis type acidic/basic sites in any kind of chemical structure
5. How to parse any kind of chemical structure into its possible synthon constituents
6. How to draw chemical structures with the most easily understood visual representation and maintaining the same aspect for all structures in writing out a complete synthesis plan
7. For any kind of reaction, how to find the starting materials in the target product structure (this is greatly facilitated if the aspects of chemical structures in reactants and products are invariant)
8. Making strong links between balancing an equation, proposing a reaction mechanism, and identifying reactant structural fragments in the target product structure for every new reaction introduced

Having said all this, my experience in launching a combined industrial and green chemistry course at York University over eight years has resulted in a win-win-win situation for students, library staff, and myself as an instructor and research scientist. However, the course has met with resistance from fellow faculty members in the chemistry department. First, from an administrative point of view, the course from the outset was not recognized as part of the core set of courses for the bachelor honors degree in chemistry. Therefore, it was considered an elective and supposed to be a general course catering to a general student chemistry audience. The department currently does not see green chemistry as a worthy subject in chemistry in both the research and teaching sense.

Additionally, although students clearly saw the superb benefits of the course in their overall education—particularly with respect to facilitating their scholastic performances in other organic chemistry courses they were taking—they were not motivated enough in filling out course evaluations as would be expected. This is mainly because science students, unlike their humanities counterparts, are often far less engaged in their educational experience. It may be that science students have significantly poorer writing skills to begin with and are therefore less likely to make the effort to put their thoughts down in coherent sentences, and so their true feelings, opinions, and voices are never known fully. Since there was little in the way of written documentation of student support, the department concluded that indeed the course was perceived to have little student benefit, and therefore reinforced their belief that it should remain as an "easy elective" type course outside of the core curriculum. Most students who realized the impact of the course on their chemistry education preferred to communicate their opinions verbally in private conversations with me, usually well after the course ended. Their latent response was not surprising given the fact that students require an incubation period to pass after the stresses of taking the course subside. This allows them time to contemplate new ideas learned and their potential consequences. Many were turned on to chemistry for the first time and asked me for recommendation letters in their quest to seek entrance in chemistry graduate programs in other universities. Almost all have succeeded and continue to apprise me of their progress.

There are other factors that need mentioning in this regard. The demography of the student body as a whole at our suburban campus is largely a first-generation immigrant population whose first language is neither English nor French. Many come from cultural backgrounds that make taking ownership of their education an unsettling experience. In addition, the science student population represents only 10% of the entire student body. In an effort to maintain sufficient enrollments and therefore to secure continued government-sponsored funding, the bar for entrance level high school marks is held significantly lower for science students (70%) than for others entering business or fine arts programs (90%), for which the university has a well-known positive reputation internationally.

A combination of all of the above factors resulted in the course being transformed, as of 2009, to a general descriptive one (under the same course title of Industrial and Green Chemistry appearing in Table 2.1), that discusses basic petrochemicals largely dealing with first-generation feedstocks from coal and crude oil. The only topics covered relevant to the chemical industry are scale-up issues and batch reactors. None of the twelve principles are mentioned. There are no problem sets or decision-making exercises of any kind. Students are evaluated on a standard protocol of two term one-hour tests, a three-hour final exam, and two descriptive term papers where they are asked to do nominal research on a chosen compound, in effect writing mini-descriptive reviews from existing published material. Since the likelihood of plagiarism is high, students are mandated to use the program Turn-It-In® to prove their academic integrity before submitting their assignments for grading. The tests are based on regurgitating memorized facts directly from course notes.

2.6 INSTRUCTIVE LECTURE CASE STUDIES

The following series of examples were among the best that were used in the course as part of the formal lecture material to showcase green chemistry principles in organic synthesis, where the relevant course sections are shown in parentheses: diazepine syntheses (application of green technologies), pravadoline (pharmaceuticals), and ibuprofen (industrial syntheses). Each synthesis plan is taken from the primary chemical literature (from either journal articles or patents). Since evaluation of green metrics plays a central role in the decision-making process for determining the green attributes of a synthesis plan, each highlighted example illustrates the ubiquitous problem that, for most of the time, no one plan has all the "right" fully optimized characteristics in every category. By being exposed to such well-documented real-world scenarios, students begin to appreciate the complexities and difficulties involved in reaction and synthesis optimization. They are able to directly link the endeavors of optimization and greening as one and the same. They also discover that because chemistry is an experimental science governed by compromise, this task of optimization inevitably involves prioritization of variables. In addition, they appreciate green metrics as a powerful tool to verify claims of greenness made in the literature and to proofread experimental procedures.

2.6.1 Application of Green Technologies: 2,2,4-Trimethyl-2,3-Dihydro-1H-Benzo[b][1,4]Diazepine

This compound is synthesized from one equivalent of *o*-phenylenediamine and two equivalents of acetone (Scheme 2.1). Since this reaction is a condensation, it produces water as the only by-product, and so its intrinsic atom economy is quite high (84%). Table 2.3 summarizes various green chemistry technologies that have been used to make this compound along with their E-factor breakdowns[18–40] (see Section 1.4 for an introduction to E-factor calculations). Entries are listed in order of increasing E-total, which can be subdivided into contributions from by-products and unreacted starting materials (E-kernel), excess reagent consumption (E-excess), and auxiliary material consumption, such as reaction solvent, workup extraction solvents, and solvents used in purification procedures (E-aux). The first entry indicates that optimization has been achieved on all counts, since it exhibits the lowest values for E-kernel, E-excess, and E-aux (the ideal value for each being zero). The most frequent green method used is the elimination of reaction solvent. This is a good example to illustrate various green technologies, and since the reaction is so simple, it can be easily implemented in the undergraduate laboratory. Figure 2.3 shows the best and worst procedures in the form of radial pentagons, which visually display a reaction performance with respect to material efficiency.[12] The five variables depicted (each expressed as a fraction between 0 and 1) are intrinsic atom economy (AE), reaction yield, excess reagent consumption (1/SF, inverse of stoichiometric factor), auxiliary material consumption due to solvents, etc. (material recovery parameter (MRP)), and reaction mass efficiency (RME). The RME is the numerical product of the four preceding factors, and hence gives a global determination of material efficiency for a chemical reaction. Clearly, in this example, it is use of reaction solvent and auxiliary materials in the workup and purification procedures that contribute the bulk of the waste, since the atom economy is high, reaction yields are high, and little excess reagents are used. All plans were advertised as green and solvent-free in the literature, but none of the authors did any kind of metrics analysis to prove the veracity of their claims. The data in the table show a wide breadth of reaction performances under the name *solvent-free*. One point that needs emphasis is that the term *solvent-free* refers to the situation that no *reaction* solvent is used—it does not imply that no solvents are used throughout the entire reaction procedure, which includes workup

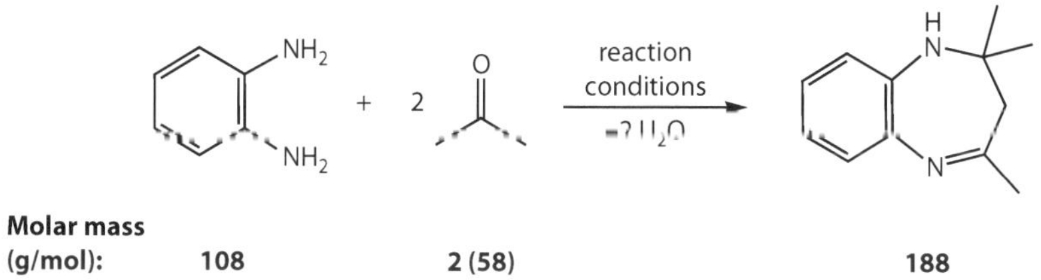

SCHEME 2.1 Reaction of *o*-phenylenediamine with acetone to form a benzodiazepine.

TABLE 2.3
Summary of E-Factor Breakdowns for Various Syntheses of 2,2,4-Trimethyl-2,3-Dihydro-1H-Benzo[b][1,4]Diazepine

E-Kernel	E-Excess	E-Aux	E-Total	Green Technology Used (reference)
0.22	0.032	0	0.26	Solvent-free, microwaves (18)
0.28	0.066	15.34	15.68	Solid-supported catalyst, microwaves (19)
0.27	0	15.9	16.16	Solvent-free (20)
0.21	0.032	22.68	22.92	Zeolite catalyst (21)
0.32	0.17	25.03	25.52	Solvent-free (22)
8.4	9.08	17.26	34.74	None—benzene solvent used (23)
0.4	0.073	41.75	42.22	Ultrasound (24)
0.28	0.034	65.89	66.2	No catalyst, [bbim]BF_4 solvent (25)
0.2	0.031	71.55	71.78	Solvent-free (26)
0.21	0.16	72.25	72.62	Solid-supported catalyst, microwaves (27)
0.21	0.16	74.97	75.34	Solid-supported catalyst, microwaves (28)
0.25	0.065	78.09	78.41	Solvent-free (29)
0.23	0.098	91.56	91.89	Solvent-free, ball milling (30)
0.28	0.083	104.92	105.29	Water solvent, nanoparticle catalyst (31)
0.29	0.17	109.71	110.17	Solvent-free (32)
0.23	0.16	120.94	121.33	Solid-supported catalyst (33)
0.25	0.16	136.48	136.89	Solid-supported catalyst, [bmim]BF_4 solvent (34)
0.28	0.033	155.54	155.86	Solid-supported catalyst, microwaves (35)
0.4	0	169.13	169.53	Solvent-free, microwaves (36)
0.23	0	307.26	307.49	Water solvent (37)
0.32	0.068	532.28	532.66	Water solvent (38)
0.27	0.033	570.93	571.23	Solvent-free (39)
0.28	0.17	678.64	679.09	Solvent-free (40)

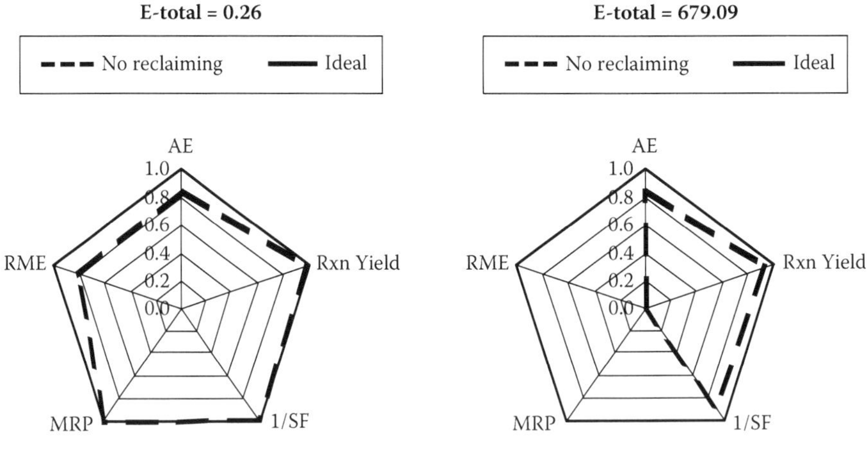

FIGURE 2.3 Radial pentagons for the best-performing[18] and worst-performing[40] syntheses of 2,2,4-trimethyl-2,3-dihydro-1H-benzo[b][1,4]diazepine.

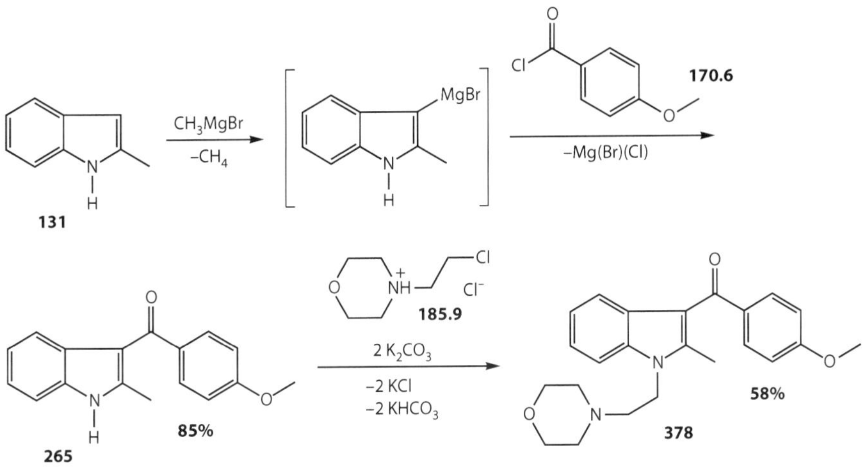

SCHEME 2.2 Sterling G1 synthetic route to pravadoline.

and purification, in addition to carrying out the reaction. This point is expanded upon in Section 3.2.

2.6.2 Pharmaceutical Synthesis: Pravadoline

Pravadoline is a nonsteroidal anti-inflammatory drug (NSAID) whose synthesis was the first pharmaceutical to be "greened up" using an ionic liquid as a reaction solvent.[41–44] Schemes 2.2–2.5 illustrate various synthetic routes, including the original medicinal chemistry approaches by the Sterling Drug Company, and another route by Cacchi.[43] In these syntheses, the molecular weights of reactants and products are included next to each structure. Table 2.4 summarizes the respective

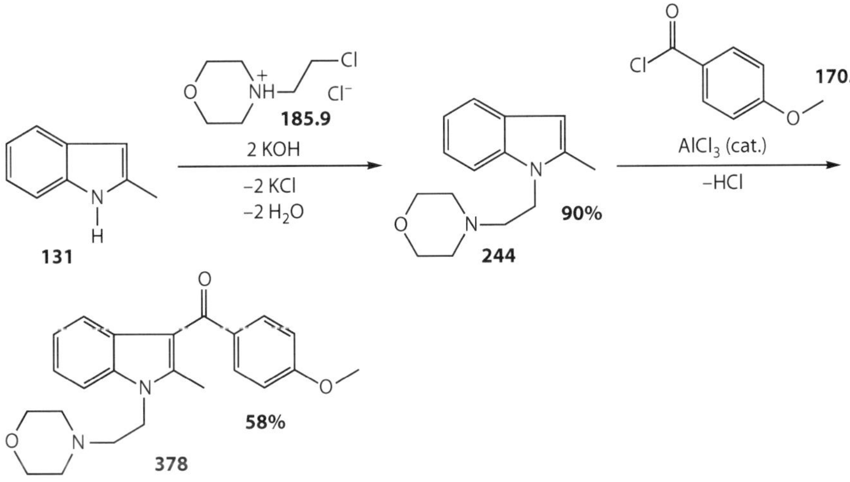

SCHEME 2.3 Sterling G2 synthetic route to pravadoline.

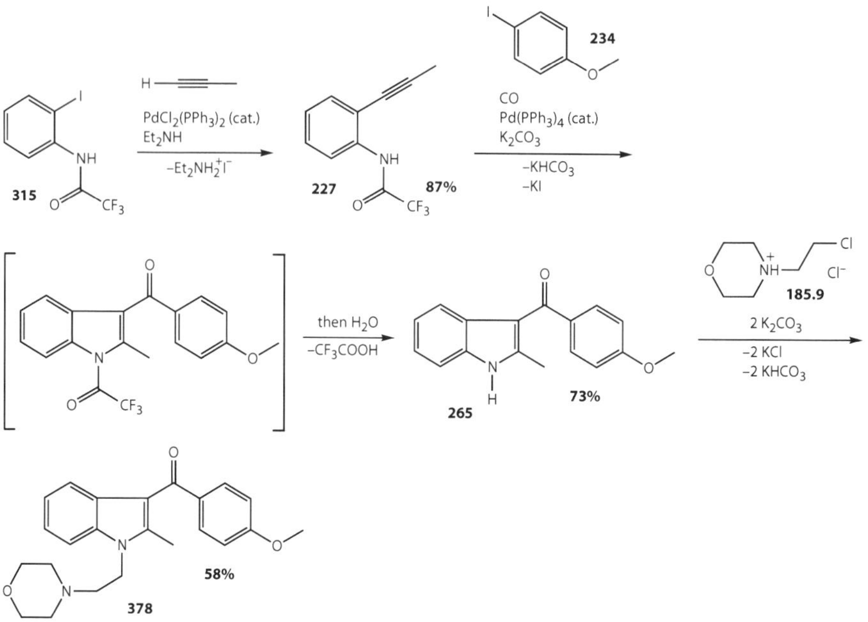

SCHEME 2.4 Cacchi synthetic route to pravadoline.

E-factor breakdowns, including that for synthesis of the ionic liquid [bmim]BF_4 (1-*n*-butyl-3-methylimidazolium tetrafluoroborate).[45] It appears from this compilation that the Sterling G1 and Cacchi plans have the best and least material efficiency performances, respectively. The target E-factor that future plans should aim for may be determined by taking the sum of the minimum values of E-kernel, E-excess, and

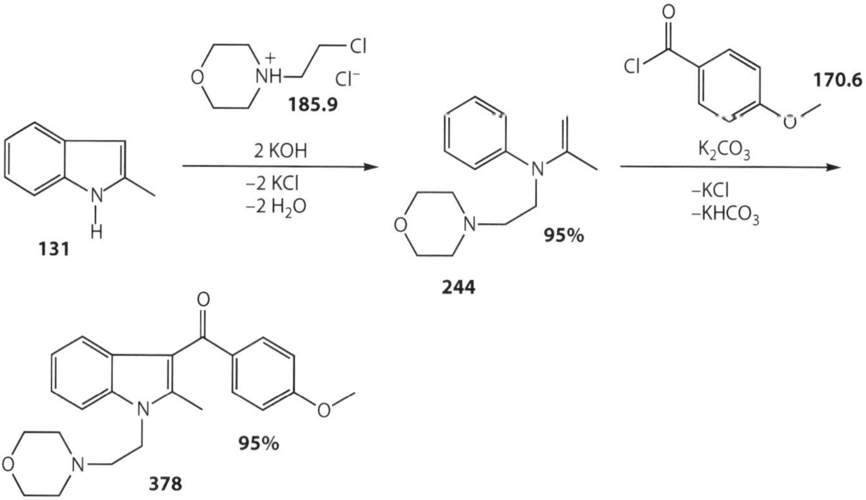

SCHEME 2.5 Seddon synthetic route to pravadoline, using ionic liquid solvent.

TABLE 2.4
Summary of E-Factor Breakdowns for Various Syntheses of Pravadoline

Plan	Year	Type	E-Kernel	E-Excess	E-Aux	E-Total	Total Mass of Waste (kg/Mole Pravadoline)
Sterling G1	1991	Linear	3.36	1.96	58.89	64.21	24.3
Sterling G2	1990	Linear	1.95	0.63	196.5	199.05	75.2
Cacchi	1994	Linear	6.79	30.4	415.3	452.53	171.1
Seddon[a]	2000	Linear	1.12	?	?	?	?
[bmim]BF_4	2002	Linear	0.56	0.15	6.88	7.58	1.7

[a] Analysis does not include synthesis of [bmim]BF_4 ionic liquid reaction solvent.

E-aux: 1.12 + 0.63 + 58.89 = 60.64, or 22.9 kg waste per mole pravadoline synthesized. The Seddon plan (Scheme 2.5) is an excellent case study for students to explore various scenarios for determining green metrics. From the data presented in Table 2.4, this plan clearly has the lowest E-kernel contribution because of its combined high atom economy (51%) and overall reaction yield (90%). It can be analyzed in two ways, depending on whether or not the synthesis of the ionic liquid solvent was included (Scheme 2.6). If the solvent is recycled and not committed to waste, as often advertised for ionic liquids used in organic synthesis, then its contribution to E-aux would be eliminated. Because of the limited data available, one can only estimate that its minimum E-total would be 1.12 + 7.58 = 8.70 if the ionic liquid is included and 1.12 + 0.56 + 0.15 = 1.83 if it is not. Both of these crude estimates are considerably lower than the best plan, having E-total = 64.2, for which all of the experimental data are known. All of this suggests that the Seddon plan may have a credible claim to being the most material efficient, and hence the greenest strategy to date.

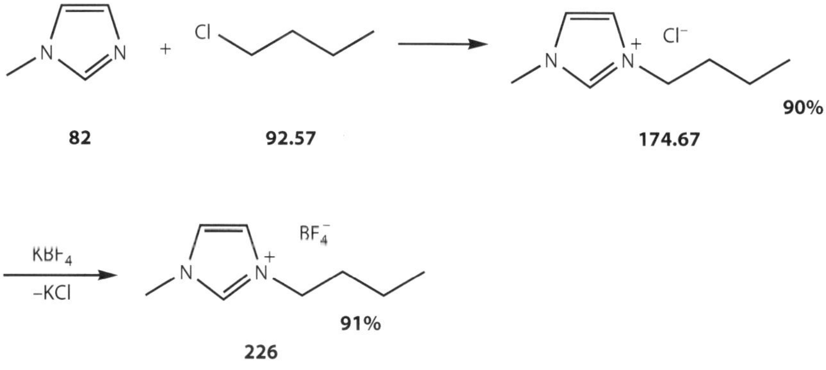

SCHEME 2.6 Synthesis of 1-*n*-butyl-3-methylimidazolium tetrafluoroborate.

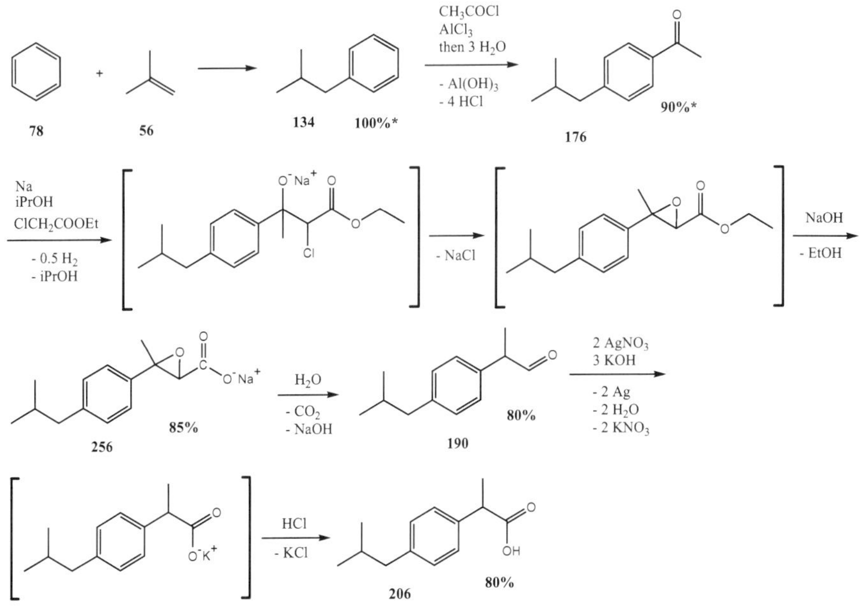

SCHEME 2.7 Boots synthetic route to ibuprofen.

2.6.3 Synthesis Evaluation of Industrial Compounds: Ibuprofen

Ibuprofen is the first pharmaceutical compound that was recognized by the U.S. Environmental Protection Agency Presidential Green Chemistry Challenge Awards in 1997.[46–57] The Boots-Hoechst-Celanese (BHC) synthesis has been used as a model of green chemistry achievement over the original Boots Drug Company plan. Since it has a special place in the evolution of green syntheses, it has been widely used in the teaching of green chemistry principles by many educators.[57] However, its analysis based on atom economy and reaction yield remains at the elementary level with respect to metrics in most textbook accounts. This compound has been a favorite target of academic and industrial researchers, as shown by the variety of strategies depicted in Schemes 2.7–2.16. For each scheme molecular weights of intermediates are given below the structures, and any reaction yields appearing with an asterisk indicate that they were not given by the authors and are taken as assumed values.

One noticeable feature about these synthesis plans is that they all begin from the same set of starting materials: benzene and isobutylene. This is a very rare situation where the starting and end points are common to all plans. Therefore, this imparts a high degree of fairness when such plans are compared with respect to waste production. Table 2.5 summarizes the E-factor breakdowns and ranks the plans in order of increasing E-total. Further improvements are still possible since the target E-total for a future plan is 1.72 + 1.03 + 30.47 = 33.22, or 6.8 kg waste per mole ibuprofen. One can see that though the Hoechst-Celanese plan has the lowest E-kernel contribution, as evidenced by its high atom economy, it does not have the lowest overall E-total. The Upjohn and DD113889 plans still win out

TABLE 2.5

Summary of E-Factor Breakdowns for Syntheses of Ibuprofen

Plan	Year	Type	E-Kernel	E-Excess	E-Aux	E-Total	Total Mass of Waste (kg/Mole Ibuprofen)
Upjohn	1977	Linear	4.78	1.03	45.9	51.7	10.7
DD113889	1975	Linear	9.17	15.55	30.47	55.10	11.1
Hoechst-Celanese	1988	Linear	1.72	1.17	111.9	114.79	23.6
Boots	1968	Linear	8.59	15.42	96.52	120.53	24.8
Ruchardt	1991	Linear	12.62	52.06	79.98	144.66	29.8
duPont	1985	Linear	4.67	61.83	150.1	216.61	44.6
Pinhey	1984	Convergent	7.8	152.2	540.3	700.33	144.3
RajanBabu	2009	Linear	3.91	21.81	1,888	1,913.8	394.2
Furstoss	1999	Linear	17.8	12.13	2,286	2,316.2	477.1
McQuade	2009	Linear	7.73	346.6	4,1968	42,322	8,718

because their E-aux contributions are smaller than those of the Hoechst-Celanese plan. Closer examination of the radial pentagons for the Hoechst-Celanese plan shows that this observation is due to the significant amount of hydrogen fluoride (HF) used in the Friedel-Crafts step, a result that was overlooked in prior published analyses of this plan. Another point to bear in mind is that the Furstoss and

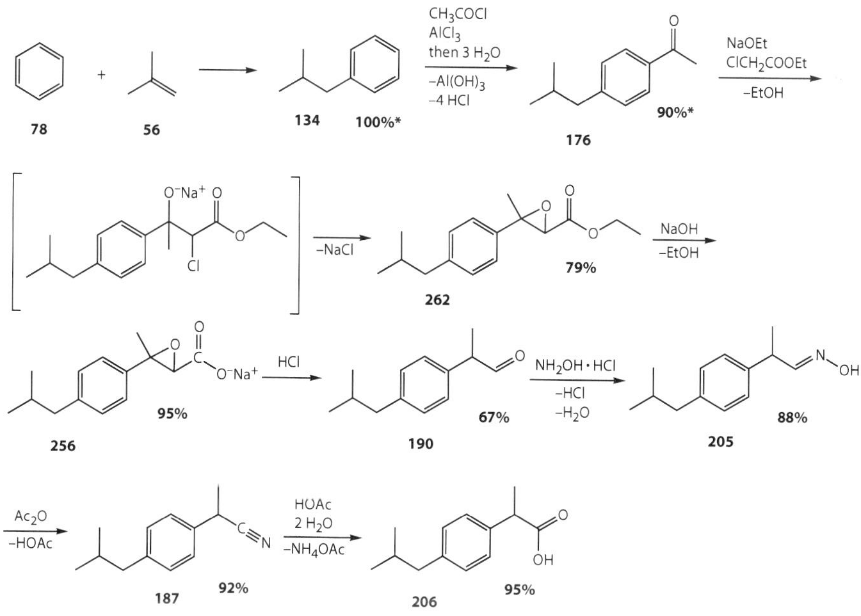

SCHEME 2.8 DD113889 synthetic route to ibuprofen.

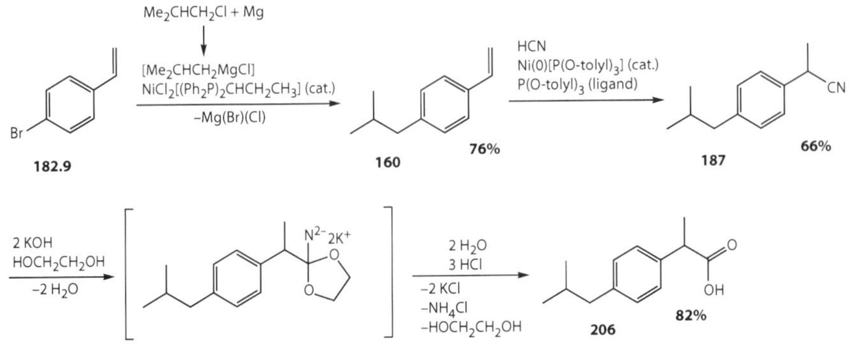

SCHEME 2.9 DuPont synthetic route to ibuprofen.

RajanBabu plans are for the (*S*)-enantiomer, whereas all others are for the racemic product. The McQuade plan utilizes flow chemistry using microchannel technology, which has several advantages over carrying out reactions in the conventional way in reaction vessels. These include safety considerations, full automation, lab space savings, built-in purification, elimination of process optimization requirements, catalysis operation under favorable kinetic conditions, on-the-fly changes in reaction conditions being possible, real-time monitoring of reaction progress by spectroscopic means, and compounds being compartmentalized until they are brought together to react. Though this has been touted as an emerging green technology in fine chemical synthesis, it does require a significant amount of reaction solvent to prevent clogging of the lines. Indeed, the E-factor entry in the E-aux

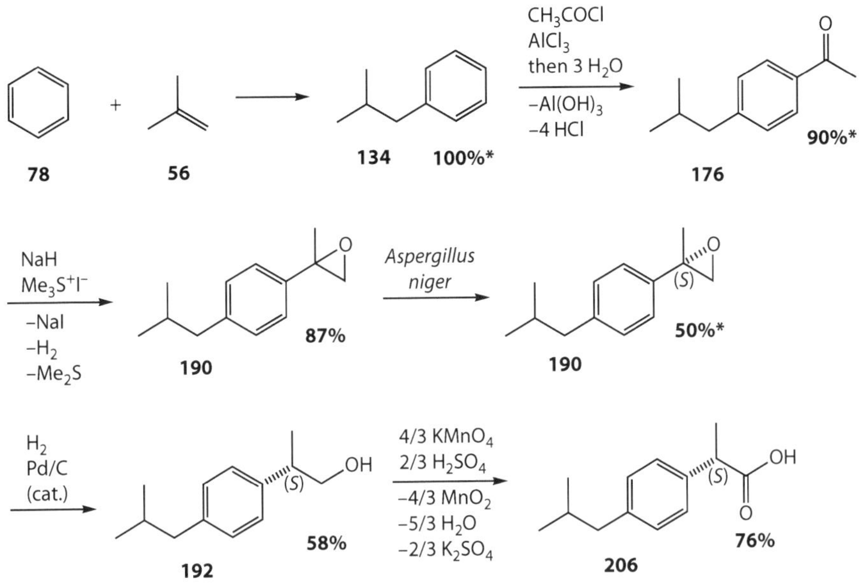

SCHEME 2.10 Furstoss synthetic route to ibuprofen.

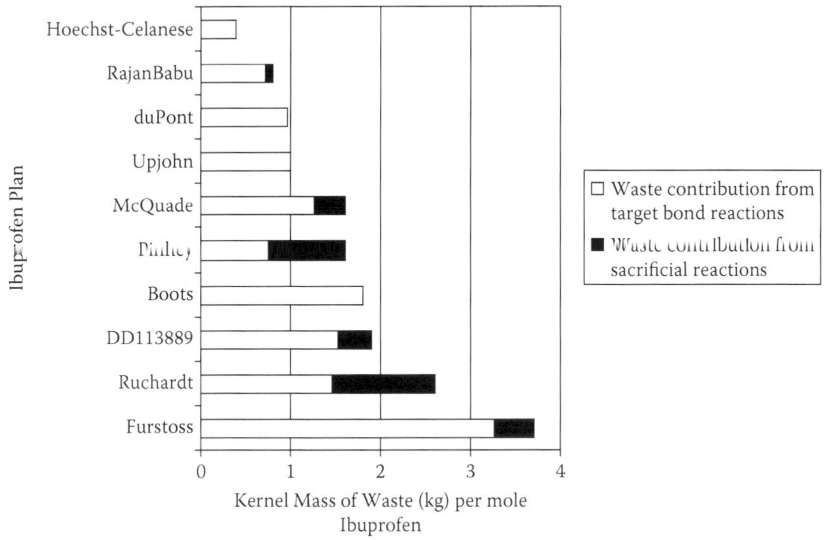

FIGURE 2.4 Bar graph showing ranking by overall kernel waste production and waste contributions from target bond and sacrificial reactions for synthesis plans of ibuprofen.

category for the McQuade process is over three orders of magnitude higher than the least E-aux value. All of these observations make for rich discussions in the classroom on the compromises and limitations of various green chemistry technologies. Figure 2.4 shows the ranking of plans according to kernel waste performance as well as target bond and sacrificial reactions. These data show that the Pinhey and Ruchardt plans have a significant proportion of sacrificial reactions that produce waste.

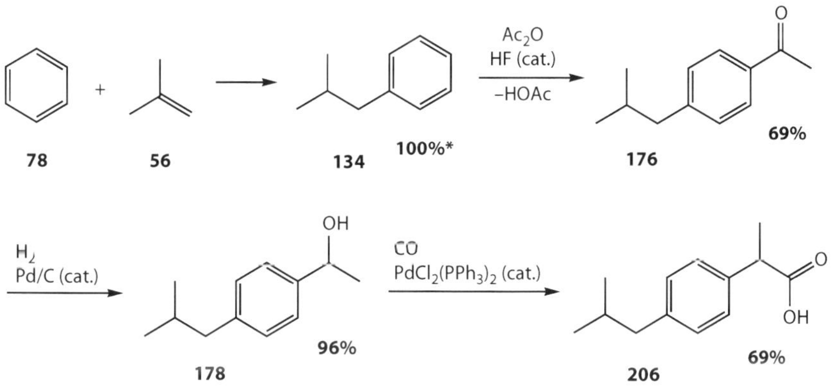

SCHEME 2.11 Hoechst-Celanese synthetic route to ibuprofen.

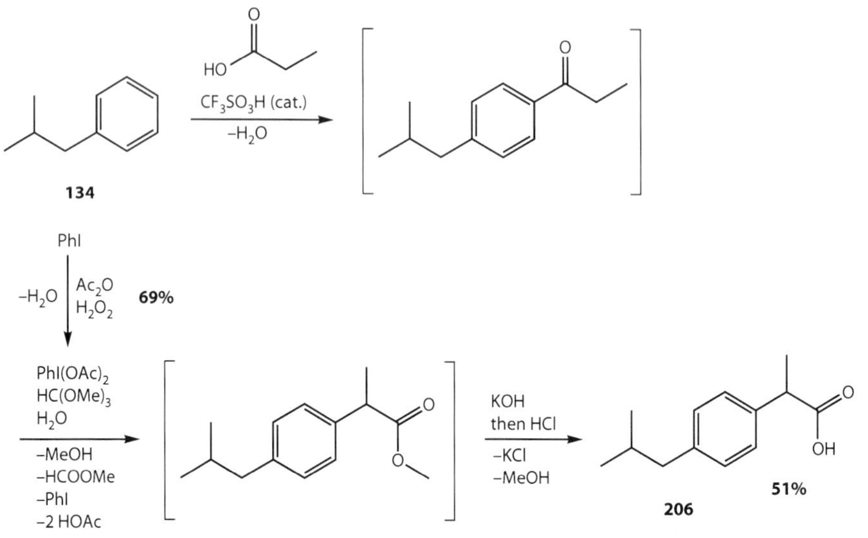

SCHEME 2.12 McQuade synthetic route to ibuprofen.

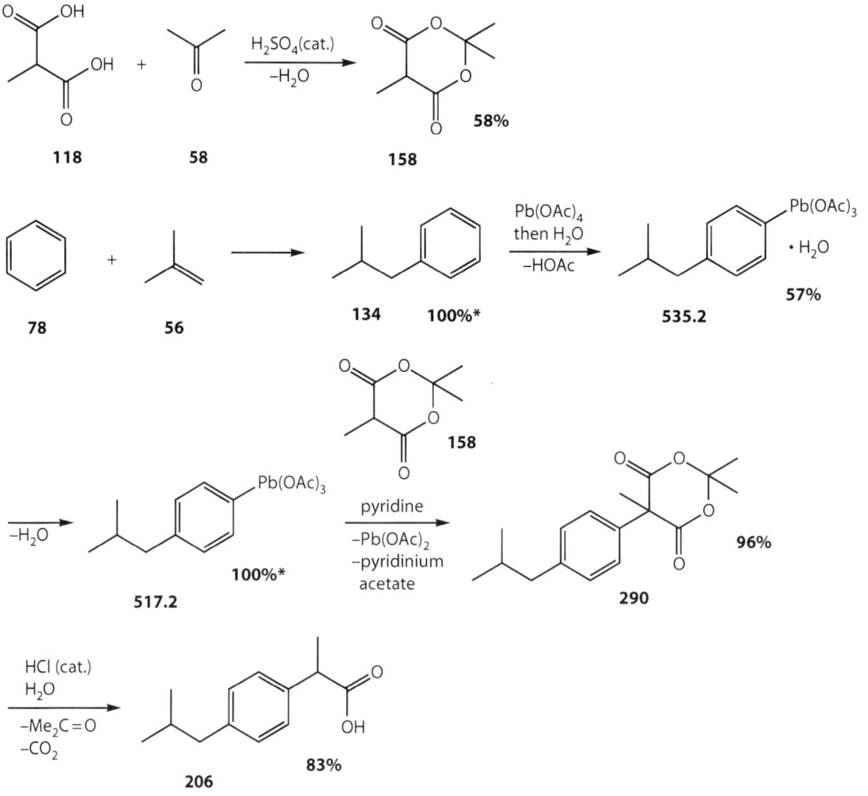

SCHEME 2.13 Pinhey synthetic route to ibuprofen.

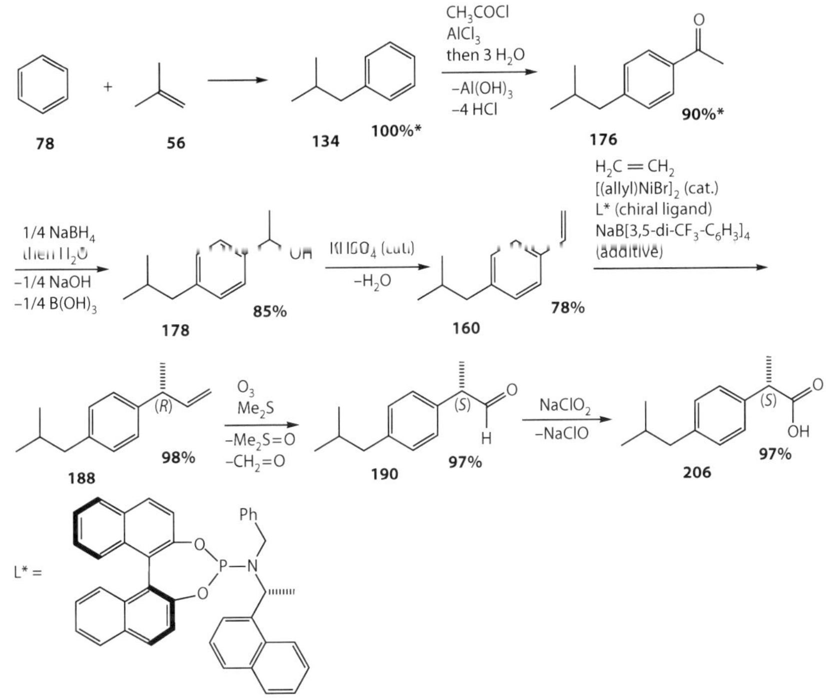

SCHEME 2.14 RajanBabu synthetic route to ibuprofen.

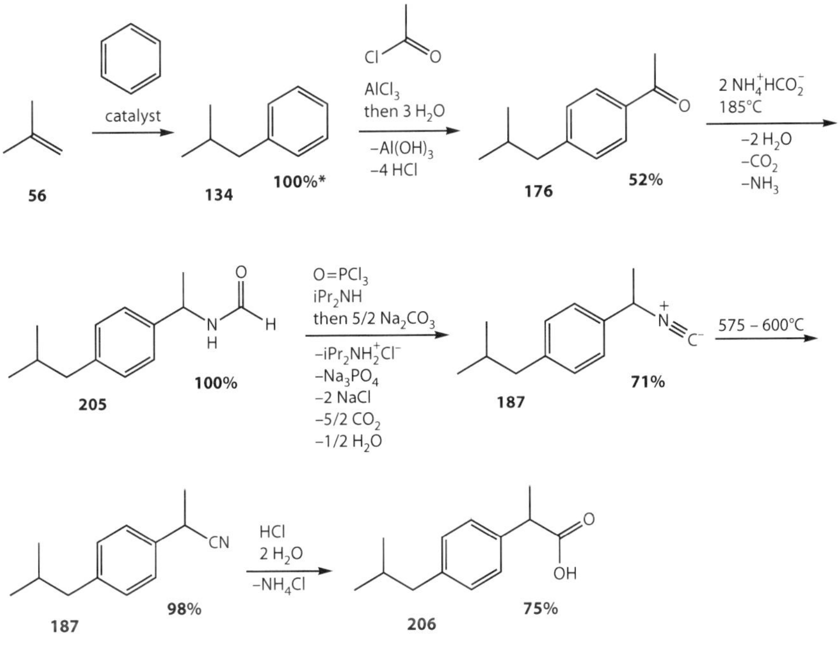

SCHEME 2.15 Ruchardt synthetic route to ibuprofen.

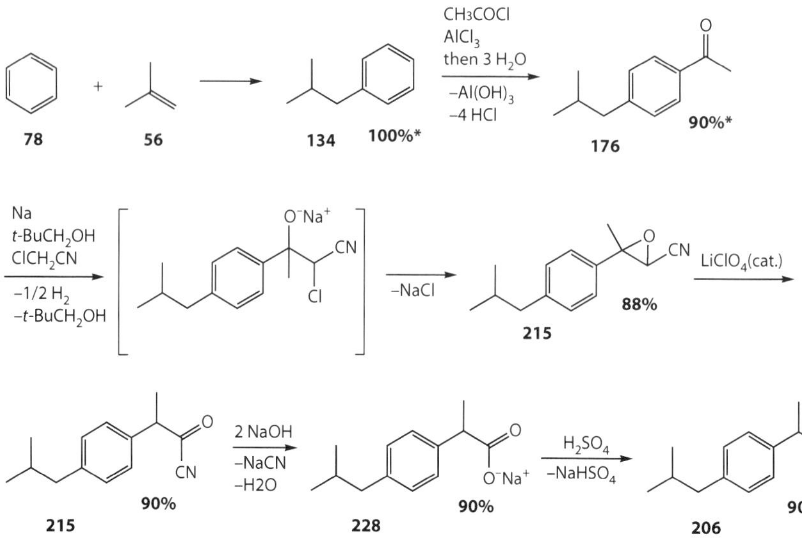

SCHEME 2.16 Upjohn synthetic route to ibuprofen.

2.6.4 Sample Problem Set Questions

The final part of this section lists ten sample questions that have been given as course problem sets. Each question is accompanied by chemical schemes and appropriate literature references. Instructors may use these as a springboard for ideas in their own course instruction. Table 2.6 summarizes the subjects of the questions and the appropriate green chemistry pedagogical skills involved in working them out.

Problem 1:[58] Three routes to 4,7-dichloroquine (Scheme 2.17) and two routes to *N,N*'-diethyl-1,4-diaminopentane (Scheme 2.18) are considered for the synthesis of the antirheumatic chloroquine (Scheme 2.19). Find the most atom-economical route to this target compound.

Problem 2:[59] A task-specific ionic liquid is used to catalyze the stereoselective reduction of acetophenone in such a way that it is recycled in the reaction (Scheme 2.20). Suggest a mechanism illustrating how this is achieved. Show the by-products of the reaction.

The task-specific ionic liquid is synthesized according to Scheme 2.21. Determine the overall reaction mass efficiency for the process.

Problem 3:[60,61] A four-step procedure to prepare the sulfa drug sulfanilamide from aniline is shown in Scheme 2.22. Costs of materials taken from the current Sigma-Aldrich® chemical catalog are also given in Table 2.7.

Reaction 1: Aniline (9.3 g), acetic anhydride (12.24 g), sodium acetate (9.8 g), and 98 wt% sulfuric acid (8 mL, d = 1.836 g/mL) are reacted in 250 mL water. The *N*-phenylacetamide product collected (12.0 g) was used directly in the next step without purification.

TABLE 2.6
Ten Green Chemistry Sample Problem Set Questions

Problem No.	Brief Description	Green Chemistry Pedagogical Aspects
1	Enumeration of synthesis routes to chloroquine from network diagram	Determination of best route based on overall atom economy for each plan
2	Stereoselective reduction of acetophenone using a task-specific ionic liquid	Mechanism of reduction; green metrics analysis of synthesis of task-specific ionic liquid combined with reduction reaction
3	Cost analysis for synthesis of the sulfa drug sulfanilamide	Radial pentagon analysis for each reaction step
4	Divergent reaction network for phthalic acid illustrating its versatility as a starting material for industrial commodity chemicals	Balancing chemical equations
5	Energy input analysis for synthesis of an antifungal agent	Synthesis using microwave technology; energy metrics
6	Solvay process for synthesis of sodium carbonate	Demonstration of potential recycling loops in process
7	Generalized Biginelli reaction	Analysis of best-case scenario for atom economy; probability analysis for greening reaction with respect to atom economy and reaction yield constraints
8	Evaluation of four plans to vanillin	Full green metrics and synthesis tree analysis to determine greenest plan according to material efficiency; comparison of biofeedstock vs. chemical synthesis routes
9	Evaluation of two plans to vitamin A	Full green metrics and synthesis tree analysis to determine greenest plan according to material efficiency; comparison of linear vs. convergent routes
10	Evaluation of two plans to DCOI: an environmentally friendly marine antifoulant	Full green metrics and synthesis tree analysis to determine greenest plan according to material efficiency

Reaction 2: *N*-Phenylacetamide (12.0 g) and chlorosulfonic acid (52.4 g) are reacted. Water (300 mL) was used to extract 4-acetamidobenzene-1-sulfonyl chloride (15.0 g) from the reaction mixture.

Reaction 3: All of the synthesized 4-acetamidobenzene-1-sulfonyl chloride was treated with a 28 wt% aqueous solution of ammonium hydroxide (60 mL, d = 0.898 g/mL). After reaction took place, the mixture was neutralized with 6 M sulfuric acid (44 wt%, 10 mL, d = 1.338 g/mL), which afforded 11.0 g of *N*-[4-(aminosulfonyl)phenyl]acetamide.

Reaction 4: All of the synthesized *N*-[4-(aminosulfonyl)phenyl]acetamide was hydrolyzed in water (60 mL) and 3 M hydrochloric acid solution (10 wt%, 26 mL, d = 1.047 g/mL). After reaction completion, the mixture was neutralized

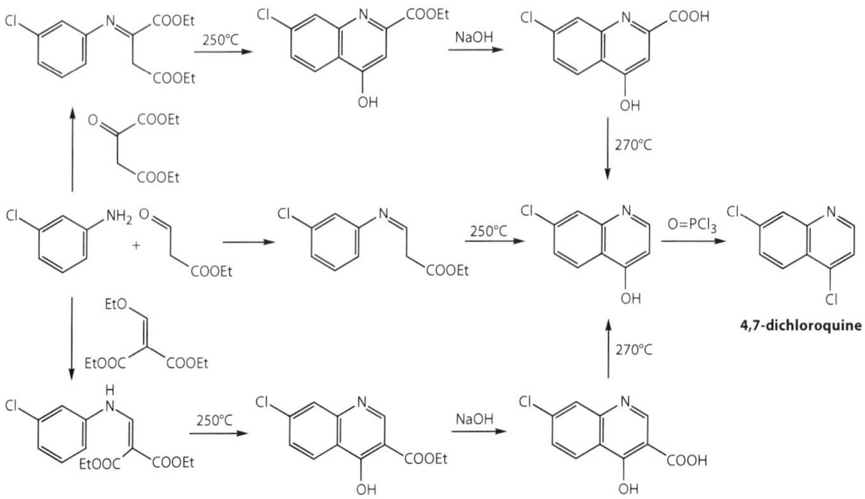

SCHEME 2.17 Three synthetic routes to 4,7-dichloroquine.

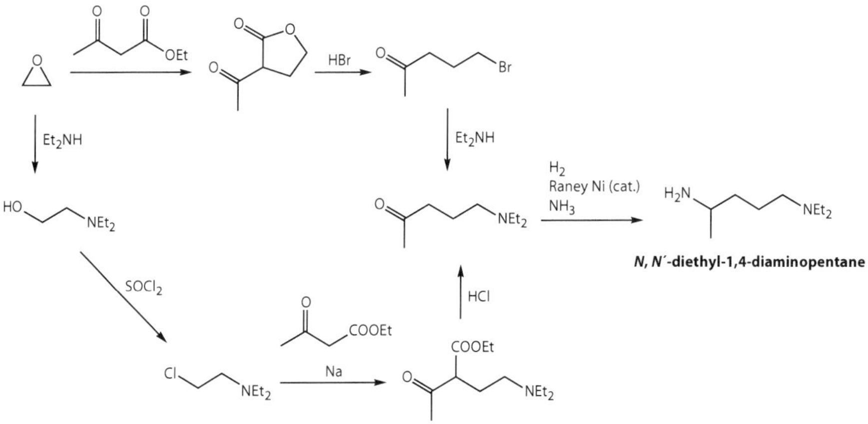

SCHEME 2.18 Two synthetic routes to *N*,*N*'-diethyl-1,4-diaminopentane.

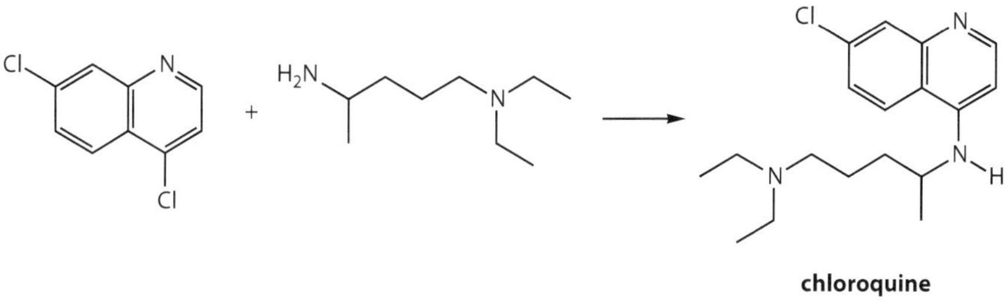

SCHEME 2.19 Synthesis of chloroquine.

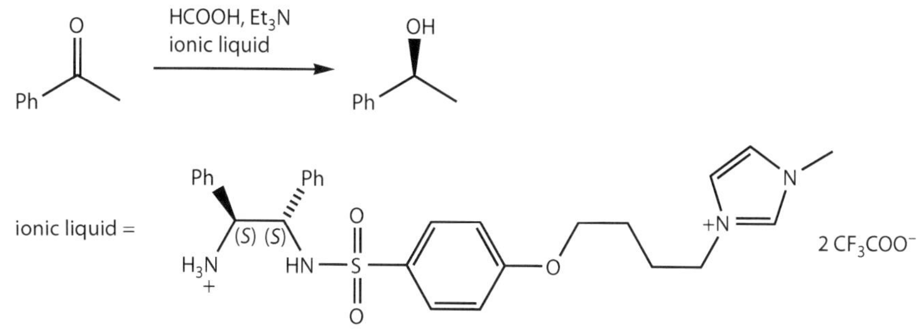

SCHEME 2.20 Stereoselective reduction of acetophenone.

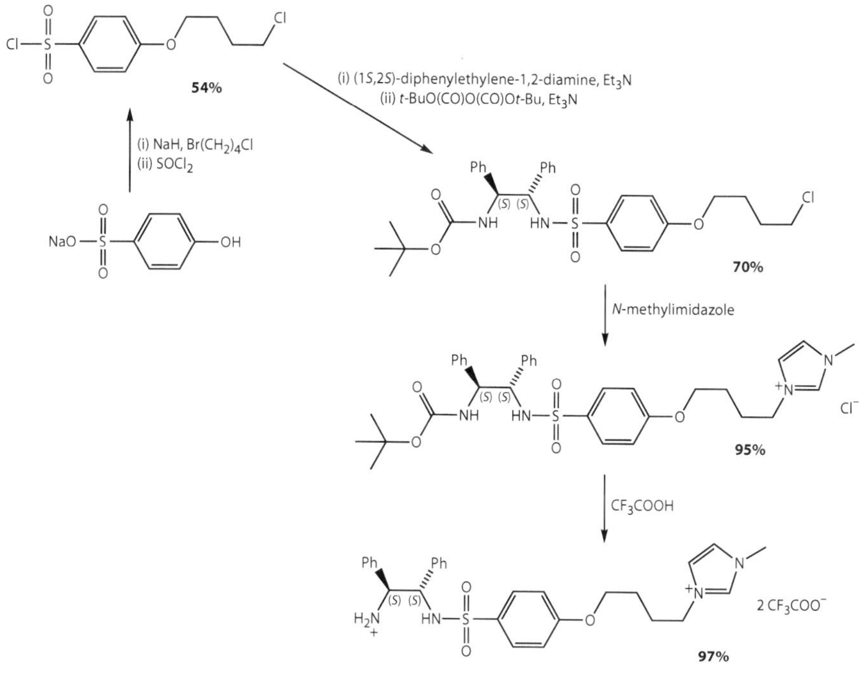

SCHEME 2.21 Synthesis of a task-specific ionic liquid.

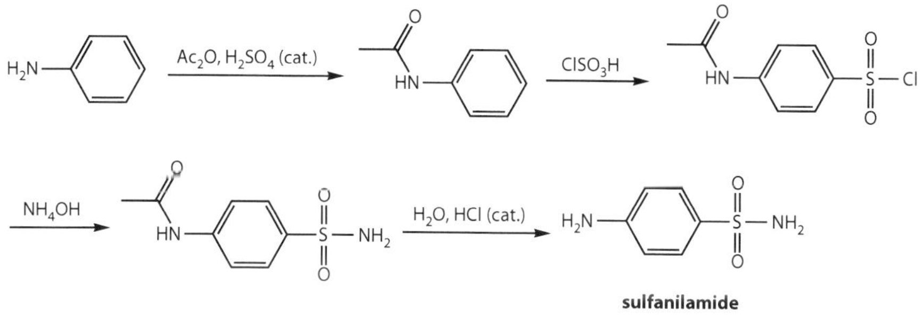

SCHEME 2.22 Synthesis of sulfanilamide from aniline.

TABLE 2.7
Cost of Chemicals for Problem 3

Material	Cost[a]
Aniline	\$537.00/18 kg
Acetic anhydride	\$554.00/18 kg
Sodium acetate	\$218.00/3 kg
98 wt% sulfuric acid	\$157.50/2.5 L
Chlorosulfonic acid	\$66.10/kg
28 wt% ammonium hydroxide	\$85.10/2 L
37 wt% hydrochloric acid	\$130.50/2.5 L
Sodium hydroxide	\$387.00/12 kg
Sodium carbonate	\$479.00/12 kg
Water	\$240.50/16 L

[a] Chemical cost from Sigma-Aldrich web site (www.sial.com), October 2, 2010.

with 6 M sodium hydroxide solution (20 wt%, 15 mL, d = 1.146 g/mL) and saturated sodium carbonate solution (14 wt%, 40 mL, d = 1.146 g/mL). The desired product sulfanilamide (8.5 g) was then collected.

a. Write out balanced chemical equations for each transformation.
b. Use the Excel radial pentagon spreadsheet to determine the atom economy (AE) and true values for reaction mass efficiency (RME) and raw material cost (RMC) for each step, assuming that all materials other than target products are destined for waste, and that all of the intermediate products are used as starting materials for the next steps.
c. Determine the overall AE and RME for the synthesis.
d. Represent this plan as a synthesis tree diagram to determine the kernel overall RME (maximum RME). Compare this value with the true overall RME determined in part (c).
e. Also from the synthesis tree determine the kernel overall RMC (minimum RMC) for this plan. Compare this value with the true overall RMC determined from the Excel analysis.
f. How much of the true overall RMC spent on all input materials is destined for waste in this synthesis plan?

Problem 4:[11] A chemical company uses phthalic anhydride as a starting material to synthesize six products (P1–P6) according to the reaction network shown in Scheme 2.23. By means of synthesis tree analysis, determine the mass of starting phthalic anhydride that is needed to prepare 1 kg of each target product. Balance all chemical equations and determine the relevant waste by-products. If all yields are optimal, determine which target product is made with the highest atom economy.

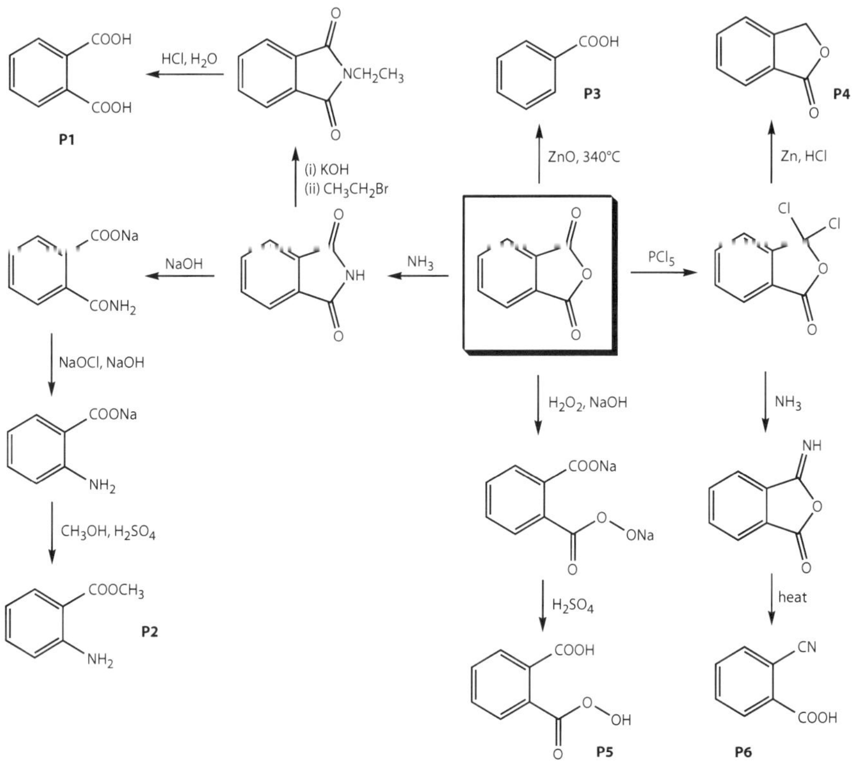

SCHEME 2.23 Synthesis of six products (P1-P6) from phthalic anhydride.

Problem 5:[62] An antifungal agent has been synthesized using microwave irradiation according to Scheme 2.24. Determine the fraction of the total input energy that is directed to make the product.

Problem 6:[63] The Solvay process to make sodium carbonate was invented in 1864. A series of reactions are shown in Scheme 2.25. Balance each chemical equation showing all by-products.

a. Determine the overall balanced chemical equation for the process.
b. Represent the entire process in the form of a synthesis tree using general reaction yields.
c. Determine the mass ratio of ammonium chloride to calcium oxide waste by-products, and determine the mass ratio of carbon dioxide input to carbon dioxide output.
d. A set of recycling reactions is added to convert the waste by-products back into ammonia.

$$CaO + H_2O \rightarrow Ca(OH)_2$$

$$Ca(OH)_2 + 2NH_4Cl \rightarrow 2NH_3 + CaCl_2 + 2H_2O$$

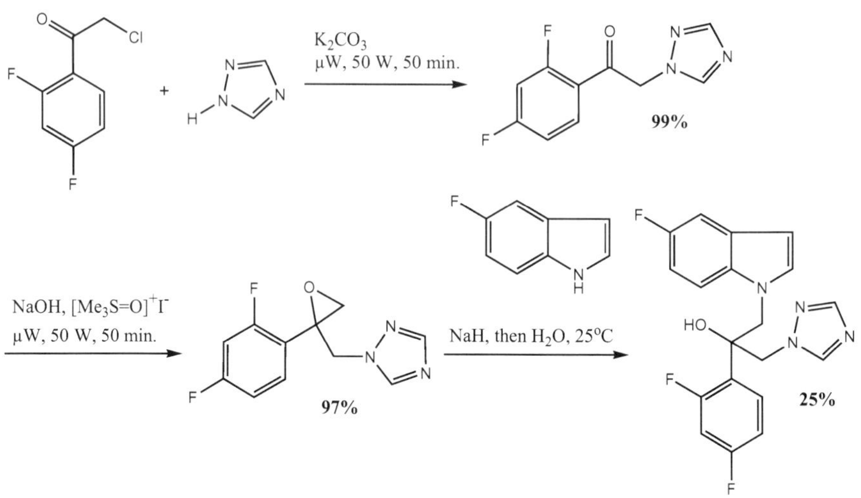

SCHEME 2.24 Synthesis of an antifungal agent.

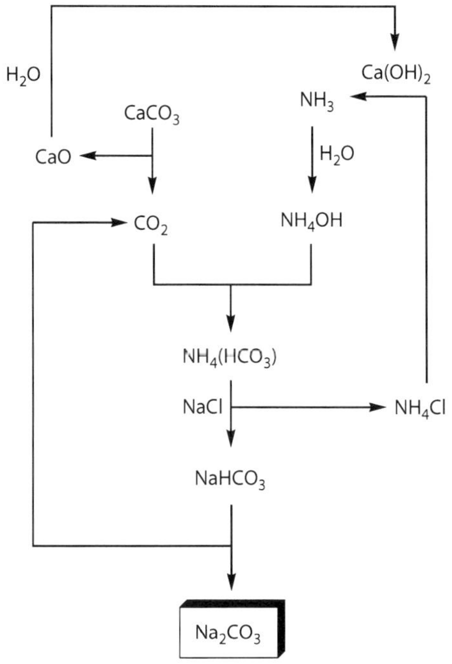

SCHEME 2.25 The Solvay process for synthesis of sodium carbonate.

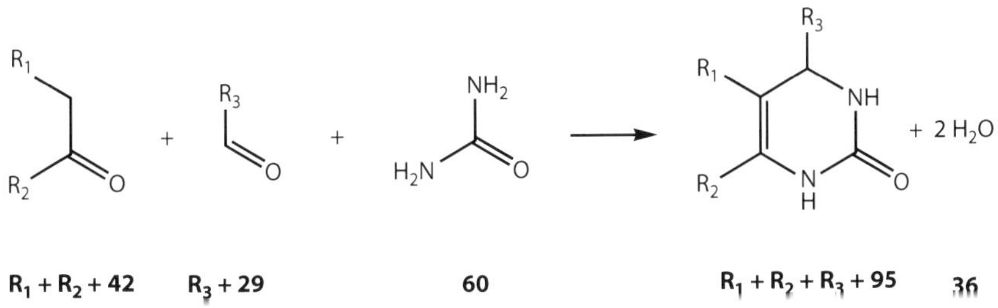

SCHEME 2.26 A generalized Biginelli reaction.

Determine the overall balanced chemical equation for the process including the recycling loop.

e. Draw a synthesis tree diagram using general reaction yields for the recycling loop only.
f. Determine the mass ratio of ammonium chloride to calcium oxide that is needed to match the mass of ammonia in the mainstream to produce sodium carbonate. Compare this value of the ratio with that found in part c. What is the condition that makes them equal?

Problem 7:[64] The Biginelli reaction is a one-pot three-component coupling reaction used to synthesize substituted 3,4-dihydro-1H-pyrimidin-2-ones, as shown in Scheme 2.26.

a. Determine the minimum AE and maximum E(mw) (E-factor based on molecular weight) for this reaction.
b. Suppose a minimum target kernel RME of 60% is imposed on this reaction. What is the probability that this target will be met given the minimum AE found in part (a)?
c. Suppose we add another constraint that the reaction yield should have a minimum value of 80%. What is the probability that the kernel RME will be at least 60% given both constraints of a minimum AE found in part a and a minimum reaction yield of 80%?

Problem 8:[65–72] A synthesis network to produce vanillin from guaiacol, eugenol, and ferulic acid feedstocks is shown in Scheme 2.27. Balance all chemical equations and rank the routes according to reaction mass efficiency.

Problem 9:[73,74] Compare and contrast the convergent (Scheme 2.28) and linear (Scheme 2.29) plans for the synthesis of vitamin A shown in the schemes using a full synthesis tree and green metrics analysis.

Problem 10:[75–79] 4,5-Dichloro-2-*n*-octyl-4-isothiazoline-3-one (DCOI) is a marine antifoulant invented at Rohm & Haas in the late 1970s. It is used as a biocide to protect the hulls of ships from unwanted accumulation of

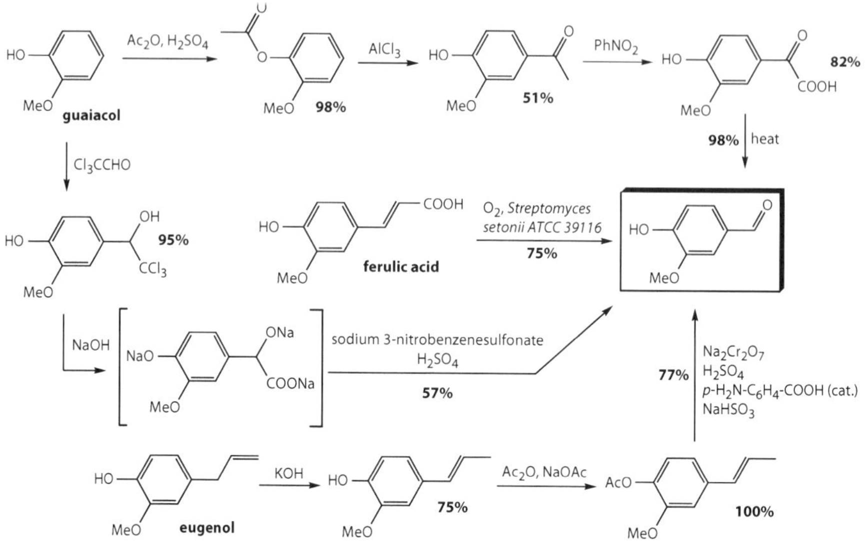

SCHEME 2.27 Synthesis of vanillin from guaiacol, eugenol, and ferulic acid.

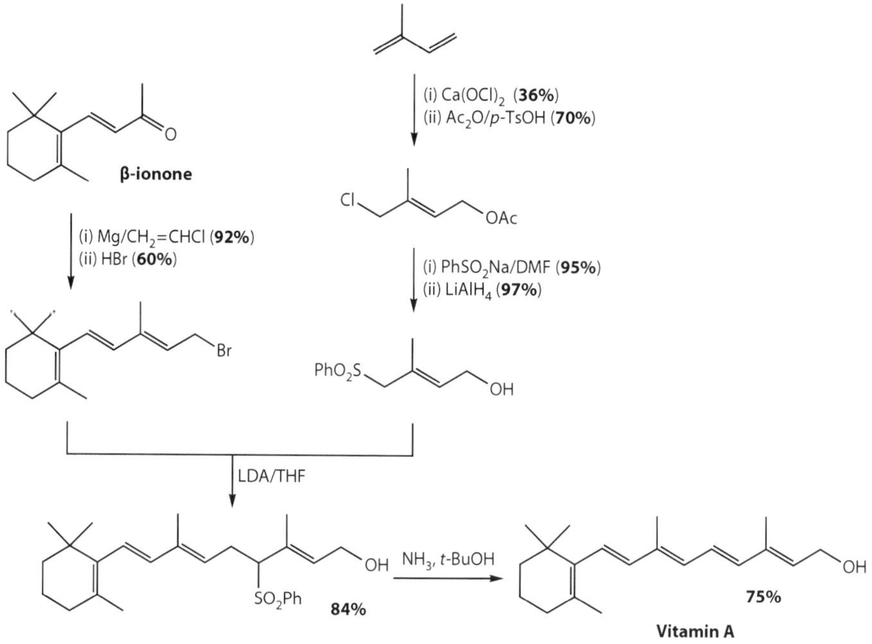

SCHEME 2.28 Convergent synthesis of vitamin A from β-ionone.

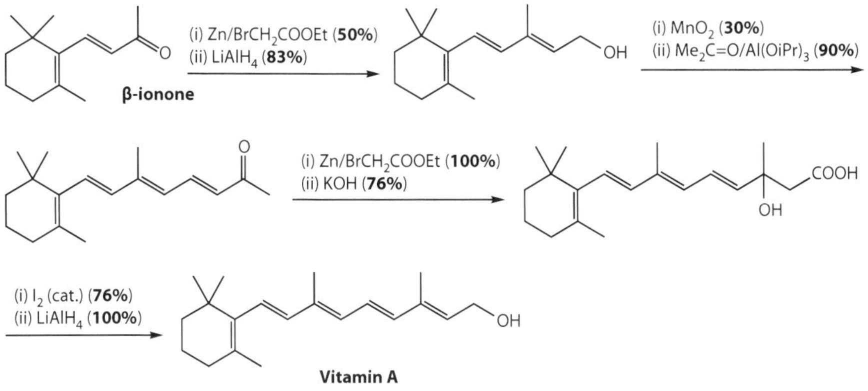

SCHEME 2.29 Linear synthesis of vitamin A from β-ionone.

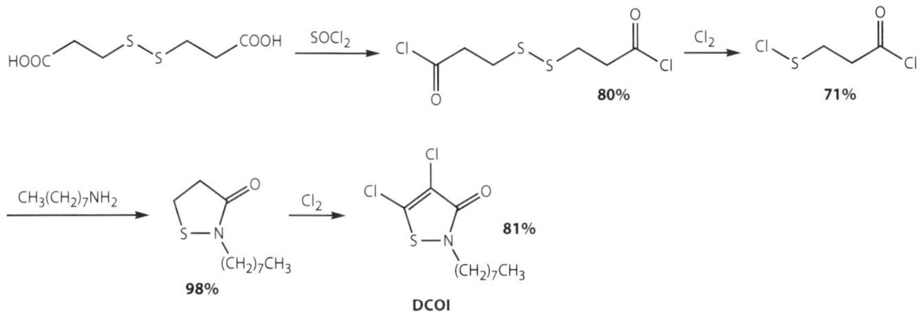

SCHEME 2.30 Plan A synthesis of 4,-5-dichloro-2-*n*-octyl-4-isothiazoline-3-one.

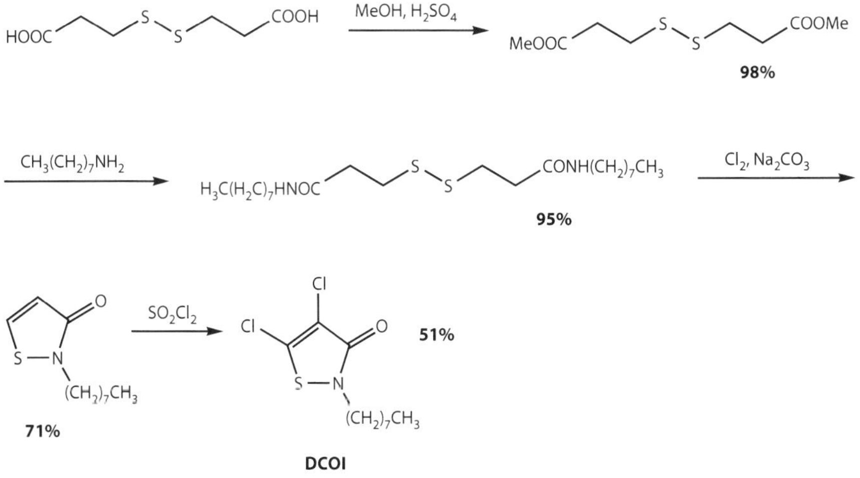

SCHEME 2.31 Plan B synthesis of 4,-5-dichloro-2-*n*-octyl-4-isothiazoline-3-one.

barnacles and other organisms that increase the drag force of ships as they travel at sea. DCOI is more environmentally friendly than tributyltin oxide that was used previously, as discussed in Section 1.4. Two synthesis plans (A and B) are shown for DCOI (Schemes 2.30 and 2.31). Analyze these using synthesis trees and a full green metrics analysis.

REFERENCES

1. Anastas, P. T., Warner J. C. *Green Chemistry: Theory and Practice*. Oxford University Press, New York, 1998.
2. McMurry, J. *Organic Chemistry*, 7th ed. Thomson Higher Education, Belmont, CA, 2008, p. 957.
3. Lee, S., Robinson, G. *Process Development: Fine Chemicals from Grams to Kilograms*. Oxford University Press, New York, 1995, p. 13.
4. Steinbach, A., Winkenbach, R. *Chem. Eng.* 2000, 107, 94.
5. Cann, M. C., Dickneider, T. A. *J. Chem. Educ.* 2004, 81, 977–980.
6. Greening the Organic Curriculum: Development of an Undergraduate Catalytic Chemistry Course. www.ched-ccce.org/confchem/2010/Spring2010/P5-Dicks_and_Batey.html (accessed December 23, 2010).
7. Marteel-Parish, A. E. *J. Chem. Educ.* 2007, 84, 245–247.
8. Andraos, J., Izhakova, J. *Chim. Oggi* 2006, 24, 31–36.
9. Andraos, J. *Org. Process Res. Dev.* 2005, 9, 149–163.
10. Andraos, J. *Org. Process Res. Dev.* 2005, 9, 404–431.
11. Andraos, J. *Org. Process Res. Dev.* 2006, 10, 212–240.
12. Andraos, J., Sayed, M. *J. Chem. Educ.* 2007, 84, 1004–1010.
13. Andraos, J. *Can. Chem. News* 2007, 59 (4), 14–17.
14. Andraos, J. *Org. Process Res. Dev.* 2009, 13, 161–185.
15. Andraos, J. Application of Green Metrics Analysis to Chemical Reactions and Synthesis Plans. In *Green Chemistry Metrics*, Lapkin, A., Constable, D. J. C., Eds. Blackwell Scientific, Oxford, 2008.
16. Anastas, P., Wood-Black, F., Masciangioli, T., McGowan, E., Ruth, L., Eds. *Exploring Opportunities in Green Chemistry and Engineering Education: A Workshop Summary to the Chemical Sciences Roundtable*. National Research Council, Washington, DC, 2007.
17. Sherlock, J.-P., Poliakoff, M., Howdle, S., Lathbury, D. *Can. Chem. News* 2009, 61 (3), 16–18.
18. Pozarentzi, M., Stephanidou-Stephanatou, J., Tsoleridis, C. A. *Tetrahedron Lett.* 2002, 43, 1755–1758.
19. Sivamurugan, V., Deepa, K., Palanichamy, M., Murugesan, V. *Synth. Commun.* 2004, 34, 3833–3846.
20. Pasha, M. A., Jayashankara, V. P. *Indian J. Chem. B* 2006, 45B, 2716–2719.
21. Hegedüs, A., Hell, Z., Potor, A. *Catal. Lett.* 2005, 105, 229–232.
22. An, L.-T., Ding, F.-Q., Zou, J.-P., Lu, X.-H. *Synth. Commun.* 2008, 38, 1259–1267.
23. Ried, W., Torinus, E. *Chem. Ber.* 1959, 92, 2902–2916.
24. Guzen, K. P., Cella, R., Stefan, H. A. *Tetrahedron Lett.* 2006, 47, 8133–8136.
25. Jarikote, D. V., Siddiqui, S. A., Rajagopal, R., Daniel, T., Lahoti, R. J., Srinivasan, K. V. *Tetrahedron Lett.* 2003, 44, 1835–1838.
26. Luo, Y.-Q., Xu, F., Han, X.-Y., Shen, Q. *Chin. J. Chem.* 2005, 23, 1417–1420.
27. Chari, M. A., Shobha, D., Syamasundar, K. *J. Heterocycl. Chem.* 2007, 44, 929–932.
28. Chari, M. A., Syamasundar, K. *Catal. Commun.* 2005, 6, 67–70.

29. Sabitha, G., Reddy, G. S. K. K., Reddy, K. B., Reddy, N. M., Yadav, J. S. *Adv. Synth. Catal.* 2004, 346, 921–923.
30. Xia, M., Lu, Y. *Heteroat. Chem.* 2007, 18, 354–358.
31. Hekmatshoar, R., Sadjadi, S., Shiri, S., Heravi, M. M., Beheshtiha, Y. S. *Synth. Commun.* 2009, 39, 2549–2559.
32. Yadav, J. S., Reddy, B. V. S., Praveenkumar, S., Nagaiah, K., Lingaiah, N., Saiprasad, P. S. *Synthesis* 2004, 901–904.
33. Shobha, D., Chari, M. A., Mukkanti, K., Ahn, K. H. *J. Heterocycl. Chem.* 2009, 46, 1028–1033.
34. Yadav, J. S., Reddy, B. V. S., Eshwaraiah, B., Anuradha, K. *Green Chem.* 2002, 4, 592–594.
35. Landge, S. M., Torok, B. *Catal. Lett.* 2008, 122, 338–343.
36. Kaboudin, B., Navaee, K. *Heterocycles* 2001, 55, 1443–1446.
37. Hazarika, P., Gogoi, P., Konwar, D. *Synth. Commun.* 2007, 37, 3447–3454.
38. Sharma, S. D., Gogoi, P., Konwar, D. *Green Chem.* 2007, 9, 153–157.
39. Kumar, S., Sandhu, J. S. *Indian J. Chem. B* 2008, 47B, 1463–1466.
40. Saini, A., Sandhu, J. S. *Synth. Commun.* 2008, 38, 3193–3200.
41. Bell, M. R., Dambra, T. E., Kumar, V., Eissenstat, M. A., Herrmann, J. L., Wetzel, J. R., Rosi, D., Philion, R. E., Daum, S. J., Hlasta, D. J., Kullnig, R. K., Ackerman, J. H., Haubrich, D. R., Luttinger, D. A., Baizman, E. R., Miller, M. S., Ward, S. J. *J. Med. Chem.* 1991, 34, 1099–1110.
42. Ward, S. J., Bell, M. R. US 4973587. Sterling Drug Inc., 1990.
43. Arcadi, A., Cacchi, S., Carnicelli, V., Marinelli, F. *Tetrahedron* 1994, 50, 437–452.
44. Earle, M. J., McCormac, P. B., Seddon, K. R. *Green Chem.* 2000, 2, 261–262.
45. Dupont, J., Consorti, C. S., Suarez, P. A. Z., de Souza, R. F. *Org. Synth.* 2002, 79, 236–243.
46. FR 1545270. Boots Pure Drug Co. Ltd., 1968.
47. Cassebaum, H., Hilger, H. DD 113889. Ger. Dem. Rep., 1975.
48. Nugent, W. A., McKinney, R. J. *J. Org. Chem.* 1985, 50, 5370–5372.
49. Cleij, M., Archelas, A., Furstoss, R. *J. Org. Chem.* 1999, 64, 5029–5035.
50. Elango, V., Murphy, M. A., Smith, B. L., Davenport, K. G., Mott, G. N., Moss, G. L. EP 284310. Hoechst-Celanese Corp., 1988.
51. Elango, V., Davenport, K. G., Murphy, M. A., Mott, G. N., Zey, E. G., Smith, B. L., Moss, G. L. EP 400892. Hoechst Celanese Corp., 1990.
52. Bogdan, A. R., Poe, S. L., Kubis, D. C., Broadwater, S. J., McQuade, D. T. *Angew. Chem. Int. Ed.* 2009, 48, 8547–8550.
53. Kopinski, R. P., Pinhey, J. T., Rowe, B. A. *Aust. J. Chem.* 1984, 37, 1245–1254.
54. Smith, C. R., RajanBabu, T. V. *J. Org. Chem.* 2009, 74, 3066–3072.
55. Wolber, E. K. A., Rüchardt, C. *Chem. Ber.* 1991, 124, 1667–1672.
56. White, D. R. US 4021478. Upjohn Co., 1977.
57. Cann, M. C., Connelly, M. E. *Real-World Cases in Green Chemistry.* American Chemical Society, Washington, DC, 2000.
58. Kleemann, A., Engel, J. *Pharmaceutical Substances: Syntheses, Patents, Applications*, 4th ed. Thieme, Stuttgart, 2001.
59. Kawasaki, I., Tsunoda, K., Tsuji, T., Yamaguchi, T., Shibuta, H., Uchida, N., Yamashita, M., Ohta, S. *Chem. Commun.* 2005, 2134–2136.
60. Galat, A. *Ind. Eng. Chem.* 1944, 36, 192.
61. Vogel, A. I. *Textbook of Organic Chemistry.* Longman, London, 1956, pp. 1005–1007.
62. Lebouvier, N., Giraud, F., Corbin, T., Na, Y. M., Le Baut, G., Marchand, P., Le Borgne, M. *Tetrahedron Lett.* 2006, 47, 6479–6483.
63. Chenier, P. J. *Survey of Industrial Chemistry*, 3rd ed. Kluwer Academic/Plenum Publishers, New York, 2002, 69.

64. Lapkin, A., Constable, D. J. C. *Green Chemistry Metrics: Measuring and Monitoring Sustainable Processes*. Wiley, Chichester, 2008, 91–96.
65. Lampman, G. M., Andrews, J., Bratz, W., Hanssen, O., Kelley, K., Perry, D., Ridgeway, A. *J. Chem. Educ.* 1977, 54, 776–778.
66. Boedecker, F., Volk, H. *Chem. Ber.* 1931, 64, 61–66.
67. Sievers, W. C. CH 89053. L. Givaudan & Cie., 1921.
68. Sievers, W. C. CH 91088. L. Givaudan & Cie., 1921.
69. Muheim, A., Müller, B., Münch, T., Wetli, M. EP 885968. Givaudan-Roure, 1998.
70. Muheim, A., Lerch, K. *Appl. Microbiol. Biotechnol.* 1999, 51, 456–461.
71. Mottern, H. O. *J. Am. Chem. Soc.* 1934, 56, 2107–2108.
72. Zasosov, V. A., Metel'kova, E. I., Onoprienko, V. S. *Meditsinskaya Promyshlennost SSSR* 1959, 13, 22–24. *Chem. Abs.* **54**: 91507.
73. Olson, G. L., Cheung, H. C., Morgan, K. D., Neukom, C., Saucy, G. *J. Org. Chem.* 1976, 41, 3287–3293.
74. Wendler, N. L., Slates, H. L., Trenner, N. R., Tishler, M. *J. Am. Chem. Soc.* 1951, 73, 719–724.
75. Weiler, E. D., Petigara, R. B., Wolfersberger, M. H., Miller, G. A. *J. Heterocycl. Chem.* 1977, 14, 627–630.
76. Lewis, S. N., Miller, G. A., Hausman, M., Szamborski, E. C. *J. Heterocycl. Chem.* 1971, 8, 571–580.
77. Hahn, S. J., Kim, J. M., Park, Y. WO 9220664. Sunkyong Industries Co., Ltd., 1992.
78. Koshiyama, T. JP 2003335763. New Japan Chemical Co., Ltd., 2003. *Chem. Abs.* **139**: 395926.
79. Morita, M., Liu, K., Yoneda, N. JP 2001181266. Chemicrea Inc., 2001. *Chem. Abs.* **135**: 61326.

3 Elimination of Solvents in the Organic Curriculum

Dr. Andrew P. Dicks

CONTENTS

3.1 INTRODUCTION

In 1994, an "In This Issue" summary published in the *Journal of Chemical Education* stated some connections the average layperson might make between chemicals, natural and artificial substances, risk, and toxicity.[1] These were listed as follows:

chemical = artificial
artificial = bad, toxic
natural = good, safe
high toxicity = high risk
low toxicity = low risk

Nearly two decades later, such perceptions are perhaps even more prevalent among the general population. Together, they have led to society demanding a reduction in the use of "chemicals." Companies have indeed been known to advertise personal care items, such as "chemical-free sunscreens," where zinc oxide and titanium dioxide are the two active ingredients![2] Conversely, "natural" products are routinely viewed in a positive light, being supposedly less toxic, greener, and safer to handle or ingest. Interestingly, with these public misconceptions existing, the chemical community at large has appreciated that much needs to be done to curb extraneous substance use in synthetic procedures. As discussed in Section 1.4, prevention of waste is one of the central tenets of green chemistry.[3] In the undergraduate laboratory, reactions can be performed on a semi-microscale or microscale level, rather than on a macroscale level (described as a waste management strategy in Section 7.3.2). Clearly, however, the worldwide demand for pharmaceuticals and other organic materials means this is not a viable option in industry. Rather, an immediate and obvious approach toward large-scale waste reduction is to minimize or eradicate auxiliary substances (those that do not appear in the desired product). A significant ancillary in many processes is the solvent, which usually provides a medium for reactant dissolution. Research into low-solvent or solvent-free reactivity is therefore currently of great interest.

Solvents have been said to account for up to 90% of utilized mass in a pharmaceutical or fine chemical operational process.[4] Recent enthusiasm has been shown in solventless oxidations by the American Chemical Society's Green Chemistry Institute Pharmaceutical Roundtable.[5] Paint manufacturers have also taken the solvent reduction notion a step further, and have changed the composition of their products, reflecting a more eco-friendly stance.[6] Organic solvents have traditionally been an integral part of many paints and varnishes, in order to provide durability and glossiness. In the European Union, legislation has meant that a reduction in the quantity of volatile organic compounds (VOCs) permitted in paint has been enforced. Certain paint companies have complied by synthesizing water-based formulations, which maintain their glossiness by having a tighter particle distribution and improved pigment dispersion. In contrast, the German company Bayer MaterialScience is currently developing polyaspartic coatings for use in hard-wearing paint designed for the construction industry. Polyaspartics are used in topcoats in order to provide weather resistance and are of low viscosity, allowing formulations to be almost absent of VOCs.[6]

Employment of solvents in chemical reactivity stretches back many centuries. In his article "Proving Aristotle Wrong," Bradley outlined how the ancient Greek philosopher stated: "No reaction proceeds without solvent."[7] Reichardt has since clarified that this is a direct translation of the quotation "*Corpora non agunt nisi fluida (or liquida) seu soluta*,"[8] and that it is better understood as, for example: "For instance, liquids are types of bodies most liable to mixing." However Aristotle is interpreted, it is clearly apparent that many reactions proceed extremely well in the absence of any medium. A comprehensive study of solventless organic reactivity has been compiled by Tanaka,[9] and recent reviews reflect research into solvent-free heterocyclic synthesis[10] and asymmetric catalysis.[11] In addition, Shearouse has written an assessment of alternative solventless methodologies in the preparation of pharmaceutical drugs.[12] With the current scope of possibilities in mind, one could in fact pose the question: How should faculty discuss solvents and their role during an undergraduate introductory organic chemistry class? The majority of instructors defer a serious mention of solvents and their purpose until a mechanistic interpretation of alkyl halide chemistry is covered. At this point, competing substitution and elimination reactions are regularly considered in the context of polar protic and polar aprotic solvents. Students typically encounter solvents earlier in an associated laboratory course component, when perfecting experimental techniques such as extractions and recrystallizations. They specifically learn that the solvent nature is very important for Grignard reactions[13] and related organometallic transformations, but are likely left with the false impression that solvents are an *essential* part of the design process for any proposed transformation.

The last ten years has seen an explosion in the publication of pedagogical articles focusing on organic reactivity in the absence of solvent. Solventless reactivity is advantageous from the perspective of both the student and course instructor, particularly if there is a desire to highlight green chemistry maxims. Besides the principle of eliminating waste, the attitude of "design for energy efficiency" is regularly reinforced.[3] Solvent-free reactions generally occur more quickly than their traditional counterparts, whether they are undertaken under conditions of grinding, microwave irradiation, or mechanical stirring in the presence or absence of heat. This is usually ascribed to there being increased reactant concentrations under solventless conditions. Many procedures that historically required lengthy heating under reflux can be performed in a three- or four-hour laboratory period, with the opportunity of accomplishing simultaneous conversions. Product workup and isolation is often simplified for solvent-free reactions, with dispensation of the need for chromatographic separations.

This chapter progresses by considering what actually constitutes a solvent-free reaction. Following this is a discussion of some significant industrial and academic case studies, where solventless reactivity has been successfully incorporated into organic transformations of note. These are chosen as examples to profile during the lecture portion of a course mounted at the second- or third-year undergraduate level. An important connection is then made between green chemistry and chemical engineering in Section 3.4. Here, an analysis is undertaken of how reactor design has evolved to accommodate issues of viscosity, inefficient mixing, and heat transfer in large-scale solvent-free processes. Some test-driven undergraduate reactions that are appropriate for inclusion across the organic curriculum are then outlined. Special attention is devoted to articles describing purely solvent-free conversions or those that use only an environmentally benign solvent or solvents during the reaction workup. A mention is made of solventless reactions that are not so green in other aspects, and the importance of providing a balanced view to students within the realm of sustainability.

3.2 SOLVENT-FREE OR NOT SOLVENT-FREE?

So what does it mean to be solvent-free? Perhaps surprisingly, there has been some debate over the use of this term during the last decade. In order to try and clarify the situation somewhat, Tom Welton wrote the following in a 2006 *Green Chemistry* editorial column: "A dry solid phase reaction is solvent-free, also a reaction where there is a liquid present, but it is not acting as a solvent (i.e., nothing is dissolved in it) is also solvent-free."[14] It has been noted that there are many reported claims of solvent-free reactivity where, in fact, one liquid reactant is present in excess. In such cases, the liquid is behaving as both reactant and solvent.[15] What authors typically mean in these instances is that no *additional* solvent needs to be added for the reaction of interest to take place. This is clearly a positive feature, but is it really solventless chemistry?

It is instructive to discuss two pedagogical laboratory examples to emphasize this point about liquid reactants. The first is one designed by a fourth-year undergraduate at the University of Toronto and concerns the microscale synthesis of a 3-acetylcoumarin derivative which is a laser dye analog[16] (Scheme 3.1). In the reaction, solid 4-(diethylamino)salicylaldehyde is combined with ethyl acetoacetate and three drops of piperidine at room temperature. This mixture is vigorously stirred, with the coumarin product being isolated as a bright yellow solid demonstrating blue fluorescence, after workup with absolute ethanol. One might make the argument that

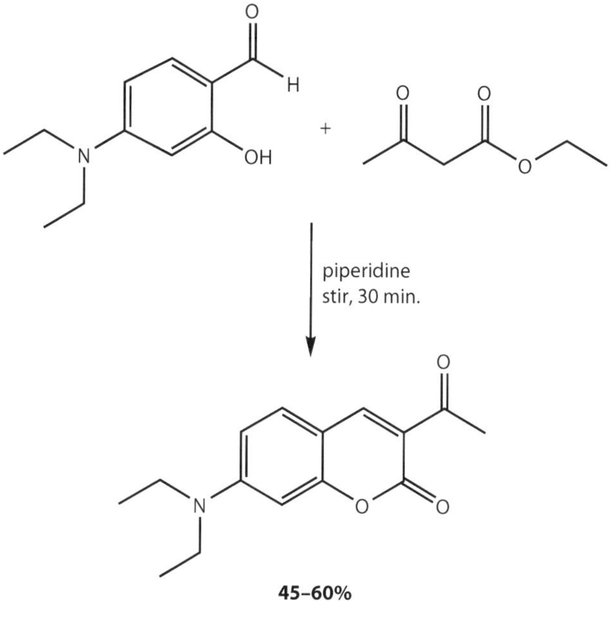

SCHEME 3.1 Coumarin synthesis from 4-(diethylamino)salicylaldehyde and excess ethyl acetoacetate.

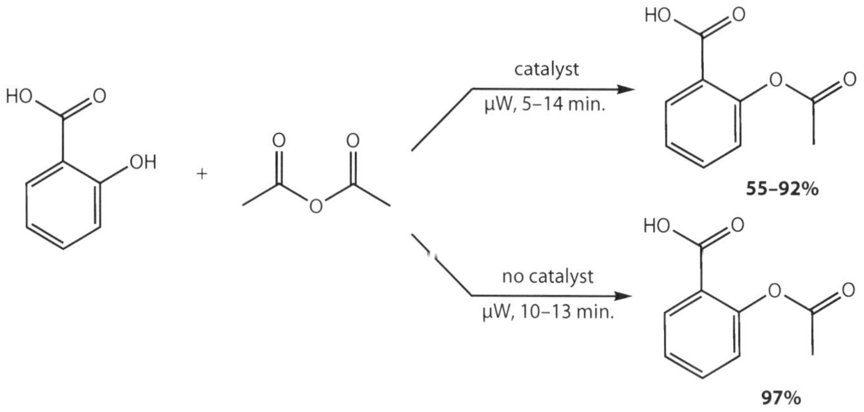

SCHEME 3.2 Aspirin synthesis under "solvent-free" conditions.

this is a solventless reaction (before workup), as no "external" medium is required. However, if attention is paid to the reactant stoichiometry, it is apparent that the β-ketoester (0.53 mL, 4.15 mmol) is present in excess of the salicylaldehyde derivative (0.4 g, 2.07 mmol). As ethyl acetoacetate is a liquid at room temperature, it is clearly participating as both reagent and solvent. (It should be noted that the authors of this article did not actually make the claim that this is a solventless process.)

In comparison, Montes et al. reported a greener synthesis of aspirin where salicylic acid and acetic anhydride are subjected to microwave irradiation, in the presence of eight different catalysts or the absence of any catalyst[17] (Scheme 3.2). Use of microwave technology represents a "greening" of many procedures, and is considered in more detail in Chapter 8. It is especially notable that aspirin can be synthesized in excellent yield without any catalyst, and the solid product does not require purification by recrystallization. However, the authors state that "by omitting the use of solvents for recrystallization in the previous procedures we can also introduce the solvent-free principle," and make several other references to the reaction taking place without a solvent. An analysis of the experimental conditions (detailed in the supporting information accompanying the article) indicates this is not the case. Solid salicylic acid (5 mmol) and liquid acetic anhydride (15 mmol) are mixed in a 50 mL beaker and heated in a domestic microwave oven, with a range of acidic and basic catalysts being analyzed. It appears a little misleading to students to call this a solvent-free conversion, when acetic anhydride is both liquid reagent and reaction medium, as per the previous example.

Another way in which reactions are somewhat erroneously labeled as solvent-free is when significant quantities of solvent (particularly those that are environmentally deleterious) are required during product isolation or purification. This is introduced in Section 2.6.1 in the context of benzodiazepine synthesis, where many literature procedures claim reactivity without any solvent use. Several such reported reactions even require a halogenated solvent for product extraction purposes. The title of a 2008 publication in *Organic Process Research and Development*, "Concise Synthesis of Vinylheterocycles through β-Elimination under Solventless Phase Transfer Catalysis

Conditions," implies that no solvent is needed for the transformation in question.[18] The abstract informs the reader that no *organic* solvent is utilized, but 50% aqueous NaOH is in fact required. Further, the experimental section indicates that more water is added on reaction completion, and then dichloromethane is used to dilute the organic phase. (It is significant to mention that the methodology described negates use of any dipolar aprotic solvents or hazardous bases, which represents a cogent green improvement on previous procedures.) The take-home message for educators and their students is that the literature must be read and interpreted very carefully before any conclusions can be drawn about the nature of solventless reactivity.

As a final point on the discussion of what constitutes a solvent-free reaction, a common approach in this venue is to mechanically grind equimolar quantities of two or more solids together. In their article "Grinding Is the New Green,"[19] Pichon and James explain how certain reactions occur effectively under so-called mechanochemical conditions. At first glance these processes would certainly seem to fit the moniker of being solventless. Interestingly, an argument has been put forward that grinding powders together could free small molecules that melt under the physical conditions, generating a liquid (solvent) that acts as a lubricant.[20] Alternatively, hygroscopic solids will remove water from the atmosphere, which could behave in a similar manner. These topics make for an interesting debate with undergraduates, who can easily perform a number of such reactions themselves.

3.3 INDUSTRIAL AND ACADEMIC CASE STUDIES

3.3.1 Synthesis of Cosmetic Ingredients

The 2009 Presidential Green Chemistry Challenge Greener Synthetic Pathways Award was granted to the Eastman Chemical Company.[21] This acknowledges Eastman's work on designing a solvent-free biocatalytic pathway to esters required by the cosmetics industry. Esters are significant components of emulsifiers and emollients, both of which are important cosmetic ingredients. For example, there is currently a great demand for new compounds that fight the onset of aging, and interest in emollients derived from a renewable resource such as rice bran oil. The traditional Fischer esterification reaction of a carboxylic acid with an alcohol suffers from several drawbacks, including the requirement of strongly acidic reaction conditions and high temperatures.[22] Organic solvents are used by other approaches that require recycling and also pose environmental hazards. Eastman has implemented enzymatic technologies to convert alcohols into esters under mild reaction conditions (Scheme 3.3). As in Fischer esterifications, the water coproduct is removed on formation, thus maximizing the isolated ester yield.

These reactions take place in the absence of solvent at relatively low temperatures and exhibit several other green features of note. The immobilized lipase enzyme can readily be collected by filtration and made available for future transformations. The benign and energy-efficient protocol reduces unwanted by-product formation, thus avoiding malodorous and colored impurities in the eventual cosmetic formulations. The reaction illustrated in Scheme 3.3 is additionally chemoselective, as esterification exclusively occurs at the benzylic

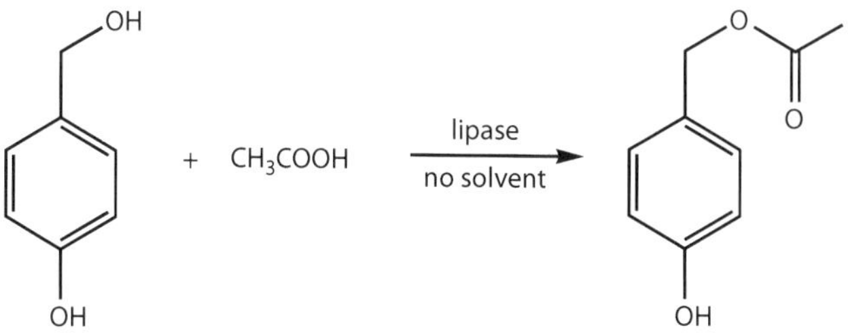

SCHEME 3.3 Solventless esterification of 4-hydroxybenzyl alcohol via biocatalysis with a lipase enzyme.

alcohol functionality, rather than at the phenol moiety. The product in this instance (4-hydroxybenzyl acetate) is a known inhibitor of the enzyme tyrosinase, which is a critical factor in melanin production. 4-Hydroxybenzyl acetate is therefore included in personal care items to promote an even skin tone and reduce unwanted coloring.

The solventless strategy designed by Eastman not only eliminates over 10 L of organic solvent needed per kilogram of ester formed, but also generates the desired product in a high degree of purity. This often reduces solvent requirements by a second route, as reaction workup protocols are simplified. The principle can be extended to more complex unsaturated fatty acids that are problematic to react under Fischer conditions due to competing oxidation chemistry occurring. Villa et al. have similarly reported preparation of long-chain aliphatic esters of cosmetic utility via solid-liquid solvent-free methodologies, utilizing either phase transfer catalysis or solid acid catalysis.[23] Some of their sample reactions are illustrated in Scheme 3.4. Butyl palmitate is synthesized in 93–97% yield upon heating palmitic acid and *n*-butanol in an oil bath or under microwave conditions. Similar results are obtained if Aliquat® 336 (a phase transfer catalyst)/potassium carbonate or *p*-toluenesulfonic acid (PTSA) is employed. As a comparison, the same ester is generated in 87% yield on refluxing the carboxylic acid and excess alcohol for three hours.[23] It is, however, important to note that diethyl ether is needed as a solvent in the workup of reactions using PTSA, and dichloromethane used similarly in the case of Aliquat 336-catalyzed reactions.

Together, these two examples make for an excellent green case study when carboxylic acid derivatives (and more specifically ester-forming reactions) are discussed with undergraduates. This typically takes place toward the end of a full-year introductory organic class, after students have learned about acid catalysis, and possibly even after having performed a Fischer esterification in the laboratory. Comparisons between traditional and more modern, greener synthetic pathways are especially valuable in a lecture setting. Some biocatalytic reactions for the undergraduate laboratory that utilize enzymes are discussed in more detail in Sections 6.2.2.1 and 6.2.7.

3.3.2 Peptide Synthesis and Preparation of Aspartame

Declerck et al. have recently outlined a general solvent-free technique to synthesize peptides via ball-milling technology.[24,25] Some well-established methods of peptide

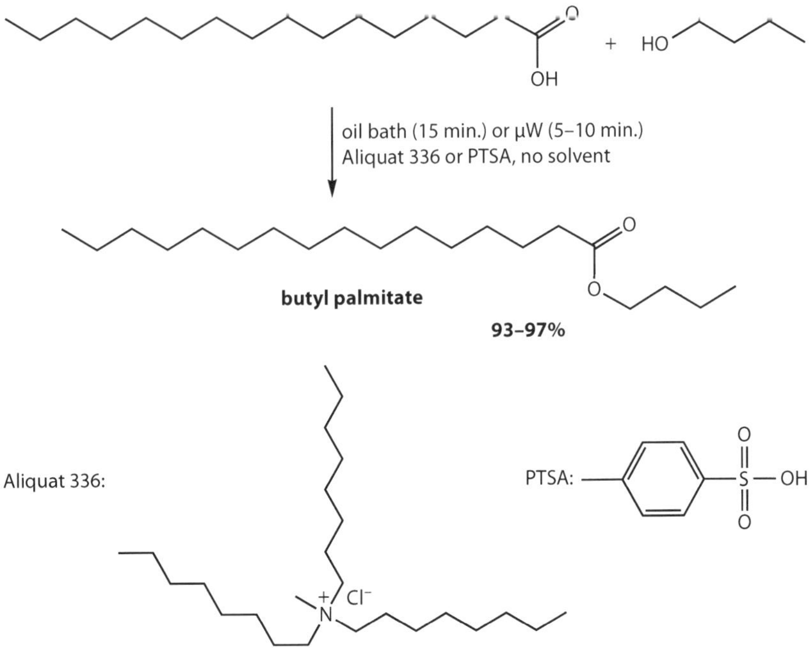

SCHEME 3.4 Synthesis of butyl palmitate under phase transfer and solid acid catalytic conditions.

bond formation[26] suffer in that large solvent quantities are required, and they are generally not as conducive to large-scale synthesis. Using a ball mill permits intimate grinding of solid reactants in the absence of solvent.[27] The mill is partially filled with both the solids to be ground and the grinding medium, which is often in the form of stainless steel balls. Upon rotation about the horizontal axis of the mill, a reaction occurs in much the same way as if solids were being ground together using a pestle and mortar. In this particular case, a urethane-protected α-amino acid *N*-carboxyanhydride (UNCA) derivative is ground with an α-aminoester hydrochloride in a ball mill for one hour to afford a dipeptide product (Scheme 3.5). The

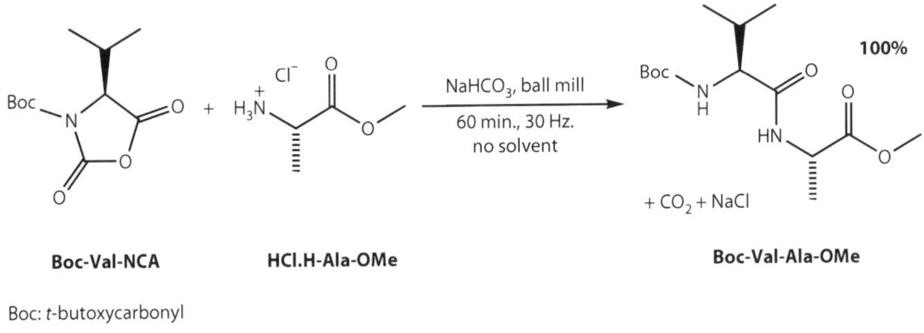

SCHEME 3.5 Ball mill solid phase dipeptide synthesis.

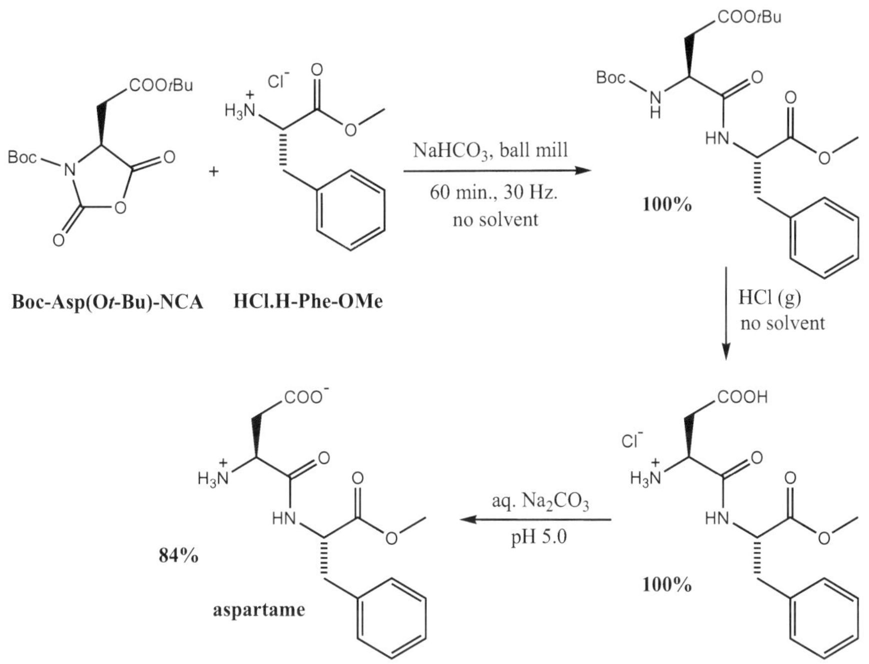

SCHEME 3.6 Aspartame synthesis avoiding use of organic solvents.

UNCA can be considered an active form of an amino acid in this reaction. Up to 500 mg of the dipeptide can be prepared in this manner.

The same authors extended their work to show how the artificial sweetener aspartame can be synthesized in the absence of organic solvents (Scheme 3.6). Aspartame is a "real-world relevant" substance of interest to students and is 150 times sweeter than sucrose.[28] The protected amino acid equivalent Boc-Asp(O*t*-Bu)-NCA is ground in a ball mill with HCl.H-Phe-OMe to generate the protected dipeptide. Acidic deprotection is then accomplished by a solventless reaction of the dipeptide with HCl gas to form aspartame hydrochloride. Aqueous dissolution and pH adjustment to the isoelectric point (pH = 5.0) yields aspartame as a white powder. This reactivity is of pertinence to a biological chemistry or upper-level synthetic organic course that deals with peptide synthesis and protection/deprotection strategies.

3.3.3 Halogen-Free Boronic Ester Synthesis

Boronic esters are key compounds used in Suzuki reactions, which represent a versatile method to make new carbon-carbon bonds under mild conditions.[29] Several undergraduate laboratory experiments showcase representative Suzuki reactions, and are discussed in Sections 4.3.1, 5.6.2, 7.5, and 8.8.4. Aryl boronic esters have historically been synthesized by the reaction of a Grignard reagent with a trialkylborate ester (Scheme 3.7).

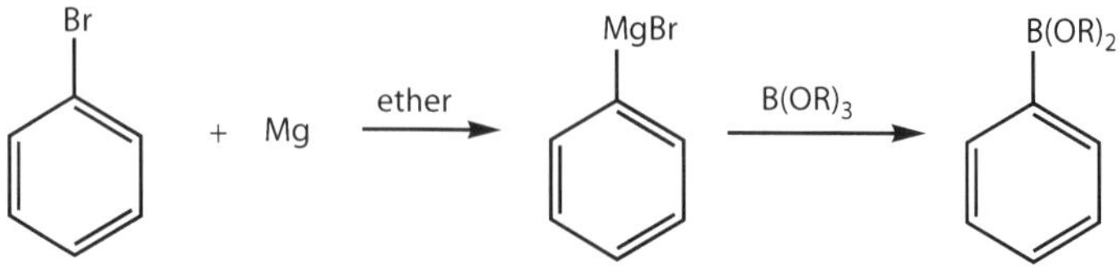

SCHEME 3.7 Historical synthesis of aryl boronic esters.

There has been much effort devoted to designing a general "halogen-free" preparative approach toward aryl and heteroaryl boronic esters. Maleczka and Smith at Michigan State University have collaborated to develop a novel catalytic C-H bond activation/borylation reaction, using iridium catalysts.[30] This means that halogenated compounds are unnecessary, and the target boronic esters are obtained under mild, solventless conditions with hydrogen gas being the only coproduct (Scheme 3.8).

Interestingly, the regiochemistry of this reaction is governed by steric rather than electronic effects. As such, 1,3-disubstituted aromatic compounds introduce a boryl group at the 5-position (*meta* to both substituents on the original ring). This is independent of the directing effects of both substituents and opens up the possibility of forming a range of compounds that are difficult to access by more conventional methods. The reaction is additionally robust, tolerating a wide variety of different functionalities. In recognition of their outstanding contribution to improving the environmental impact of how a fundamental group of compounds is produced, Maleczka and Smith were honored with a 2008 Presidential Green Chemistry Challenge Award in the Academic category.[31]

3.3.4 Enantioselective Synthesis of (*S*)-3-Aminobutanoic Acid

Synthesis of enantiomerically pure β-amino acids provides a challenge either with or without incorporation of green principles into the methodology. (*S*)-3-aminobutanoic acid and other short-chain β-amino acids have been manufactured by a short procedure that employs cheap and easily accessible starting compounds, thus fulfilling several sustainability criteria (Scheme 3.9).[32] However, a drawback of this approach is the need for column chromatography as a purification technique for the first solvent-free chemoenzymatic transformation, with associated organic solvent requirements. In addition, although (*S*)-3-aminobutanoic acid is produced with excellent enantioselectivity, the overall yield is 25%, and the intermediate

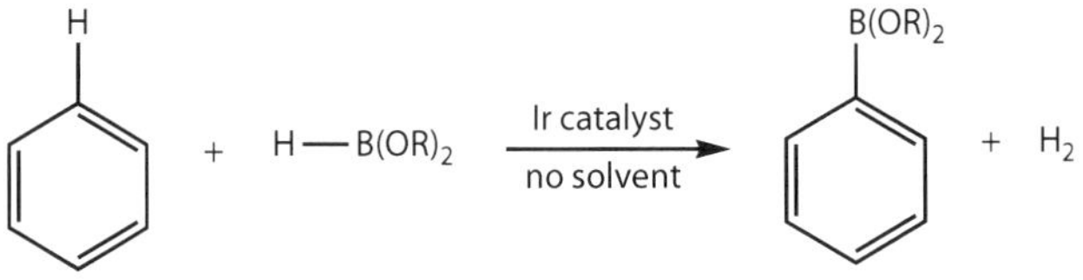

SCHEME 3.8 Iridium-catalyzed synthesis of aryl boronic esters.

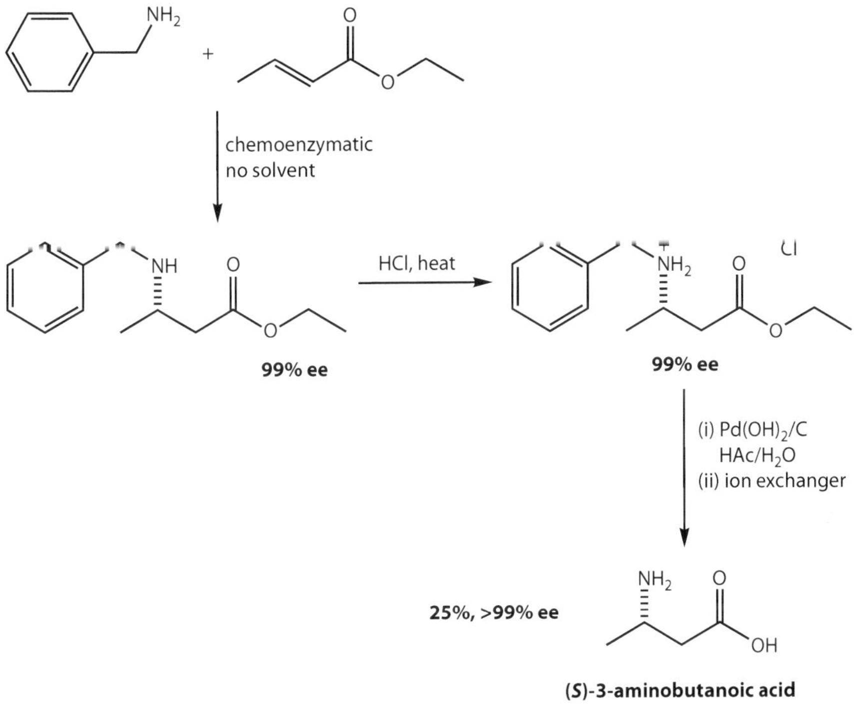

SCHEME 3.9 Partially green synthesis of (*S*)-3-aminobutanoic acid.

hydrochloride salt has to be isolated. These disadvantages produce a very high E-factor of 359 (Section 1.4) when all the reaction and isolation procedures are accounted for.

The same authors have recently designed a greener preparation of (*S*)-3-aminobutanoic acid with a slight improvement in overall yield to 28%.[33] Again, the first chemoenzymatic reaction is performed sans solvent (a thermal aza-Michael reaction between ethyl crotonate and benzylamine, followed by a biocatalytic resolution with a lipase enzyme). The immobilized enzyme is then recovered by filtration and washed with methyl *t*-butyl ether (MTBE) to remove the desired β-aminoester. A number of amide by-products are also present in the MTBE filtrate at this point, one of which is challenging to separate from the β-aminoester without resorting to chromatographic means. A chemoselective ester hydrolysis with NaOH/MTBE is subsequently required to isolate the sodium salt of (*S*)-3-(benzylamino)butanoic acid. After passing this product through an ion exchanger, deprotection with H_2 and Pd/C under acidic conditions leads to the hydrochloride salt of (*S*)-3-aminobutanoic acid in 99% enantiomeric excess (ee) (Scheme 3.10).

Because this synthesis avoids column chromatography and simply utilizes extractions and ion exchange chromatography, the E-factor (including all isolations) falls from 359 to 41, which is in the range of those calculated for the industrial preparation of pharmaceuticals. The lipase enzyme can additionally be recycled and reused several times, albeit with a slight decrease in activity. It should be noted, however,

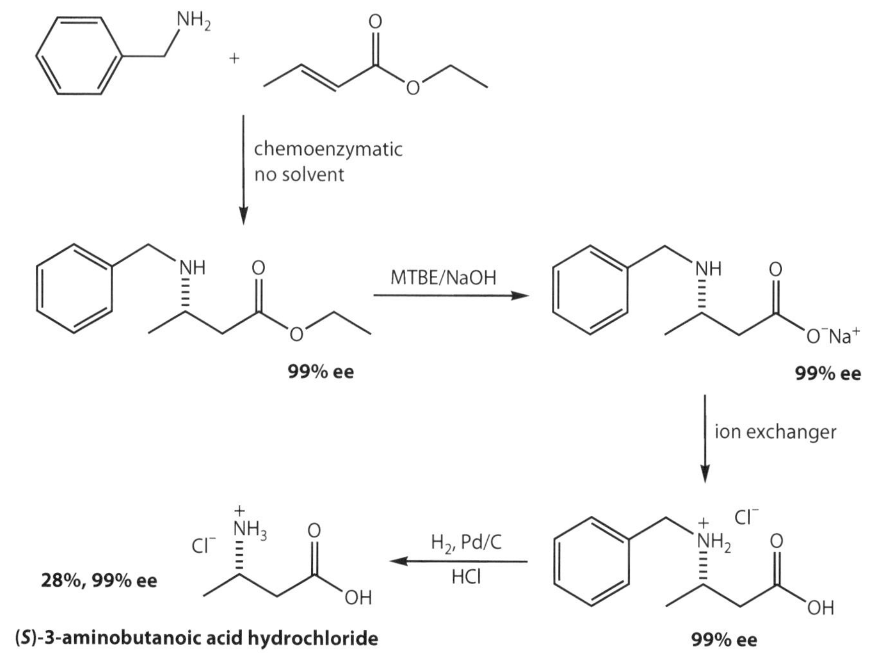

SCHEME 3.10 Greener synthesis of (*S*)-3-aminobutanoic acid hydrochloride.

that both thermal aza-Michael reactions in Schemes 3.9 and 3.10 require an excess of benzylamine (2.2 equivalents compared with 1.0 equivalent of ethyl crotonate). As such, some benzylamine is acting as a solvent in the first transformation, even though no "extra" solvent is added.

3.4 SOLVENT-FREE REACTOR DESIGN

Having undergraduates learn about the possibility of performing reactions in the absence of solvent facilitates a first-rate opportunity to couple principles of organic synthesis, thermodynamics, and chemical engineering. An important industrial focus is the design and manufacture of new reactors that can deal with the challenges associated with solvent-free processes. These include the lack of a heat sink for exothermic reactions and stirring problems/inefficient mixing associated with liquid viscosity. This section presents several examples of how large- and small-scale reactor design has developed to solve such issues.

3.4.1 Spinning Tube-in-Tube Reactors for Ionic Liquid Synthesis

Ionic liquids are covered in detail in Section 5.5, and represent reusable, nonvolatile, and polar replacements for more conventional organic solvents. These compounds are commonly synthesized by the S_N2 reaction of a heterocycle with an alkyl halide, in either the presence or absence of solvent (Section 5.5.2.1). However, an organic solvent such as diethyl ether or dichloromethane is usually

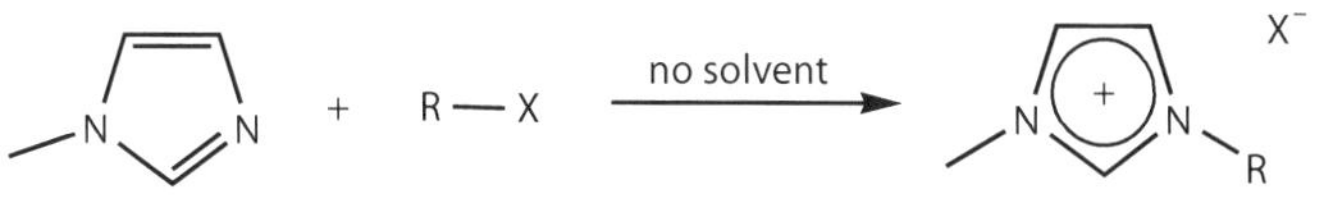

R = Et, iPr, *t*-Bu, *n*-Bu, hexyl, octyl, benzyl
X = Cl, Br, I, OTs, OTf

SCHEME 3.11 Ionic liquid synthesis using a spinning tube-in-tube reactor.

required in the reaction workup steps as part of product isolation and purification. In addition, these reactions often do not go to completion, with batch syntheses demanding high temperatures and lengthy reaction times. If ionic liquid syntheses are run solvent-free using regular laboratory glassware and heating apparatus, ineffective stirring often leads to product separation from reactants. Many ionic liquids are very viscous in nature, so this is a common problem associated with their preparation.

Researchers have developed both ultrasound and microwave-assisted approaches to ionic liquid synthesis.[34,35] These both have limitations in terms of the reactant quantities that can be utilized in a given time. Gonzalez and Ciszewski have reported use of a spinning tube-in-tube (STT) reactor, which affords both good temperature control and continuous flow, leading to an improvement in mixing effectiveness.[36] A thin film of reactants is generated by allowing a heterocycle (e.g., 1-methylimidazole) and an alkyl halide to flow into a small gap between a rotating cylinder and a static shell. This gap is typically less than 0.5 mm in width. Extremely rapid rates of rotation aid film formation (up to 12,000 rpm). The high-purity ionic liquid formed flows along the cylinder surface to an outlet where it is collected. Scheme 3.11 illustrates the reactions studied using an STT reactor.

When 1-methylimidazole is combined with 1-bromohexane (1.05 equivalents) at 131°C in an STT reactor, over 99% conversion of reactants is observed. Over 18 kg of ionic liquid product can be synthesized by this strategy per day. Unsurprisingly, alkyl chlorides and more hindered alkyl halides exhibit lower reactivity and require higher temperatures. The reaction rate is dependent on the shear rate, and altering the spin rate and gap size allows scale-up in the reactor to be straightforward. The continuous flow methodology allows for a ready change of reagents (e.g., from one alkyl halide to another) and facile fine-tuning of reaction conditions. It should not be overlooked that the STT reactor also permits real-time monitoring by analytical techniques—one of the twelve principles stated in Section 1.3. In this instance, ^{1}H and ^{13}C nuclear magnetic resonance (NMR) spectra can be obtained only six minutes after reaction commencement in order to analyze for product purity.

In a related process, a solvent-free synthesis of ionic liquids using a domestic microwave oven with water moderation has been reported.[37] This is in an attempt to address a significant challenge associated with the absence of reaction solvent, which is the possibility of overheating and even thermal "runaways." The latter occurs when the rate of heat removal from a reaction does not correlate with the rate of heat production. Although undergraduates often consider solvents to be merely "there to dissolve reactants," they provide a valuable heat sink in many cases.[38] In this instance, 1-bromobutane and 1-methylimidazole are mixed in a flask equipped with a drying tube and placed inside a household microwave oven.[37] The flask and tube are immersed in a beaker containing 500 mL of water at 60°C and irradiated for a few minutes. During this time the water temperature rises to 80°C and the ionic liquid forms quantitatively. General research into reaction calorimetry is currently of great interest so that transformations can be appropriately undertaken on an industrial scale.[39] Reactants can now be mixed in extremely narrow channels (diameters of down to 10 μm) in so-called microreactors to manage significantly exothermic solvent-free reactions.[40]

3.4.2 A Spinning Disc Reactor for Polymerizations under Ultraviolet Irradiation

On an industrial scale, addition polymers are often formed by free radical mechanisms involving initiation by heating. Many of these reactions are undertaken under low concentrations in an organic solvent. High heats of polymerization and rapid propagation rates mean that special reactors are necessary. Dunk and Jachuck have disclosed details of a novel spinning disc reactor that has been used for large-scale polymerizations of ethene derivatives using ultraviolet irradiation.[41] The key to this technology is formation of extremely thin reaction films (100–300 μm) by reactor surface rotation, which allows for efficient mixing and effective heat dissipation. This is similar in approach to the STT reactor described in Section 3.4.1. The thin film also facilitates consistent irradiation characteristics. Ultraviolet irradiation is considered an effective technique, as it allows polymerization control with a rapid initiation rate, which forms a polymer with low polydispersity. It is possible to synthesize up to 40 kg of polymer using this type of reactor in one hour, even though the diameter of the apparatus is only 180 mm.

3.4.3 Organic Reaction Scale-Up Using Focused Microwaves

Chapter 8 describes how commercial microwave reactors can be incorporated into an undergraduate curriculum in terms of microscale or semi-microscale chemistry. Cléophax et al. have dealt with the challenge of solvent-free reaction scale-up by utilizing a microwave batch reactor. This permits preparation of several hundred grams of product, and can be adapted to a range of reaction types, including alkylations, heterocyclic phenacylations, halogenations, epoxidations, and dealkylations.[42] The Prolabo Synthwave 1000 reactor has a 1 L reaction vessel, mechanical stirrer, and dropping funnel for reagent addition.[43] Using this setup, reactions are irradiated at a maximum power of 800 W for very short time periods. As an example, the S_N2 reaction of a potassium carboxylate with a primary alkyl bromide proceeds smoothly under microwave heating for only five minutes on a 2 M reaction scale (Scheme 3.12). Here, the phase transfer catalyst Aliquat 336 (Section 3.3.1) is required at an imposed temperature of 160°C. The total reaction mass (carboxylate salt + alkyl bromide + phase

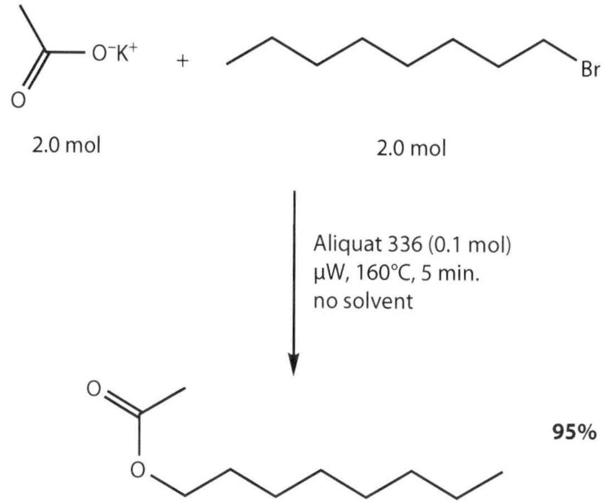

SCHEME 3.12 Preparation of *n*-octyl acetate via batch reactor microwave irradiation.

transfer catalyst) is over 620 g. However, the ester product is extracted with diethyl ether to separate it from the catalyst.

3.4.4 Synthesis of Polyglycerol-3 Laurate in a Bubble Column Reactor

The preparation of polyester surfactants under biocatalytic conditions presents a challenge due to the highly viscous conditions, among other factors.[44,45] The reaction between lauric acid and polyglycerol-3 using Novozym 435 (a lipase enzyme) as a catalyst has been studied in detail[44] (Scheme 3.13). The polyglycerol-3 laurate surfactant cannot be generated via this route using conventional fixed-bed technology, as this is limited to reactants and products having a low enough viscosity to be pumped through an immobilized enzyme bed. A stirred tank reactor is additionally impractical due to degradation of the enzyme by mechanical stresses. Correspondingly, a new type of approach (known as a bubble column reactor) has been utilized that facilitates synthesis of a wide range of surface active agents. Bubble columns have typically been employed in two-phase systems where interaction between a liquid and a gas phase is difficult.[46] For surfactant preparation, the main principle is development of an alternative mixing process that forms a reactant emulsion while avoiding enzyme breakdown. In this bubble column reactor, generation of polyglycerol-3 laurate commences as a four-phase system (polyol, carboxylic acid, enzyme, and pressurized air). The air serves two important functions: it removes water formed during the condensation and aids mixing of the lauric acid and polyglycerol-3 without significantly affecting lipase stability. The rate-limiting step during synthesis of polyglycerol-3 laurate is reactant mixing, but water removal is also a necessity for a complete reaction to take place. At 75°C the Novozym 435 enzyme can be reused up to nine times, although the reaction half-life is 19.9 hours.[44] Characterization of the system shows that the viscosity increases by a factor of 20 as the reaction proceeds, with pilot plant scale-up being achieved.

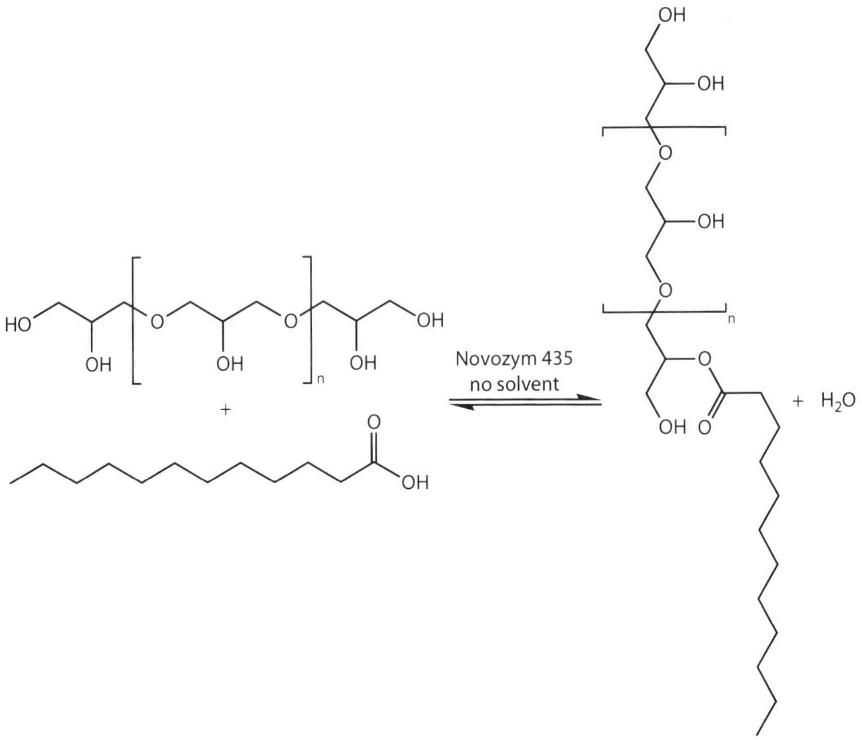

SCHEME 3.13 Enzyme-catalyzed polyglycerol-3 laurate synthesis from polyglycerol and lauric acid.

3.4.5 Solventless Reactor Cleaning

An underappreciated role of solvents in an industrial venue is for cleaning of large-scale reaction vessels, following target compound synthesis. Indeed, at the pilot plant level, the amount of solvent needed for this purpose is approximately two or three times that employed in the synthesis itself.[47] Solvents used for cleaning include methanol and acetone, which are relatively green, and are recycled for further use. Recycling of waste solvent is possible via distillation (especially if the solvent is low boiling), but this comes with a concomitant energy cost. A significant problem exists with simply rinsing vessels with the reaction solvent, as unwanted by-products are often structurally different than the desired compound, and therefore insoluble, meaning they cannot be removed. Alternative approaches that have found success are use of water jets under elevated pressure[48] or cleaning with aqueous hydrogen peroxide.[49] Total elimination of organic solvents in this regard would have a significant impact on the environmental profile of industrial synthetic chemistry.

3.5 ELIMINATING SOLVENTS IN THE INTRODUCTORY ORGANIC LABORATORY

3.5.1 Introduction

The wide range of solventless reactivity designed for the undergraduate organic laboratory has been recently documented.[50] In this paper, twenty-seven reactions from the chemical education literature are highlighted from 2000 until the beginning of 2009. Reactions are primarily included from laboratory textbooks and peer-reviewed journal articles. The procedures are grouped into three general types according to the experimental technique employed in each one. These are microwave/visible light irradiation, conventional heating, and room temperature mechanical mixing (including reactant grinding and ambient stirring). Each technique is divided further into three reaction types where appropriate: C-C bond formers, C-N bond formers, and functional group conversions involving C-H or C-O bond-forming processes.

It is not the intention of this chapter section to include a comprehensive review in the same style of that article. Rather, Section 3.5 focuses on a smaller number of reactions that are all appropriate for inclusion in the introductory organic curriculum. Some of these have been considered previously,[50] but others have been published during the last two years and are very worthy of being highlighted. Reactions are broadly delineated into two general categories: (1) those not requiring any solvent at all during the procedure and (2) those using an environmentally benign solvent for product isolation or purification. Processes have been selected that are practically straightforward, and that proceed by a range of different mechanisms under different experimental conditions. Appropriate theory regarding reactions taking place sans solvent is included where applicable. This background information will be of interest to both instructors and students in combination with the reactions themselves. For organizational reasons, solvent-free reactions performed under microwave irradiation are elaborated upon in Chapter 8. Finally, in an attempt to provide the bigger picture, Section 3.5 concludes with an analysis of several solvent-free reactions that are somewhat less green in other key areas.

3.5.2 Transformations Avoiding Any Solvent Use

3.5.2.1 Carbonyl Condensation Reactions

A number of pedagogical solventless condensation reactions have been designed that are straightforward to undertake, as they simply involve solid-solid or solid-liquid grinding. Cave and Raston combined two such processes sequentially by linking a mixed aldol condensation with a Michael reaction[51,52] (Scheme 3.14). In the first step, liquid acetophenone (9.42 mmol) is ground with one equivalent of solid sodium hydroxide using a pestle and mortar, and then one equivalent of liquid benzaldehyde is added. Further grinding for fifteen minutes forms the yellow chalcone product as a free-flowing powder, after the viscosity of the mixture increases. The chalcone can be characterized and reacted further with a heterocyclic ketone to form a Michael

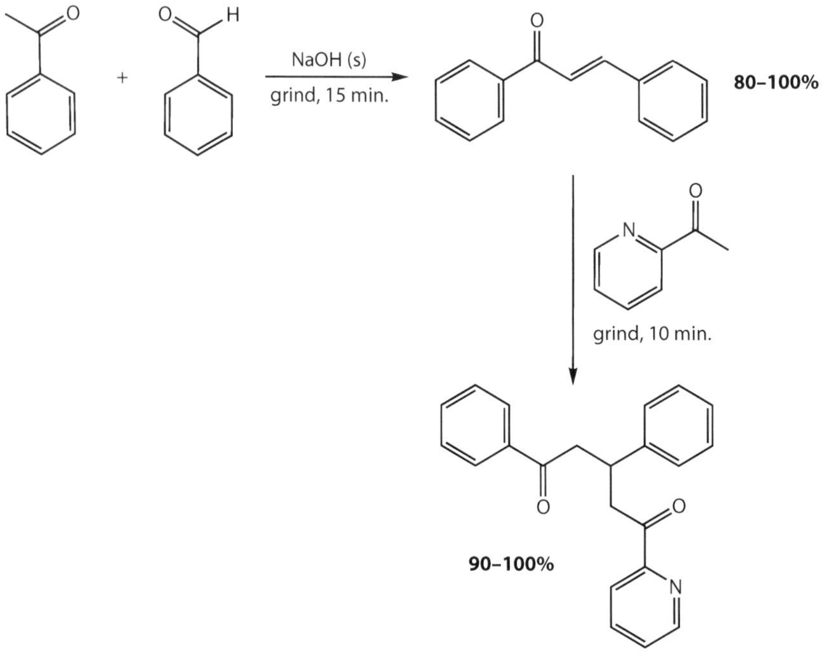

SCHEME 3.14 Consecutive mixed aldol and Michael reactions.

addition product. This is easily achieved by adding 2-acetylpyridine (9.42 mmol) to the mortar and grinding for ten minutes. The 1,5-diketone generated is sufficiently pure enough to be heated with ammonium acetate in acetic acid solvent to form a Kröhnke pyridine.[51]

The same authors have written an excellent review on the nature of several types of solventless reactions.[38] Aldol condensations of the type described here are considered, particularly ones where a solid aldehyde bearing no α-hydrogen atoms is ground together with a solid ketone in the presence of a base (rather than a liquid aldehyde reacting with a liquid ketone). In the former instances, the assertion is made that a *eutectic mixture* (with a melting point lower than room temperature) is formed. This means the reaction mixture is best described as a mutual solution of the carbonyl compounds in which the condensation product precipitates. Instantaneous dehydration seems to be important for this to occur smoothly. Indeed, it has been argued that "the existence of a liquid phase is a prerequisite for reaction in these systems."[38]

Palleros has also described chalcone synthesis under similar conditions in the undergraduate laboratory.[53] Twenty different products can be formed in very good yields by employing a range of *meta-* and *para-*substituted benzaldehyde and acetophenone derivatives (Scheme 3.15). Of these twenty, seventeen are formed as pure solids within ten minutes of grinding. At least one liquid reactant was used in most of the combinations, and during reactions where two solids were ground together, a eutectic mixture formed before the addition of base. The three chalcone products that did not form pure solids generally had lower melting points (<80°C). These

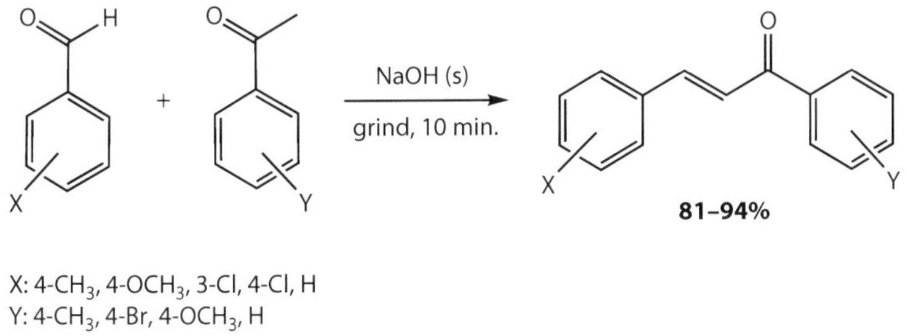

SCHEME 3.15 Chalcone syntheses via mixed aldol condensations.

compounds (e.g., X = 4-OCH_3, Y = H) tended to remain as liquids, as any by-products lowered their melting points. From a theoretical perspective, Gálvez et al. have recently attempted to predict the times of solvent-free aldol reactions undertaken by conventional stirring or ball milling.[54] Their approach invokes molecular topology principles, where reactant structures are characterized by subsets of topological indices.

Reactions such as these clearly reduce the amount of waste formed in the laboratory, and two related experiments are highlighted in this context in Section 7.3.4. One of these is the aldol condensation of two solid reactants: 1-indanone and 3,4-dimethoxybenzaldehyde.[55] The other is solventless imine formation by reaction between *p*-toluidine and *o*-vanillin.[56]

3.5.2.2 Diels-Alder Reactions and Their Analysis

Diels-Alder reactions are routinely used to introduce pericyclic mechanisms to undergraduate organic students. They have the green advantage of being 100% atom economic and produce good yields if the correct types of diene and dienophile are selected. A range of aqueous Diels-Alder reactions and an aqueous hetero-Diels-Alder reaction are profiled in Section 4.3.2, and a similar process in polyethylene glycol solvent is outlined in Section 5.6.2. Microwave Diels-Alder reactions are discussed in Section 8.8.5. McKenzie et al. have taken the green Diels-Alder reactivity one step further by mixing an allylic dienol with maleic anhydride in the absence of any reaction or workup solvent[55] (Scheme 3.16). This is a variation on a more traditional strategy, where the same reactants are refluxed in toluene for five minutes to afford

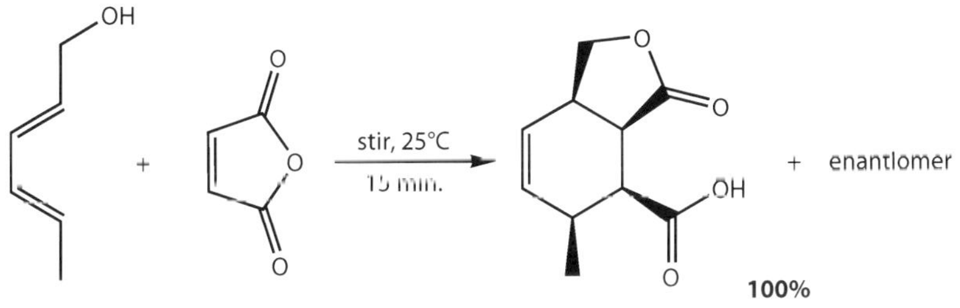

SCHEME 3.16 Diels-Alder reaction of (2*E*,4*E*)-hexa-2,4-dien-1-ol with maleic anhydride followed by intramolecular nucleophilic acyl substitution.

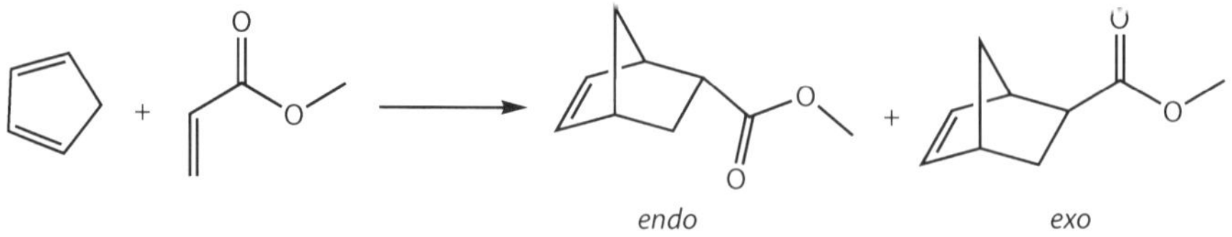

SCHEME 3.17 Diels-Alder reaction between cyclopentadiene and methyl acrylate.

the lactone product, after intramolecular nucleophilic acyl substitution occurs on the initial adduct.[57] In the solventless microscale approach, (2*E*,4*E*)-hexa-2,4-dien-1-ol (0.444 mmol) and maleic anhydride (0.449 mmol) are simply combined in a beaker and stirred with a spatula at room temperature for fifteen minutes. The solid product precipitates and the presence of a carboxylic acid can be determined by the reaction of a small amount with aqueous sodium bicarbonate. Although sixteen possible stereoisomers could be formed in this reaction, only two are in fact generated, and product formation is quantitative.

The greenness of alternative reaction media has been assessed within the framework of another Diels-Alder reaction.[58] Reinhardt et al. have studied the transformation of cyclopentadiene and methyl acrylate into the corresponding endo and exo adducts (Scheme 3.17). They analyzed the reaction taking place in different solvents (an ionic liquid, methanol, methanol/water, acetone, cyclohexane, and a citric acid/*N*,*N*'-dimethyl urea melt). For comparison purposes they also monitored the solventless reaction. Each alternative was analyzed by a simplified life cycle assessment (SLCA) approach, where reactant and solvent production is accounted for, along with synthesis, product workup, waste disposal, and recycling, where appropriate.

The solvent-free version of this Diels-Alder reaction fared very well in comparison with its competitors during the analysis. At room temperature and over a forty-eight-hour reaction time, 98% of the methyl acrylate was converted into product (only bettered by the citric acid/*N*,*N*'-dimethyl urea melt at 25°C). It was noted, however, that the endo:exo ratio was slightly lower than that obtained for reactions occurring in solvents. The authors evaluated the reaction energy factor (EF) for each system, which calculates the cumulative energy demand resulting from the following: (1) supply of reagents, solvents, and any auxiliaries; (2) reaction performance; (3) workup procedures; (4) product application; and (5) waste disposal. Interestingly, the EF was smallest for the solventless reaction by a significant margin compared to most alternatives, apart from methanol and methanol/water. The EF for reagent supply is higher for the solventless process than for methanol and aqueous methanol due to the lower observed endo:exo product ratio. However, the EF per kilogram endo product is relatively low, as no extra energy for solvent supply is needed, and the product workup does not require any additional solvent removal (e.g., by distillation).

The environmental and human health factor (EHF) for each condition was analyzed by comparing the reagents, solvents, and auxiliaries used, in terms of their human and environmental risk. In general, chemical acute toxicity, chronic toxicity, and water-mediated impact are evaluated in this approach. The solvent-free system had slightly higher EHF values than other solvents due to greater starting material consumption. Finally, the cost factor (CF) represents the sum of chemical prices and energy, disposal, equipment, and process costs and was calculated in each case. Unsurprisingly, the solventless reaction exhibited lower costs, in keeping with the more conventional solvents studied. In summary, the authors concluded that "the solvent system methanol/water or the solvent-free synthesis seem to be the most ecological sustainable alternatives, yet."[58]

3.5.2.3 Cocrystal Controlled Solid-State Synthesis

Cocrystal controlled solid-state synthesis (abbreviated C^3S^3) is an approach to reactivity that can easily be implemented in the organic laboratory.[59] Cocrystals are formed when two solid materials become crystallized together (in either the presence or absence of solvent). In doing this a new crystal lattice structure is formed, which has demonstrably different characteristics than the lattice of either initial solid. The cocrystal lattice exhibits intermolecular attractive forces (typically hydrogen bonds) between functional groups associated with the original two solids. These functionalities can sometimes be held in such close proximity to one another that it is possible for a reaction to take place. Cocrystal formation between hydroquinones and bis-(*N*,*N*-dialkyl)terephthalamides has been of great interest to researchers at Polaroid (Section 1.4). It is important to note that the C^3S^3 technique differs from more classical solid-state synthesis. In the latter, the reactants do not have their lattice structure disrupted and reaction occurs via a less well-defined system.

Cheney et al. have published a student experiment where a dianhydride (0.11 mmol) is ground together with *m*-aminobenzoic acid (0.22 mmol) in the absence of any solvents to generate a pale yellow cocrystal.[60] After sand bath heating for over two hours, the cocrystal is converted into a diimide, which can be characterized by standard spectroscopic techniques, as well as x-ray diffraction and differential scanning calorimetry (Scheme 3.18). Other aldehydes and amines can be employed to provide experimental variety among an undergraduate class. Alternatively, a solvent-drop grinding method allows a tiny amount of solvent (typically *N*,*N*-dimethylformamide) to be introduced to the solid mixture.

In a partly related procedure, cocrystal formation between resorcinol and *trans*-1,2-bis(4-pyridyl)ethene forms the basis of a solvent-free [2 + 2] photodimerization, forming a tetrasubstituted cyclobutane derivative[61] (Scheme 3.19). Equimolar amounts of the diol and alkene derivative are combined in ethanol solvent to form a cocrystal, which is collected by vacuum filtration and characterized. The cocrystal is composed of four molecules that are linked to one another by four hydrogen bonds between the pyridine N atoms and the H atoms of the aromatic alcohol groups. The cocrystal system is set up to efficiently undergo photodimerization in the solid state. The two alkene double bonds are aligned in parallel and separated by 3.65 Å, which is within the 4.2 Å requirement previously reported.[62] The cocrystal is ground to a fine powder, placed between two transparency films, and irradiated in a photoreactor for eighteen hours. Alternatively, the sample can be exposed to sunlight for a period of six hours during either winter or summer months. After this time the cyclobutane product is detected by proton NMR spectroscopy and the template molecules

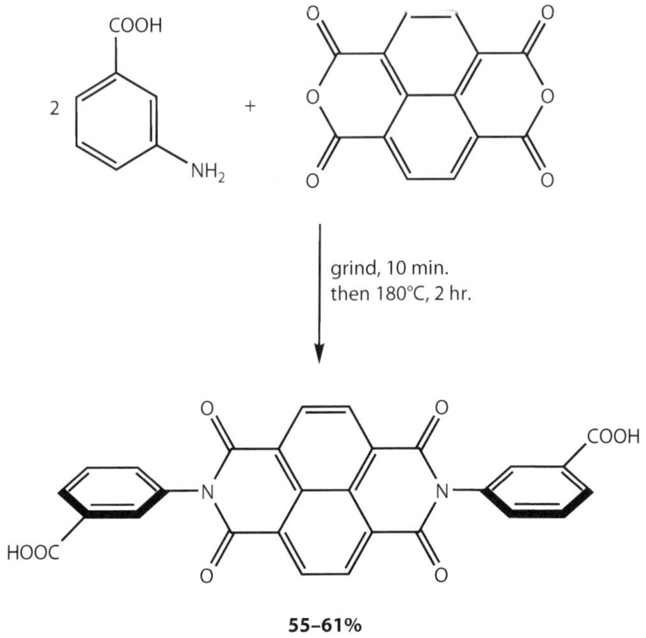

SCHEME 3.18 Cocrystal controlled solid-state synthesis (C^3S^3).

recycled. This gives students an insight into aspects of supramolecular and solid-state chemistry in the context of more familiar topics, such as hydrogen bonding and pericyclic mechanisms.

3.5.3 Solvent-Free Reactions Requiring an Environmentally Benign Solvent during Workup

3.5.3.1 Semicarbazone Derivative Preparation

A classical method for qualitative characterization of aldehydes and ketones (both aliphatic and aromatic) is via reaction with semicarbazide hydrochloride. This is usually achieved in methanol solvent in the presence of pyridine (to form the semicarbazide free base), followed by product recrystallization from ethanol or ethanol/water.[63] The semicarbazone product is identified by its melting point, providing indirect confirmation of the carbonyl compound structure. Pandita et al. have reported an alternative convenient procedure that takes advantage of grinding methodology.[64] Semicarbazide hydrochloride (1 mmol) is ground with sodium acetate to form a liquid, followed by the addition of a carbonyl compound (1 mmol). Further grinding for five minutes, followed by product isolation with water and recrystallization from ethanol, leads to the pure semicarbazone derivative (Scheme 3.20). This strategy has led to the characterization of fourteen different semicarbazone products.

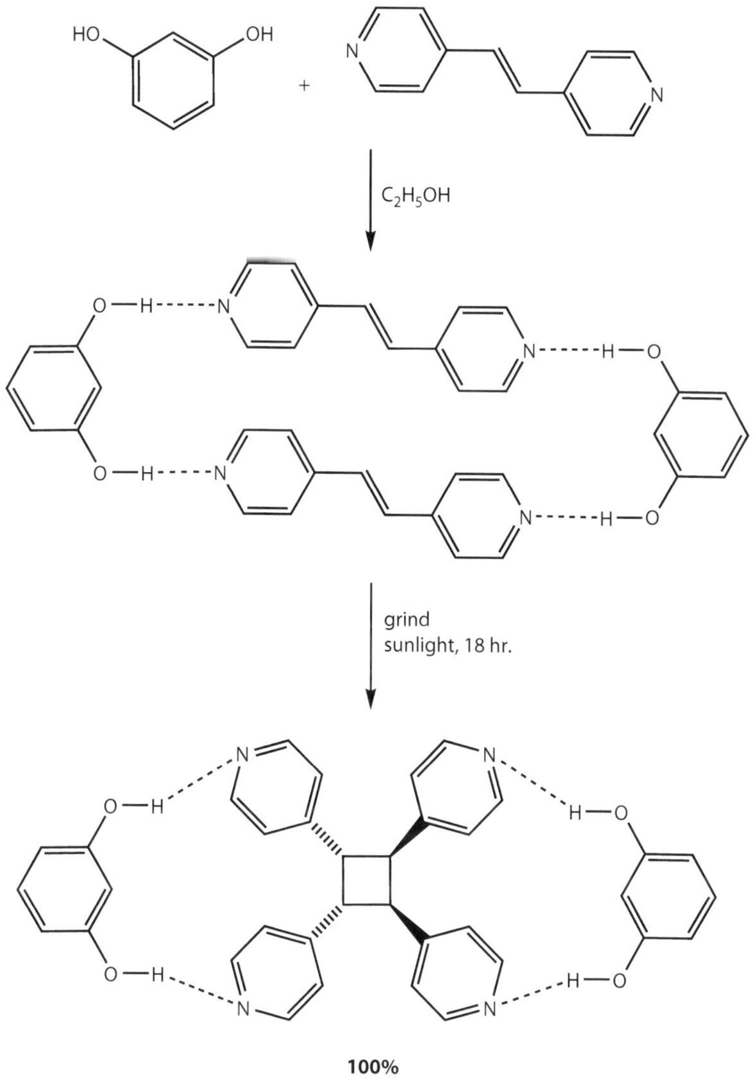

SCHEME 3.19 Solid-state [2+2] photodimerisation of *trans*-1,2-bis(4-pyridyl)ethene:resorcinol cocrystals.

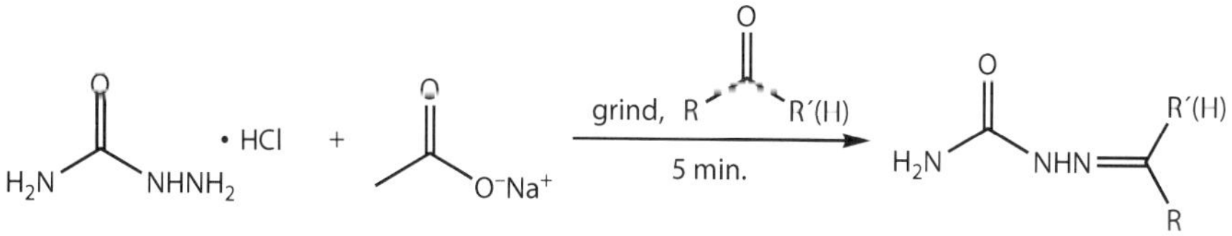

SCHEME 3.20 Semicarbazone derivative synthesis from aliphatic and aromatic carbonyl compounds.

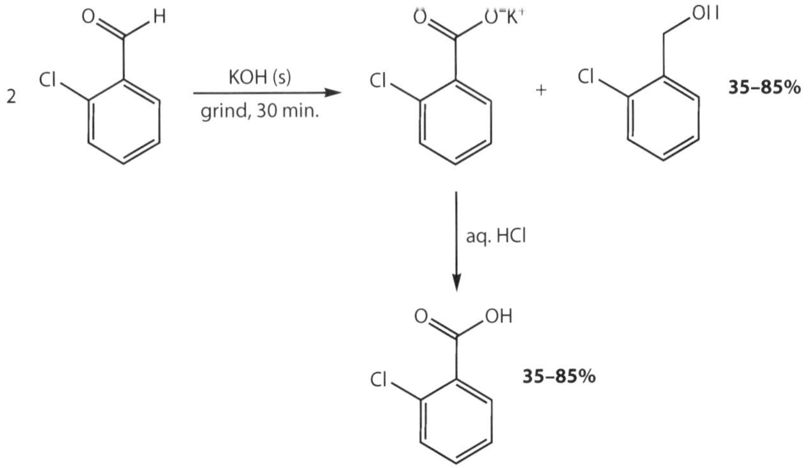

SCHEME 3.21 Cannizzaro disproportionation of 2-chlorobenzaldehyde.

3.5.3.2 Cannizzaro Disproportionation Reactions

The Cannizzaro reaction of an aromatic aldehyde with an hydroxide ion to form a carboxylate anion and a primary alcohol has been studied as a pedagogical tool. 1-Naphthaldehyde can be used as the reactant and heated with potassium hydroxide in a solventless procedure, forming 1-naphthalenemethanol and 1-naphthoic acid.[65] The former product is isolated by extraction with ether. In comparison, disproportionation of 2-chlorobenzaldehyde can be performed under grinding conditions using potassium hydroxide (KOH) at room temperature.[66] The liquid aldehyde and solid base are ground together using a pestle and mortar for thirty minutes, during which time a thick paste is formed, and water is added (Scheme 3.21). The solid 2-chlorobenzyl alcohol is insoluble in water and collected by vacuum filtration. In contrast, potassium 2-chlorobenzoate is soluble in water and 2-chlorobenzoic acid precipitated by acidification. As such, organic solvents are eliminated during the workup protocol. The experiment can be easily tailored to adopt a guided inquiry approach. Here, students are empowered to design their own experimental procedure and to use the twelve principles as an analytical guide. Although the model Cannizzaro disproportionation may not specifically be taught during the lecture component of an introductory organic class, it represents a useful teaching reaction from both mechanistic and green chemistry perspectives.

3.5.3.3 3-Carboxycoumarin Synthesis

As outlined in Section 3.2, certain coumarins are compounds of industrial interest as laser dyes. The coumarin nucleus is also present in a wide array of biologically relevant substances, from the anticoagulant Warfarin® to plant growth regulators, fungistats, and bacteriostats.[67,68] 3-Carboxycoumarins can be synthesized by the condensation of Meldrum's acid with salicylaldehyde or an appropriately substituted derivative, in the presence of catalytic ammonium acetate[69,70] (Scheme 3.22). On grinding *o*-vanillin (R = OCH_3) with Meldrum's acid in the absence of a catalyst,

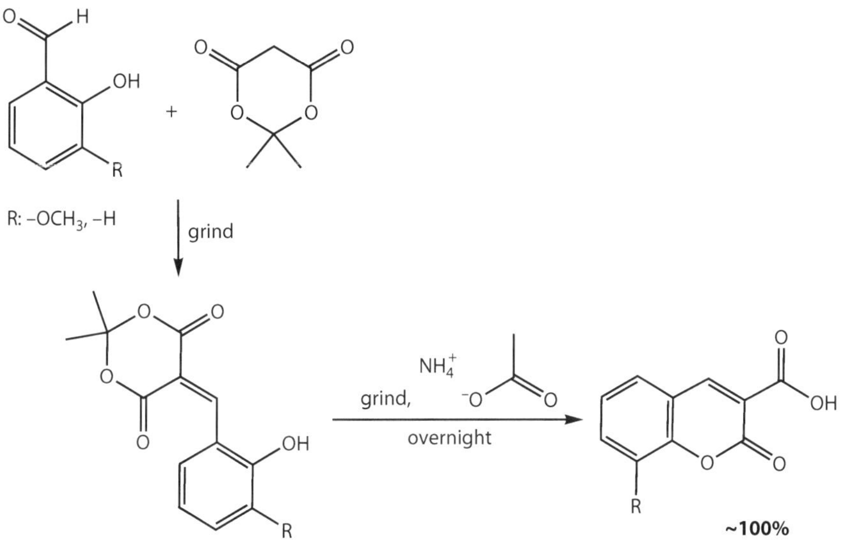

SCHEME 3.22 Coumarin synthesis from two salicylaldehydes and Meldrum's acid.

the conjugated benzylidene intermediate product is formed via a Knoevenagel condensation. Grinding at room temperature in the presence of ammonium acetate, followed by overnight standing, leads to the 3-carboxycoumarin product in quantitative yield. The reaction is not quite 100% atom efficient, as one molecule of acetone is eliminated on coumarin formation. The product is purified by an aqueous wash that removes the catalyst. This solution can be reused in subsequent reactions. Indeed, although Meldrum's acid and the salicylaldehyde are insoluble in water, these reactions also proceed very effectively under aqueous conditions. This reactivity has been incorporated into the teaching of green chemistry in a final-year undergraduate course at Monash University, Australia.[70]

3.5.3.4 Two Multicomponent Named Reactions

Multicomponent reactions (ones where three or more reactants are combined in one pot to form a product) are currently creating a lot of interest from a green perspective. These processes usually occur efficiently with high atom economies, and are often utilized to make substances of medicinal interest. An example is the Biginelli reaction between a β-ketoester, an aldehyde, and urea, as highlighted in Section 2.6.4.[71] This reaction dates from 1893, and was largely underused by researchers until the discovery that the 3,4-dihydropyrimidin-2(1H)-ones formed are biologically active as calcium channel blockers.[72] In this regard, Biginelli products often exhibit vasodilatory activity and cause blood vessels to relax (dilate). The Biginelli reaction was originally performed in refluxing ethanol with an HCl catalyst for several hours to achieve moderate yields.[73] In 2001, Holden and Crouch published a microscale synthesis of the target compound in Scheme 3.23. This reduced the reaction time to ninety minutes, making it amenable to a three- or four-hour undergraduate laboratory period.[74] This particular transformation (a traditional

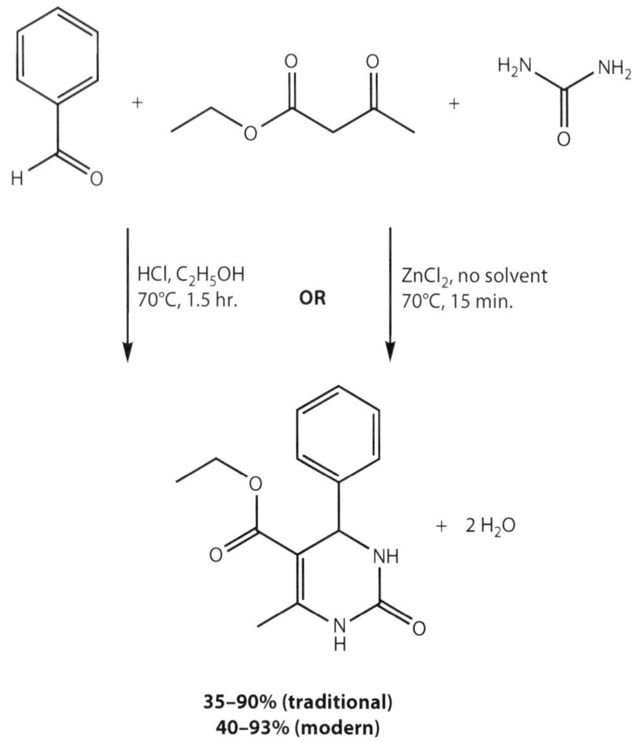

SCHEME 3.23 Traditional and modern Biginelli syntheses.

synthesis) has recently been coupled with a modern solvent-free reaction. In the latter strategy, the same starting materials are combined with a Lewis acid catalyst (zinc(II) chloride) and heated for fifteen minutes[75] (Scheme 3.23). One can see that the range of isolated yields for the traditional vs. the modern method is almost identical. Students are able to individually perform and analyze both reactions during a three-hour laboratory period.

From a green perspective, the solvent-free modern Biginelli synthetic method affords a number of advantages. A sixfold rate acceleration is evident (fifteen-minute reaction time contrasting with ninety minutes for the traditional approach), leading to improved energy efficiency. The intrinsic atom economy is 88% (two water molecules are eliminated during the transformation), which clearly represents a benign by-product. However, when the experimental atom economy for each method is compared, an important difference arises. Accounting for reactant and product masses indicates the modern method is 80% atom efficient, whereas the traditional approach is only 72% atom efficient. Students are surprised when calculating overall reaction efficiencies (Section 1.4), as they discover that only around 40%–50% of the reactant atoms actually make their way into the isolated 3,4-dihydropyrimidin-2(1H)-one product! Both methods utilize environmentally benign solvents (water and ethanol) during purification, and recrystallization is not required in either instance, so energy savings are achieved. Students are encouraged to adopt the asymptotic approach

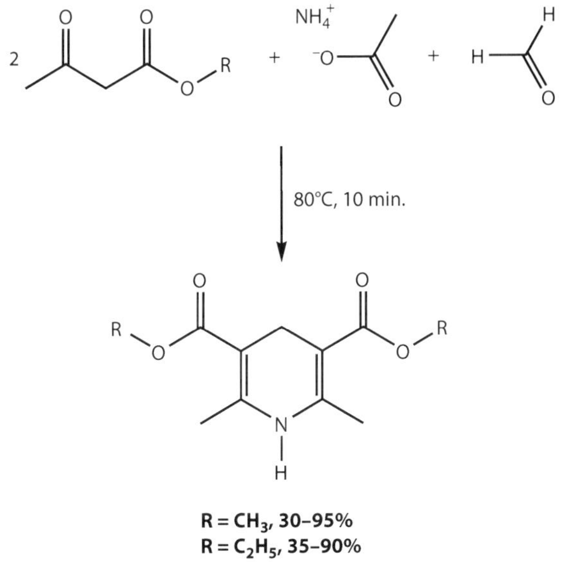

SCHEME 3.24 Hantzsch synthesis of two 1,4-dihydropyridines.

espoused by Goodwin, where the following three postlaboratory questions are posed for experimental work:[76]

- What was green about the experiment?
- What was not green?
- How could the experiment be made greener?

Regarding the third question, students appreciate that neither catalyst (HCl or $ZnCl_2$) is recycled during each reaction and, therefore, can design how to adapt their synthesis to change this.

In a related reaction, the Hantzsch synthesis of 1,4-dihydropyridines can be undertaken without any additional organic solvent needed.[77] One of two β-ketoesters is combined with ammonium acetate as the nitrogen source and aqueous formaldehyde, without any added catalyst (Scheme 3.24). On heating for ten minutes, a bright yellow product precipitates, which is collected by vacuum filtration and recrystallized from ethanol. The isolated products are structurally analogous to the commercially available vasodilators nifedipine and lacidipine[78] and showcase the 1,4-dihydropyridine ring as a "privileged structure."[79] It should be noted that a small amount of water is present during the reaction that comes from the aqueous formaldehyde.

3.5.3.5 Polysuccinimide Synthesis

Thermal polyaspartate (TPA) is a biodegradable, eco-friendly, and nontoxic polymer that has several uses in agricultural venues to improve the uptake of nutrients by plants. It can also be used as part of a water treatment program, and to stimulate

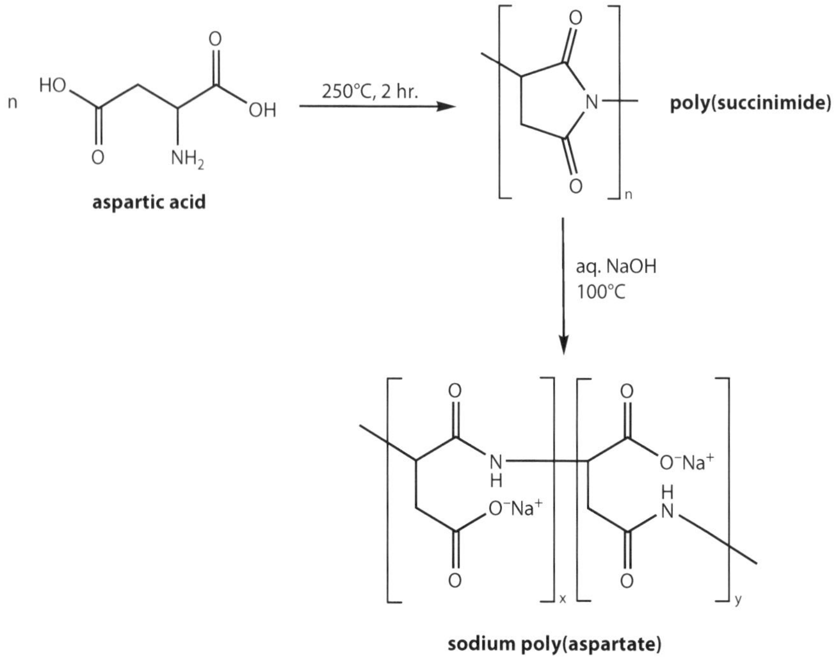

SCHEME 3.25 Polymerization of aspartic acid to polysuccinimide and sodium polyaspartate.

the proficiency of gas and oil manufacturing. The biodegradability of TPA affords an advantage over other polymers, such as polyacrylic acid, which does not readily break down. In 1996, Donlar Corporation (now NanoChem Solutions, Inc.) received a Presidential Green Chemistry Challenge Small Business Award for its synthesis and development of TPA.[80] Donlar invented two processes for TPA preparation, both of which typify industrial green chemistry. The first method involves polymerization of a naturally occurring nonessential amino acid (L-aspartic acid) under solventless conditions to form polysuccinimide, in a routine yield of over 97%. Aqueous basic hydrolysis of polysuccinimide subsequently forms TPA. The second method utilizes a polymerization catalyst, allowing the reaction to be performed at a lower operating temperature. The TPA product formed has many positive characteristics, including enhanced biodegradability.

An undergraduate protocol for polymerization of L-aspartic acid to form polysuccinimide (the precursor for TPA) has been developed, which is adapted from the work of the Donlar Corporation.[81] The amino acid (10 mmol) is heated in a beaker using a sand bath at 250°C for two hours, followed by treatment with aqueous acid and base (Scheme 3.25). The hydrophobic polysuccinimide formed is hydrolyzed with 0.1 M aqueous NaOH to form the sodium salt of TPA (sodium polyaspartate). The molecular weight of the polymer can be calculated by titration. Sodium polyaspartate is an example of a step growth polymer, and is also a good example of a polypeptide, both of which are polymeric materials that students are familiar with. The formation of polysuccinimide takes place in a solventless environment, and no organic solvents are used throughout the whole experiment. A superb connection is made between a real-world industrial process and green chemistry classroom issues.

3.5.4 Solvent-Free, but ...

As mentioned in the introduction to this chapter, there have been many educational experiments devised in the last ten years that feature solvent-free reactivity. That this has taken place is laudable, as undergraduates can appreciate that solvents are not essential for organic reactions, and that eliminating them is a worthwhile green chemistry goal. However, it is important to place solventless reactions into context with all twelve principles. More specifically, instructors should train their students to perform a comprehensive assessment of reaction greenness, rather than focusing on a single metric or axiom. The laboratory is an excellent venue to begin to do this.

The Wittig olefination of an aldehyde or ketone is a process that has been modified for the undergraduate organic curriculum in several different ways. The regioselective nature of the reaction is of great importance, as less highly substituted alkenes (which are often problematic to generate by alcohol and alkyl halide eliminations) can be formed. Recently, several solvent-free green procedures for Wittig reactions have been reported, including via microwave heating[82] (Section 8.8.1.2), conventional heating[83] (Scheme 3.26), room temperature stirring[83] (Scheme 3.27), or grinding solid reactants[84] (Scheme 3.28). These reactions are operationally straightforward to undertake, avoid use of potentially hazardous strong bases, and bring to life an otherwise abstract transformation taught in the lecture. Nevertheless, for the

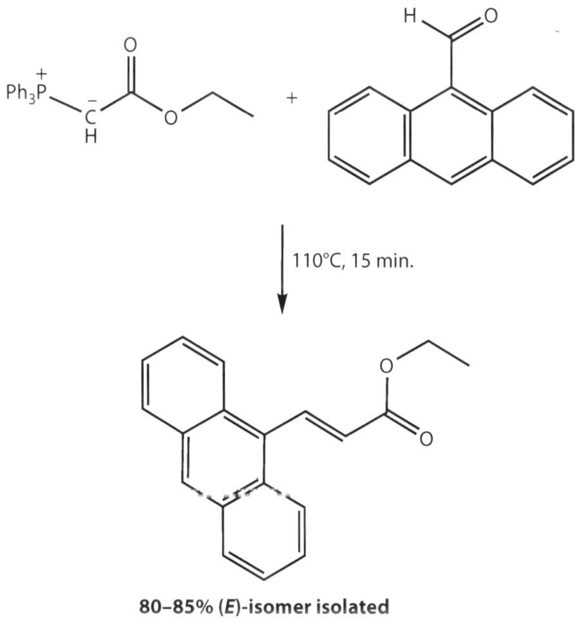

SCHEME 3.26 Wittig synthesis of (*E*)-ethyl 3-(9-anthracenyl)propenoate.

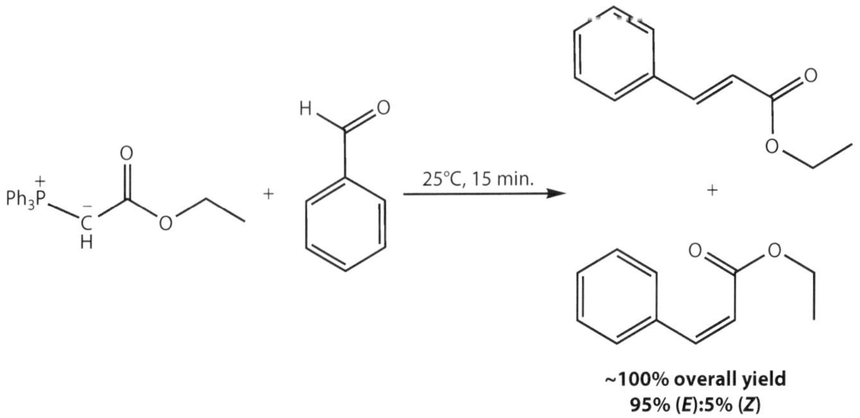

SCHEME 3.27 Wittig synthesis of diastereomeric ethyl cinnamate esters.

sake of balance, it is important to apply a broader perspective to the Wittig reaction. Cann and Dickneider have demonstrated that for the Wittig synthesis of methylenecyclohexane from cyclohexanone, the intrinsic atom economy is only 18%.[85] The primary reason for this is the formation of triphenylphosphine oxide as a by-product. If one calculates the intrinsic atom economies for the reactions as represented in Schemes 3.26, 3.27, and 3.28, they are 50%, 39%, and 33%, respectively, without accounting for generation of the Wittig reagent in each case. This means that when the experimental atom economy and overall efficiency is computed for each reaction, less than half the reactant atoms have their fate in the isolated alkene products. This important observation does not devalue the pedagogical benefits of these reactions in any way, but educators are doing students a disservice if they allow them to think that making a reaction solvent-free automatically makes it

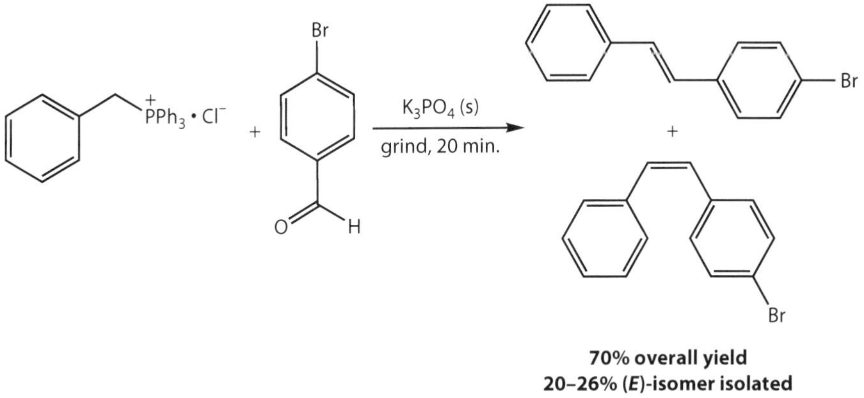

SCHEME 3.28 Wittig synthesis of diastereomeric stilbene derivatives.

completely green. Chapter 2 illustrates that there are many complex issues at play when teaching green chemistry. The approach taken by Goodwin[76] (described in Section 3.5.3.4) regarding questions to ask at the end of a laboratory experiment has special pertinence again here.

3.6 CONCLUSION

This chapter has described the importance and utility of solventless reactivity from both an instructional and a commercial perspective. Reactions sans solvent are now commonplace in teaching, academic, and industrial laboratories, with more examples reported daily in the primary literature. Undertaking organic chemistry without a solvent seems counterintuitive to many students, but is in keeping with several of the twelve principles, most notably those addressing waste prevention, auxiliary substances, and energy efficiency. A significant opportunity exists to integrate lecture and practical examples into a preexisting course or a stand-alone green chemistry offering. Indeed, one could imagine combining a variety of published experiments to form a laboratory curriculum where no solvents are used! A strong argument could, of course, be mounted against this. Students would lack exposure to particular practical techniques (e.g., separatory funnel extractions) and would be poorly prepared for graduate or industrial work. Regardless of this, introducing at least one or two solventless reactions into undergraduate laboratories is a simple and very worthwhile move, which will have an impact on student perceptions of green chemistry and organic reactivity.

REFERENCES

1. (a) Anon. *J. Chem. Educ.* 1994, 71, 906. (b) Blumberg, A. A. *J. Chem. Educ.* 1994, 71, 912–918.
2. Halford, B. *Chem. Eng. News* 2007, 85 (28), 56.
3. Anastas, P. T., Warner J. C. *Green Chemistry: Theory and Practice*. Oxford University Press, New York, 1998.
4. Constable, D. J. C., Jimenez-Gonzalez, C., Henderson, R. K. *Org. Process Res. Dev.* 2007, 11, 133–137.
5. Andrews, I., Cui, J., Dudin, L., Dunn, P., Hayler, J., Hinkley, B., Hughes, D., Kaptein, B., Lorenz, K., Mathew, S., Rammeloo, T., Wang, L., Wells, A., White, T. D. *Org. Process Res. Dev.* 2010, 14, 770–780.
6. Houlton, S. *Chem. World* 2010, 7 (3), 46–49.
7. Bradley, D. *Chem. Br.* 2002, 38 (9), 42–45.
8. Reichardt, C. *Org. Process Res. Dev.* 2007, 11, 105–113 and references therein.
9. Tanaka, K. *Solvent-free Organic Synthesis*. Wiley-VCH, Weinheim, Germany, 2003.
10. Martins, M. A. P., Frizzo, C. P., Moreira, D, N., Buriol, L., Machado, P. *Chem. Rev.* 2009, 109, 4140–4182.
11. Walsh, P. J., Li, H., Anaya de Parrodi, C. *Chem. Rev.* 2007, 107, 2503–2545.
12. Shearouse, W. C. *Curr. Opin. Drug Discovery Dev.* 2009, 12, 772–783.
13. McMurry, J. *Organic Chemistry*, 7th ed. Thomson Higher Education, Belmont, CA, 2008, 613–615.
14. Welton, T. *Green Chem.* 2006, 8, 13.

15. Blackmond, D. G., Armstrong, A., Coombe, V., Wells, A. *Angew. Chem. Int. Ed.* 2007, 46, 3798–3800.
16. Aktoudianakis, E., Dicks, A. P. *J. Chem. Educ.* 2006, 83, 287–289.
17. Montes, I., Sanabria, D., García, M., Castro, J., Fajardo, J. *J. Chem. Educ.* 2006, 83, 628–631.
18. Albanese, D., Ghidoli, C., Zenoni, M. *Org. Process Res. Dev.* 2008, 12, 736–739.
19. Pichon, A., James, S. L. *Chem. World* 2007, 4 (7), 35.
20. Solvent-Free Synthesis? I Don't Think So. http://prospect.rsc.org/blogs/cw/2007/07/03/solvent-free-synthesis-i-dont-think-so/ (accessed December 23, 2010).
21. The Presidential Green Chemistry Challenge Award Recipients 1996–2009, 2009 Greener Synthetic Pathways Award. U.S. Environmental Protection Agency, Office of Pollution Prevention and Toxics, Washington, DC, 2009, 6–7.
22. McMurry, J. *Organic Chemistry*, 7th ed. Thomson Higher Education, Belmont, CA, 2008, 795–796.
23. Villa, C., Mariani, E., Loupy, A., Grippo, C., Grossi, G. C., Bargagna, A. *Green Chem.* 2003, 5, 623–626.
24. Declerck, V., Nun, P., Martinez, J., Lamaty, F. *Angew. Chem.* 2009, 121, 9482–9485.
25. Martinez, J., Lamaty, F., Declerck, V. WO 2008125418, 2008.
26. McMurry, J. *Organic Chemistry*, 7th ed. Thomson Higher Education, Belmont, CA, 2008, 1033–1038.
27. Polshettiwar, V., Varma, R. S. Environmentally Benign Chemical Synthesis via Mechanicochemical Mixing and Microwave Irradiation. In *Eco-Friendly Synthesis of Fine Chemicals*, Ballini, R., Ed. Royal Society of Chemistry Green Chemistry Series, Vol. 3. Royal Society of Chemistry, Cambridge, 2009.
28. Mazur, R. H., Schlatter, J. M., Goldkamp, A. H. *J. Am. Chem. Soc.* 1969, 91, 2684–2691.
29. Miyaura, N., Suzuki, *A. Chem. Commun.* 1979, 866–867. (b) Miyaura, N., Suzuki, A. *Chem. Rev.* 1995, 95, 2457–2483.
30. Cho, J.-Y., Tse, M. K., Holmes, D., Maleczka, Jr., R. E., Smith III, M. R. *Science* 2002, 295, 305–308.
31. The Presidential Green Chemistry Challenge Award Recipients 1996–2009, 2008 Academic Award. U.S. Environmental Protection Agency, Office of Pollution Prevention and Toxics, Washington, DC, 2009, 12–13.
32. Weiß, M., Gröger, H. *Synlett* 2009, 8, 1251–1254.
33. Weiß, M., Brinkmann, T., Gröger, H. *Green Chem.* 2010, 12, 1580–1588.
34. Namboodiri, V. V., Varma, R. S. *Org. Lett.* 2002, 4, 3161–3163.
35. Deetlefs, M., Seddon, K. R. *Green Chem.* 2003, 5, 181–186.
36. Gonzalez, M. A., Ciszewski, J. T. *Org. Process Res. Dev.* 2009, 13, 64–66.
37. Law, M. C., Wong, K-Y., Chan, T. H. *Green Chem.* 2002, 4, 328–330.
38. Cave, G. W. V., Raston, C. L., Scott, J. L. *Chem. Commun.* 2001, 2159–2169.
39. Correa, W. H., Edwards, J. K., McCluskey, A., McKinnon, I., Scott, J. L. *Green Chem.* 2003, 5, 30–33.
40. Anastas, P. T., Beach, E. S. *Green Chem. Lett. Rev.* 2007, 1, 9–24.
41. Dunk, B., Jachuck, R. *Green Chem.* 2000, 2, G13–G14.
42. Cléophax, J., Liagre, M., Loupy, A., Petit, A. *Org. Process Res. Dev.* 2000, 4, 498–504.
43. CEM Corporation Prolabo Support Page. www.cemservice.us/prolabo/prolabo.htm (accessed December 23, 2010).
44. Hilterhaus, L., Thum, O., Liese, A. *Org. Process Res. Dev.* 2008, 12, 618–625.
45. Korupp, C., Weberskirch, R., Müller, J. J., Liese, A., Hilterhaus, L. *Org. Process Res. Dev.* 2010, 14, 1118–1124.
46. Jakobsen, H. A. *Chemical Reactor Modeling—Multiphase Reactive Flows.* Springer-Verlag, Berlin, 2008, 757–806.

47. Constable, D. J. C., Dunn, P. J., Hayler, J. D., Humphrey, G. R., Leazer, Jr., J. L., Linderman, R. J., Lorenz, K., Manley, J., Pearlman, B. A., Wells, A., Zaks, A., Zhang, T. Y. *Green Chem.* 2007, 9, 411–420.
48. Meng, P., Geskin, E. S., Leu, M. C., Tismenetskiy, L. Waterjet In-Situ Reactor Cleaning. *Proceedings at the 13th International Conference on Jetting Technology*, Vol. 21. BHR Group Conference Series Publications, 1996, 347–358.
49. Brant, F. R., Cannon, F. S. *J. Environ. Sci. Health A* 1996, 31, 2409–2434.
50. Dicks, A. P. *Green Chem. Lett. Rev.* 2009, 2, 87–100.
51. Cave, G. W. V., Raston, C. L. *J. Chem. Educ.* 2005, 82, 468–469.
52. Raston, C. L., Scott, J. L. *Green Chem.* 2000, 2, 49–52.
53. Palleros, D. R. *J. Chem. Educ.* 2004, 81, 1345–1347.
54. Gálvez, J., Gálvez-Llompart, M., García-Domenech, R. *Green Chem.* 2010, 12, 1056–1061.
55. McKenzie, L. C., Huffman, L. M., Hutchison, J. E., Rogers, C. E., Goodwin, T. E., Spessard, G. O. *J. Chem. Educ.* 2009, 86, 488–493.
56. Touchette, K. M. *J. Chem. Educ.* 2006, 83, 929–930.
57. McDaniel, K. F., Weekly, R. M. *J. Chem. Educ.* 1997, 74, 1465–1467.
58. Reinhardt, D., Ilgen, F., Kralisch, D., König, B., Kreisel, G. *Green Chem.* 2008, 10, 1170–1181.
59. Cheney, M. L., McManus, G. J., Perman, J. A., Wang, Z., Zaworotko, M. *J. Cryst. Growth Des.* 2007, 7, 616–617. (b) Trask, A. V., Jones, W. *Top. Curr. Chem.* 2005, 254, 41–70.
60. Cheney, M. L., Zaworotko, M. J., Beaton, S., Singer, R. D. *J. Chem. Educ.* 2008, 85, 1649–1651.
61. Friščić, T., Hamilton, T. D., Papaefstathiou, G. S., MacGillivray, L. R. *J. Chem. Educ.* 2005, 82, 1679–1681.
62. Schmidt, G. M. J. *Pure Appl. Chem.* 1971, 27, 647–678.
63. Williamson, K. L., Minard, R. D., Masters, K. M. Semicarbazones. In *Macroscale and Microscale Organic Experiments*, 5th ed. Houghton Mifflin, Boston, 2007, 511–512.
64. Pandita, S., Goyal, S., Arif, N., Passey, S. *J. Chem. Educ.* 2004, 81, 108.
65. Esteb, J. J., Gligorich, K. M., O'Reilly, S. A., Richter, J. M. *J. Chem. Educ.* 2004, 81, 1794–1795.
66. Phonchaiya, S., Panijpan, B., Rajviroongit, S., Wright, T., Blanchfield, J. T. *J. Chem. Educ.* 2009, 86, 85–86.
67. Wong, T. C., Sultana, C. M., Vosburg, D. A. *J. Chem. Educ.* 2010, 87, 194–195.
68. Murray, R. D. H., Méndez, J., Brown, S. A. *The Natural Coumarins: Occurrence, Chemistry and Biochemistry*. Wiley, New York, 1982.
69. Scott, J. L., Raston, C. L. *Green Chem.* 2000, 2, 245–247.
70. Raston, C. L., Scott, J. L. *Pure Appl. Chem.* 2001, 73, 1257–1260.
71. Biginelli, P. *Gazz. Chim. Ital.* 1893, 23, 360–413.
72. Kappe, C. O. *Eur. J. Med. Chem.* 2000, 35, 1043–1052.
73. Folkers, K., Harwood, H. J., Johnson, T. B. *J. Am. Chem. Soc.* 1932, 54, 3751–3758.
74. Holden, M. S., Crouch, R. D. *J. Chem. Educ.* 2001, 78, 1104–1105.
75. Aktoudianakis, E., Chan, E., Edward, A. R., Jarosz, I., Lee, V., Mui, L., Thatipamala, S. S., Dicks, A. P. *J. Chem. Educ.* 2009, 86, 730–732.
76. Goodwin, T. E. *J. Chem. Educ.* 2006, 83, 287–289.
77. Cheung, L. L. W., Styler, S. A., Dicks, A. P. *J. Chem. Educ.* 2010, 87, 628–630.
78. Loev, B., Goodman, M. M., Snader, K. M., Tedeschi, R., Macko, E. *J. Med. Chem.* 1974, 17, 956–965. (b) Ali, S. L. *Anal. Profiles Drug Subst. Excipients* 1989, 18, 221–288.
79. Evans, B. E., Rittle, K. E., Bock, M. G., DiPardo, R. M., Freidinger, R. M., Whitter, W. L., Lundell, G. F., Veber, D. F., Anderson, P. S., Chang, R. S. L., Lotti, V. J., Cerino, D. J., Chen, T. B., Kling, P. J., Kunkel, K. A., Springer, J. P., Hirshfield, J. *J. Med. Chem.* 1988, 31, 2235–2246.

80. The Presidential Green Chemistry Challenge Award Recipients 1996–2009, 1996 Small Business Award. U.S. Environmental Protection Agency, Office of Pollution Prevention and Toxics, Washington, DC, 2009, 138–139.
81. Bennett, G. D. *J. Chem. Educ.* 2005, 82, 1380–1381.
82. Martin, E., Kellen-Yuen, C. *J. Chem. Educ.* 2007, 84, 2004–2006.
83. Nguyen, K. C., Weizman, H. *J. Chem. Educ.* 2007, 84, 119–121.
84. Leung, S. H., Angel, S. A. *J. Chem. Educ.* 2004, 81, 1492–1493.
85. Cann, M. C., Dickneider, T. A. *J. Chem. Educ.* 2004, 81, 977–980.

4 Organic Reactions under Aqueous Conditions

Dr. Effiette L. O. Sauer

CONTENTS

4.1 INTRODUCTION

4.1.1 HISTORICAL USE OF WATER AS A SOLVENT

For generations, chemistry students have been taught to regard water as an inappropriate solvent for carrying out organic reactions. However, this has not always been the case. The synthesis of urea by Friedrich Wöhler in 1828,[1] generally regarded as the birth of synthetic chemistry,[2] was carried out in water by heating an aqueous solution of ammonium cyanate. Use of aqueous reactions continued into the early part of the twentieth century, and many of the named reactions known today owe their discoveries to reactions performed in water. Some examples include the Baeyer-Villiger oxidation,[3] the Curtius rearrangement,[4] the Hofmann degradation,[5] the Sandmeyer reaction,[6] and the Wolff-Kishner reduction.[7]

Given the historical significance of water as a reaction medium, its avoidance today merits discussion. There were two significant factors that led to a shift toward organic solvents in the early twentieth century: the advent of organometallic chemistry and the expansion of the petroleum industry.[8] Early organometallic species, such as alkylzinc[9] and Grignard[10] reagents, opened the door to new methods of carbon-carbon bond formation. The impact of these reactions was enormous, and chemists quickly came to rely on them for synthetic transformations. Unfortunately, these highly reactive organometallic reagents were extremely sensitive to hydrolysis and decomposed violently in the presence of water. Thus, chemists were forced to abandon the use of water in favor of dry, aprotic solvents. At about the same time, the petroleum industry was undergoing its own rapid expansion as society moved from coal to oil to meet its energy needs. A by-product of this industry was the sudden availability of new, inexpensive hydrocarbon-based solvents—perfect for use with water-sensitive reagents. However, reports of aqueous reactions continued to appear, with a particularly influential article in 1980 by Rideout and Breslow on the Diels-Alder reaction.[11] Inspired by nature's use of water as a solvent, they investigated the reaction between various dienes and dienophiles under aqueous conditions. Not only did the reactions work, but their rates and selectivities were improved compared to the same processes in organic solvents (Section 4.2.1). Extensive research into the breadth and origin of these effects followed, and the resulting body of work has become the foundation for much of today's renewed interest in aqueous phase chemistry.[12]

4.1.2 Water as a Green Solvent

With the awakening of green chemistry, chemists have been forced to take a critical look at their reactions and the amount of waste they generate. A useful metric for doing so is the E-factor, which compares the mass of waste generated to the mass of desired product, as presented in Section 1.4.[13] For the fine chemical and pharmaceutical industries, this ratio generally lies between 5 kg of waste and more than 100 kg of waste per kg of product.[13a] Of the waste generated, it has been estimated that up to 90% by mass comes from solvents[14] (Section 3.1). A major focus of green chemistry has therefore become reduction (or even elimination) of solvent use. Unfortunately, this is not always possible, leaving chemists to search for solvents that are safe, nontoxic, and hazard-free.

Among the many alternative solvents available, water has been identified as one of the most attractive.[15] Its obvious advantages include its low cost, relative abundance, low toxicity, and inflammability. In addition, water's unique physical properties lend themselves to further, less obvious advantages. Its high specific heat provides a sink for exothermic reactions; its immiscibility with nonpolar substances can facilitate isolation of insoluble products; and the existence of water's OH groups obviates the need for hydroxyl protection-deprotection sequences. Therefore, adoption of water as a solvent has the potential to address several green chemistry principles, including use of innocuous solvents, reduction of hazardous substances, increased safety, and in some cases, reduced derivatization.

4.1.3 Overcoming Solubility and Reactivity Issues with Aqueous Phase Chemistry

Despite its many advantages, water is still used for relatively few reactions in both industry and academia. One likely reason for this is the misconception that water is an inappropriate solvent for organic reactivity, for both solubility reasons and concerns over functional group compatibility. While it is true that these issues require consideration, there are many examples of how such challenges can be overcome. Given that organic reactants are generally much less polar than water, many substrates and reagents will indeed exhibit poor solubility. The result of this has often been highly dilute reactions, which are impractical for large-scale preparations.[16] There are, however, numerous approaches available to address this problem. The addition of cosolvents or auxiliary agents such as surfactants[17] and phase transfer catalysts[17,18] are proven methods for overcoming limited solubility. Alternatively, substrates, reagents, and catalysts can be modified to increase their water solubility. This approach typically involves incorporation of water-solubilizing sulfonate or phosphonate groups, and has been met with great success in the area of aqueous phase catalyst design.[19]

The notion that solubility is a requirement for reactivity has itself come under challenge recently. A 2005 report by Narayan et al.[20] describes several examples of aqueous reactions where the reactants remain insoluble and are stirred as a suspension or emulsion. In cases where these heterogeneous reactions are also enhanced by their biphasic nature (with respect to either their rate or selectivity), the authors suggest using the term *on water* to distinguish them from other aqueous reactions taking place in water or in the presence of water.[21] In one example, the $[2\sigma + 2\sigma + 2\pi]$ cycloaddition between dimethyl azodicarboxylate (DMAD) and quadricyclane was carried out as either a neat reaction, in toluene, or an aqueous suspension (Scheme 4.1). The reaction rate was dramatically increased when carried out on water relative to the same reaction in toluene or under solvent-free conditions.

The possibility of carrying out reactions on water makes aqueous phase chemistry possible with even the least polar reactants. This eliminates the need for cosolvents, auxiliary agents (surfactants, phase transfer catalysts), or high dilution

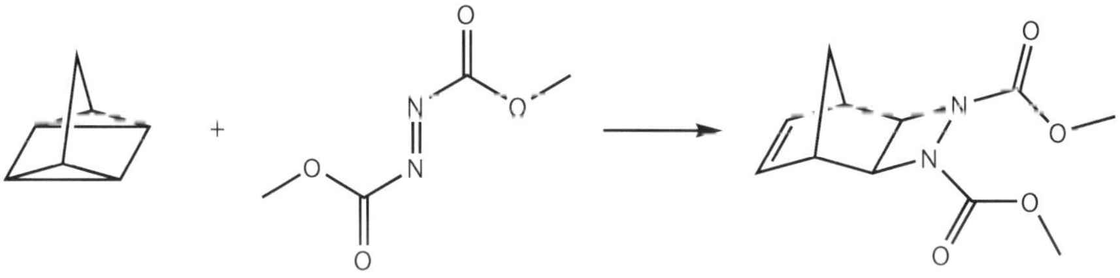

SCHEME 4.1 Reaction between dimethyl azodicarboxylate and quadricyclane.

conditions. Since this initial report,[20] numerous other examples of on-water reactions have appeared in the literature, highlighting the broad applicability of the approach.[22]

A second hindrance to water adoption is its incompatibility with certain reagents and reactive intermediates. For some transformations, such as the Grignard reaction, this is a very real problem. However, potential benefits to using aqueous conditions have prompted research into alternatives to traditionally anhydrous reactions. One success story is the development of the aqueous Barbier-Grignard type reaction, involving nucleophilic additions to carbonyl compounds.[23] Prompted by early reports of water-compatible organometallic species,[24] researchers have developed metal-mediated nucleophilic additions that can take place in water. Examples have expanded to include allyl, vinyl, propargyl, alkynyl, aryl, and alkyl halides, which, in the presence of a metal such as zinc, tin, or indium, can be added to aldehydes or ketones. Reactions such as these defy conventional wisdom and challenge the notion of what can and cannot be achieved in an aqueous environment. While it is true that some reactions will likely never be possible in water, it is important that there is a continuing search for alternative methodologies.

4.1.4 Ongoing Challenges with Aqueous Reactivity

While many of the limitations of aqueous phase chemistry can be overcome with a little creativity, there are some very real challenges that remain.[25] Among these are difficulties associated with isolation of water-soluble products and the treatment of wastewater. In contrast to reactants and reagents where water solubility is generally desired, products become easier to collect if they are *insoluble* in aqueous media. In such cases, isolation can be readily achieved by simple phase separation or filtration. Unfortunately, many aqueous reactions yield water-soluble products, or products that are soluble under the reaction conditions where cosolvents and other solubilizing agents may be present. In these cases, isolation typically involves extraction with another organic solvent, thus negating much of the advantage of using water in the first place.

Perhaps the biggest concern with using aqueous phase chemistry, however, relates to waste treatment. While few would disagree that pure water is an environmentally friendly solvent, the wastewater produced from aqueous phase chemistry is rarely, if ever, pure. Dissolved reactants and auxiliary agents all need to be removed before release into the environment. These treatment processes can be both costly and energy intensive, and need consideration when evaluating the "greenness" of water as a solvent. The management of wastewater produced from the synthesis of aromatic sulfonic acids (Section 7.2.3) shows what is possible in terms of recycling.

Given the associated complexities of aqueous phase chemistry, it is important that students are encouraged to think critically when evaluating a new reaction. Many aqueous phase reactions offer significant advantages over their traditional counterparts (Section 4.4), while others suffer from serious limitations. Rather than equating water with green, students should be trained to evaluate reactions on a case-by-case basis.

4.2 STUDIES ON THE ORIGIN OF ENHANCED REACTIVITY UNDER AQUEOUS CONDITIONS

The remarkable rate accelerations and superior selectivities observed for certain aqueous phase reactions provide an opportunity to present students with some very interesting chemistry. Extensive research has been devoted to understanding the origin of these effects, and several theories exist in the literature.[26] For homogeneous aqueous systems, the improvements are generally attributed to some combination of polarity effects, hydrophobic interactions, and hydrogen bonding. In heterogeneous systems, hydrogen bonding interactions across the water-organic phase boundary become important. An introduction to each of these effects is presented here.

4.2.1 Polarity Effects

Students of organic chemistry learn early on about the importance of solvent polarity as part of their studies of substitution and elimination reactions. A reaction that proceeds via a transition state that is more polar than its initial state will be faster in polar solvents that can stabilize the developing charges (e.g., an S_N1 reaction of *t*-butyl bromide). Conversely, a reaction that passes through a less polar transition state will be slower in polar solvents owing to preferential stabilization of the initial state (e.g., an S_N2 reaction with an anionic nucleophile). Water is a highly polar solvent with a dielectric constant (ε_r) of 80.1[27] and a Reichardt $E_T(30)$ value of 61.3.[28] Consequently, its ability to influence reaction rates is to be expected.

Rideout and Breslow were the first to quantify the rate enhancements observed in aqueous reactions of small molecules.[11] Kinetic studies on the Diels-Alder reaction between cyclopentadiene and butenone (Scheme 4.2) revealed a reaction rate more than 700 times greater in water than in isooctane. Diels-Alder reactions normally exhibit a slight increase in polarity at their transition state.[29] Thus, an increased reaction rate is consistent with moving to a more polar solvent. To explore whether solvent polarity alone was behind this rate enhancement, Rideout and Breslow repeated the reaction in methanol. Based on Reichardt's $E_T(30)$ parameters, the polarity of methanol lies roughly halfway between that of isooctane and water,[28] predicting a reaction rate ~350 times greater than in isooctane. Experimentally, the reaction between cyclopentadiene and butenone in methanol was faster than in isooctane, although only by a factor of 12. This rate enhancement is only a fraction of what was observed in water, suggesting that while solvent polarity plays a role in enhanced aqueous reactivity, the magnitude of these changes cannot be attributed to polarity effects alone.

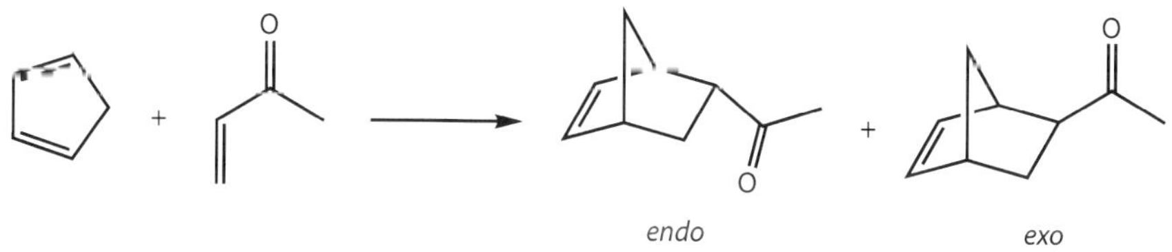

SCHEME 4.2 Diels-Alder reaction between cyclopentadiene and butenone.

4.2.2 The Hydrophobic Effect

In general terms, the hydrophobic effect describes the tendency of nonpolar species to aggregate when placed in an aqueous environment.[30] The driving force for this aggregation is the minimization of the hydrocarbon-water interface in order to maintain the thermodynamically favored network of hydrogen bonds that exists between water molecules. This effect is well documented in biological systems, where it is responsible for the three-dimensional structure of proteins and nucleic acids, as well as the selective binding between many enzymes and their substrates. Noting that the observed rate enhancements for the aqueous Diels-Alder reaction could not entirely be explained by polarity effects, Rideout and Breslow suggested that hydrophobic effects could be influencing these aqueous reactions.[11] They proposed that rate acceleration was due to the aqueous environment forcing together the nonpolar reactants (so-called hydrophobic packing), thereby increasing their local concentrations and bimolecular reaction rate.

Following Rideout and Breslow's evocation of the hydrophobic effect, Engberts et al. went on to make a subtle but important distinction between hydrophobic packing and what was termed "enforced hydrophobic interactions."[31] The former describes the aggregation of the diene and dienophile in a geometry that may or may not lead to the activated complex. In comparison, the latter acknowledges that transition state formation during the Diels-Alder reaction is accompanied by a reduction in the hydrophobic surface area. Through detailed kinetic studies, it was demonstrated that it is the compact nature of the transition state that is instrumental to the hydrophobic effect, yielding significant rate enhancements.[31]

In subsequent studies, research by both Breslow et al.[32] and Grieco et al.[33] revealed that the endo/exo selectivity of the Diels-Alder reaction was also affected by solvent changes. For example, the reaction between cyclopentadiene and butenone gave an endo/exo ratio of 3.85 when carried out with excess cyclopentadiene as the solvent, and 8.5 when the reaction was performed in ethanol.[32a] In water, however, the ratio increased to 22.5. Again, the magnitude of this difference did not match changes in solvent polarity, prompting the researchers to suggest the hydrophobic effect as the root cause. In this case, the more compact endo transition state reduces hydrophobic interactions compared to the larger exo transition state, and is therefore favored under aqueous conditions.

A common test for the hydrophobic effect involves addition of different salts to aqueous reactions. Salts with small cations and anions (e.g., LiCl, NaCl) are known to decrease the solubility of organic species in water (an effect known as salting out).[34] An explanation for this behavior is believed to be contraction of the solvent volume accompanying dissolution of these salts. As water molecules rearrange themselves to surround the dissolved ions, the network of hydrogen bonds contracts, leaving less empty space to accommodate the dissolved hydrocarbons. The end result is an effective increase in the hydrophobic effect. Conversely, salts where one or both of the ions are large (e.g., $C(NH_2)_3Cl$, $LiClO_4$, $C(NH_2)_3ClO_4$) have the opposite effect. These salts increase the water solubility of nonpolar species and are said to act as "salting in" agents. The molecular cause of this effect is not fully understood.[35] One explanation is that the larger ions of these salting in agents break up the hydrogen bonding

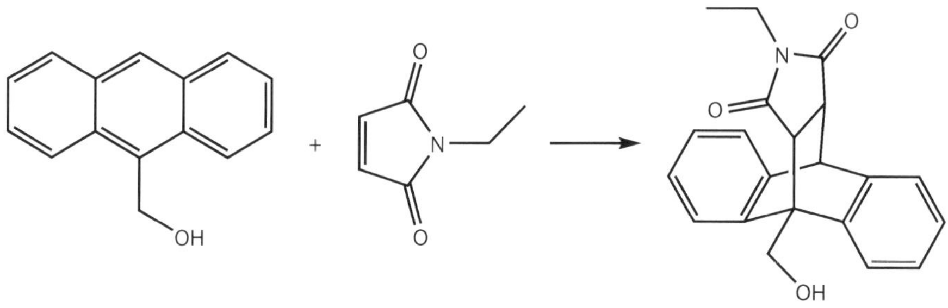

SCHEME 4.3 Diels-Alder reaction between anthracene-9-carbinol and *N*-ethylmaleimide.

network, making it easier to form a cavity for the nonpolar substance. Alternatively, the large ions could be directly improving the solubility of nonpolar substances by acting as a bridge between the hydrocarbon and water molecules. Regardless of the rationale, addition of these salting in agents effectively decreases the hydrophobic effect in aqueous solutions.

To further distinguish between simple polar effects and the hydrophobic effect, Breslow et al.[11,32] investigated the effect of adding various salts to aqueous Diels-Alder reactions. For the reaction between anthracene-9-carbinol and *N*-ethylmaleimide (Scheme 4.3), addition of lithium chloride was found to increase the reaction rate relative to the reaction in pure water, while addition of guanidinium perchlorate decreased the rate.[32a] Similar salt effects were also observed with endo/exo selectivity for the cyclopentadiene/butenone reaction.[32b]

4.2.3 Hydrogen Bonding

The hydrophobic effect is now generally regarded as a major contributing factor to the accelerated reactivity and increased selectivity observed with some aqueous phase reactions. Significantly though, evidence suggests that hydrophobic effects alone are not enough to explain such enhancements. Theoretical studies by several groups have shown that for Diels-Alder reactions such as that between cyclopentadiene and butenone, the comparatively polarized transition state is preferentially stabilized by hydrogen bonding with water molecules compared to the reactants.[36–38] The net result is a decrease in the activation energy of these reactions in aqueous media. The extent to which this stabilization is predicted to occur depends on the nature of the reacting species. For example, the activation energy for the reaction between cyclopentadiene and 1,4-naphthoquinone has been calculated to decrease by 3.2 kcal/mol on moving from methanol to water as the solvent.[36e] By comparison, acrylonitrile as the dienophile leads to a predicted decrease in activation energy of just 1.2 kcal/mol. Both the hydrophobic effect and hydrogen bonding were invoked in these models. However, it is the enhanced hydrogen bonding capabilities of 1,4-naphthoquinone compared to acrylonitrile that appears responsible for the different sensitivities of these reactions to aqueous rate enhancements. Importantly, these calculated changes in activation energies are in excellent agreement with those determined experimentally.[11,39] More recent studies have shown similar results for 1,3-dipolar cycloadditions[40] and Claisen rearrangements,[41] suggesting that the relative contributions from hydrogen bonding and hydrophobic effects vary according to both the reaction and substrates under study.

4.2.4 Heterogeneous Systems

When determining the effect of water on heterogeneous systems (so-called on-water reactions), there are two possibilities to consider. Either one assumes that some nominal amount of each reactant is soluble in the bulk water (rendering these systems dilute homogeneous solutions), or there is something unique happening at the phase boundary that is responsible for the observed rate enhancements. In reality, it is not trivial to distinguish between these two scenarios, and for many reactions, it may be that both are taking place. However, it has become apparent that in at least some biphasic systems, reactions are taking place at the aqueous-organic interface rather than in solution. Most early aqueous phase chemistry was carried out at low concentrations in order to ensure solubilization of reactants. In some cases, however, researchers noted that reactions proceeded even when reactant concentrations rendered the systems heterogeneous.[32a,33a] Detailed kinetic studies were generally not done to compare these reactions. However, Breslow et al. noted the time to reach 50% completion for the reaction between cyclopentadiene and butenone was approximately three times greater for the homogeneous system than for the heterogeneous reaction.[32a] This result provided the first piece of evidence that heterogeneous systems may benefit from some additional effect not present in homogeneous aqueous reactions.

In the 2005 report by Narayan et al.,[20] heterogeneity was found to be crucial for the increased rate enhancements of the on-water reaction between DMAD and quadricyclane (Scheme 4.1). The cycloaddition was carried out in aqueous methanol solutions under both homogeneous and heterogeneous conditions. As long as the reaction was heterogeneous, the reaction time remained constant at ten minutes, regardless of whether methanol was present. However, the reaction time dropped to four hours when enough methanol was added to make the system homogeneous. Importantly, heterogeneity alone was not enough to induce a rate enhancement, as the reaction *on* C_6F_{14} was slower than any of the other cases. These results, along with an observed deuterium kinetic isotope effect for the reaction on D_2O, led the authors to suggest that unique properties at the water-organic interface were behind the enhanced reactivity.

Earlier work by Shen et al.[42] showed that 25% of the water molecules at an oil-water interface have free OH groups (not hydrogen bonded) that protrude into the hydrophobic boundary. Jung and Marcus[43] proposed that these "dangling" OH groups could be responsible for increased activity observed with the on-water reaction between DMAD and quadricyclane.[20] Since these OH groups are not hydrogen bonded to the bulk water, they are free to bind with reactants or transition states at the water-hydrophobic interface. In contrast, OH groups in the bulk solvent would first need to break free from their network of hydrogen bonds before coordinating to any reactants or transition states. This process adds an associated energetic cost to homogeneous aqueous reactions that is not present with on-water reactions. Jung and Marcus used this model to calculate theoretical rate constants for the reaction between DMAD and quadricyclane under neat, homogeneous, and heterogeneous conditions.[43] By incorporating preferential binding of dangling OH groups to transition states in the on-water reaction, along with an energetic penalty for breaking hydrogen bonds in the homogeneous model, they were able to calculate rate constants that closely match those found by experiment.[20] Similar studies based on this model have since been published for the aromatic Claisen rearrangement.[44]

4.3 AQUEOUS CHEMISTRY IN THE UNDERGRADUATE ORGANIC LABORATORY

In keeping with heightened research interest in aqueous phase chemistry, the pedagogical literature has seen a similar increase in publications of undergraduate reactions taking place in water. A 2009 article by Dicks presents a comprehensive review of aqueous phase reactivity for the teaching laboratory.[45] The remainder of this section presents new experiments not included in this review, as well as a selection of experiments that were reviewed. Preference is given to examples that showcase the breadth of chemistry possible in an aqueous environment or build connections to examples of aqueous phase success stories in an industrial venue. Examples are also included that, although carried out in water, are not necessarily much greener than their traditional counterparts due to solvent use in purification or poor atom economy. Such reactions represent an excellent opportunity to introduce green chemistry concepts while encouraging students to think critically while evaluating the greenness of a process. To facilitate integration with course content, reactions have been divided by mechanism into the following six categories: transition metal catalyzed, pericyclic, reduction-oxidation, nucleophilic additions and substitutions, electrophilic additions and substitutions, and radical reactions. Examples of biocatalyzed reactions in water are discussed in Chapter 6.

4.3.1 Transition Metal-Catalyzed Reactions

Extensive research has been devoted to development of water-compatible transition metal catalysts,[19] and several instances have been incorporated into industrial processes. The Suzuki coupling is one such example (Section 4.4.2). In a typical Suzuki reaction, a palladium catalyst is used to couple an aryl or vinyl halide with a boronic acid or ester. Several undergraduate experiments for aqueous phase Suzuki reactions have been reported, including a recent example by Pantess and Rich.[46] Using 1 mol% of palladium(II) acetate as catalyst and potassium carbonate as the base, phenylboronic acid is coupled with an aryl bromide (Scheme 4.4). Although water is the only solvent used, tetra *n*-butylammonium bromide is required to solubilize the reactants and the product is purified by column chromatography.

In a related example, Aktoudianakis et al. have outlined a Suzuki coupling between phenylboronic acid and 4-iodophenol that requires no solubilizing agent, uses inexpensive palladium on carbon as the catalyst, and produces an insoluble product that is isolable by simple filtration[47] (Scheme 4.5). Purification by recrystallization from

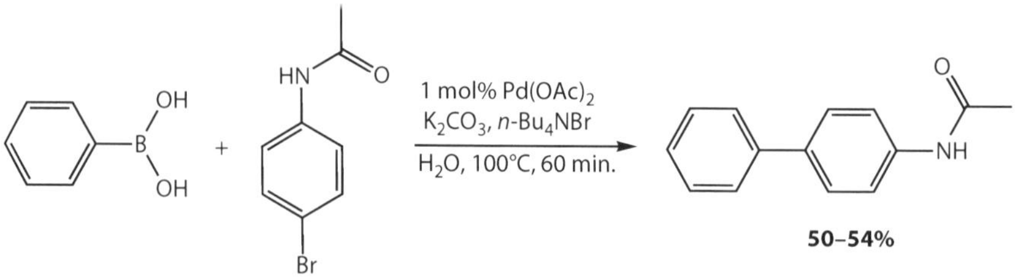

SCHEME 4.4 Suzuki cross-coupling between phenylboronic acid and an aryl bromide.

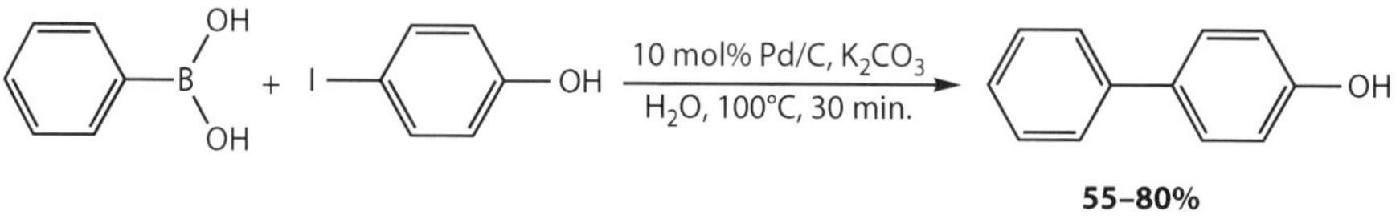

SCHEME 4.5 Suzuki cross-coupling between phenylboronic acid and an aryl iodide.

aqueous methanol affords 4-phenylphenol, a structural analog to several nonsteroidal anti-inflammatory drugs (NSAIDs). Additional greener undergraduate Suzuki reactions are covered in Sections 5.6.2, 7.5, and 8.8.4.

Other examples of palladium-catalyzed cross-coupling experiments include the Heck reaction[48] and Sonogashira coupling.[49,50] In an interesting application of the latter, a greener approach to benzofuran synthesis is described. Hutchison et al.[50] reacted a terminal alkyne with an aryl iodide using palladium(II) acetate, a water-solubilizing ligand (TPPTS), and *N*-methylpyrrolidine (NMP) as a cosolvent (Scheme 4.6). During the one-week reaction time, the intermediate cyclizes to yield a benzofuran ring.

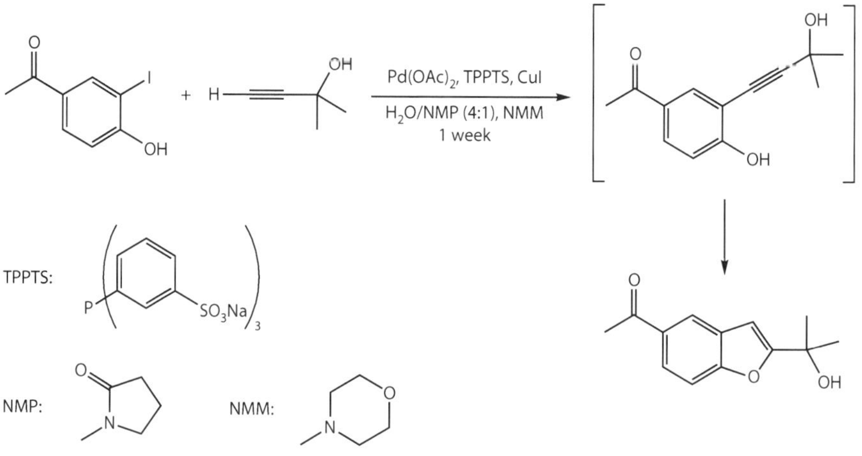

SCHEME 4.6 Sonogashira coupling followed by cyclization to yield a benzofuran.

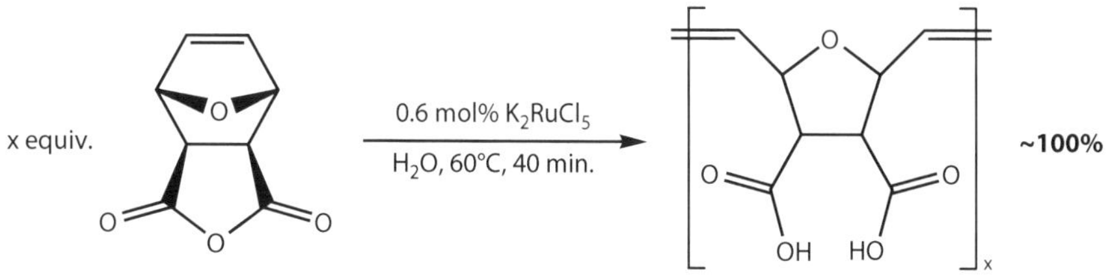

SCHEME 4.7 Ring opening metathesis polymerization of a Diels-Alder adduct.

In addition to palladium-catalyzed reactions, ruthenium-based chemistry has also found use in the undergraduate laboratory. Viswanathan and Jethmalani[51] have reported a ring-opening metathesis polymerization (ROMP) of the bicyclic Diels-Alder adduct formed from furan and maleic anhydride (Scheme 4.7). Heating an aqueous solution of the adduct and 0.6 mol% of K_2RuCl_5 yields the polymeric product in near-quantitative yield after just forty minutes. The water in this reaction is believed to act not only as a solvent, but also as a cocatalyst to lower the reaction induction time.[52]

4.3.2 Pericyclic Reactions

Given the extensive research into Diels-Alder reactions in water, it is not surprising that the educational literature contains several excellent examples of undergraduate experiments that highlight this aqueous reaction. A recent example by Hutchison et al. examines the reaction between anthracene-9-carbinol and *N*-methylmaleimide[53] (Scheme 4.8). It is worth noting that this reaction is virtually identical to one originally studied by Breslow et al.[32a] (Scheme 4.3), thus providing an excellent opportunity to introduce this seminal work to the undergraduate laboratory. In this experiment, the reagents are refluxed in water for one hour and the product is collected as a precipitate at the end of the reaction. An excess of *N*-methylmaleimide is required, although the excess remains in the aqueous phase and can be collected/reused if desired. In another recent aqueous Diels-Alder experiment, microwave conditions are used to accelerate the reaction.[54]

The hetero-Diels-Alder reaction has also been found to exhibit aqueous rate enhancements.[55] An undergraduate experiment that incorporates this reaction was published by Sauvage and Delaude in 2008.[56] For this experiment, an iminium ion is

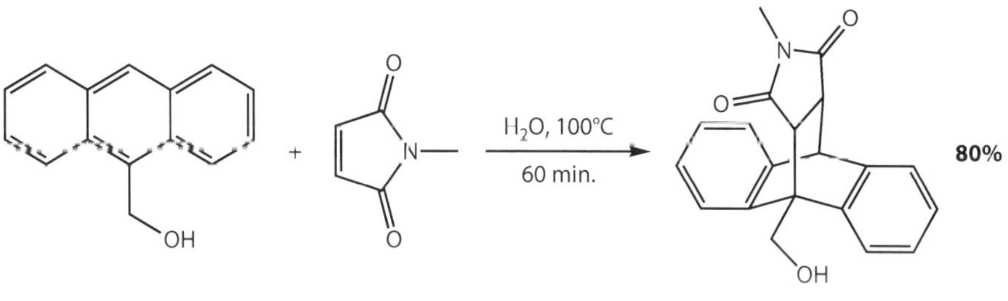

SCHEME 4.8 Diels-Alder reaction between anthracene-9-carbinol and *N*-methylmaleimide.

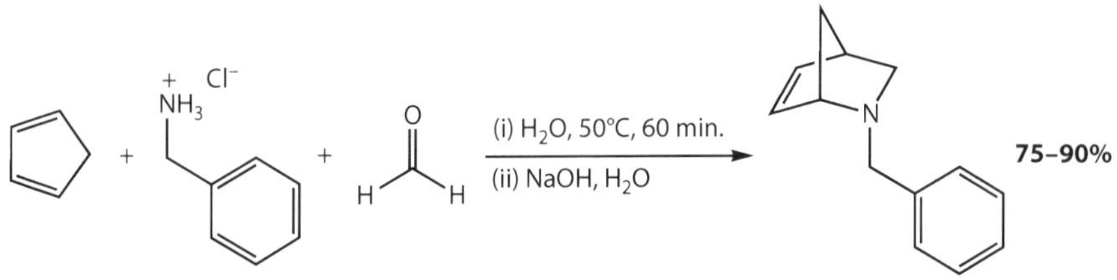

SCHEME 4.9 Hetero-Diels-Alder reaction between cyclopentadiene and an iminium ion.

formed in situ between formaldehyde and benzylamine hydrochloride (Scheme 4.9). Also present in the reaction mixture is cyclopentadiene. After thirty minutes at ambient temperature followed by thirty minutes of heating to 50°C, the cycloadduct is formed as an oil, so isolation is achieved by extraction with ether.

"Click" reactions are so named for their ability to quickly and reliably generate new substances by joining smaller units together.[57] To qualify as a click reaction, the conditions must be operationally simple, utilize no solvents (or use benign solvents where necessary), involve simple purification methods such as filtration, and form products in near-quantitative yields. As such, click chemistry incorporates many of the principles of green chemistry. Copper-catalyzed 1,3-dipolar cycloadditions between an alkyne and an azide were one of the first click reactions to be discovered.[58] An undergraduate experiment involving an aqueous version of this click reaction has recently been reported.[59] Various terminal alkynes are coupled to benzyl azide in a 1:1 mixture of water and *t*-butanol (Scheme 4.10). Copper(II) sulfate (5 mol%) is reduced in situ by sodium ascorbate to produce the active copper(I) catalyst. After two hours at 60°C, the products precipitate from solution and are isolated by simple filtration. An alternative version of the reaction is also available where the azides are produced in situ.

4.3.3 Reduction and Oxidation Reactions

Reduction and oxidation reactions are estimated to make up 15% of transformations undertaken in the pharmaceutical and fine chemicals industry.[60] Unfortunately, many classical methods for performing these transformations require harsh reaction conditions and an excess of hazardous reagents. There are alternative methods available, some of which can be performed in water. The following examples highlight aqueous redox reactions that have been adapted for the undergraduate laboratory.

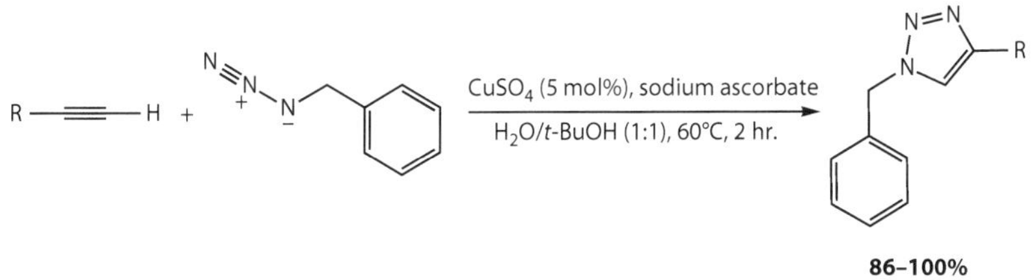

SCHEME 4.10 1,3-Dipolar cycloaddition between benzyl azide and a terminal alkyne.

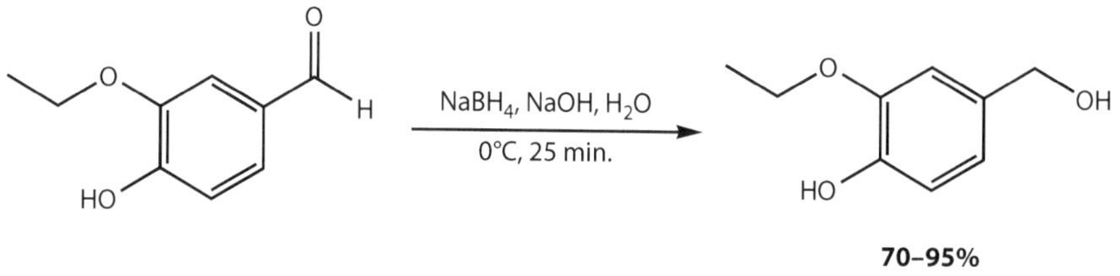

SCHEME 4.11 Sodium borohydride reduction of vanillin in aqueous sodium hydroxide.

Sodium borohydride is a relatively mild hydride donor and is often used in alcoholic solvents for reduction of ketones and aldehydes (Section 6.2.2.2). Under basic conditions, this reagent can also be used in water, and several experiments have been reported for its use as such.[61–64] For example, the reduction of vanillin to vanillyl alcohol has been described using sodium borohydride in a 1 M sodium hydroxide solution[64] (Scheme 4.11). After thirty minutes stirring at 0°C, the product precipitates out of solution. If desired, the product can be further transformed into Methyl Diantilis, a commercially important fragrance, as discussed in Section 7.5. Other common aqueous reductions involve use of biocatalysis. Several such examples have been reported for use in undergraduate experiments;[65–67] reactions of this type are developed in Section 6.2.2.1.

A common oxidation performed in the undergraduate laboratory involves epoxide formation from an alkene. The typical procedure, however, uses *m*-chloroperbenzoic acid (*m*CPBA) in chlorinated solvents. An alternative protocol has been outlined by Broshears et al.[68] using Oxone® (potassium peroxymonosulfate: $2KHSO_5 \cdot KHSO_4 \cdot K_2SO_4$) in a 1:1 solution of acetone and water (Scheme 4.12). Oxone is a strong oxidizer and, in the presence of acetone, generates dimethyldioxirane, a good epoxidizing agent. The reaction takes thirty minutes and proceeds with near-quantitative yields for epoxidation of cyclohexene, norbornylene, and β-pinene. While the reaction works well under these aqueous conditions, it is worth noting that use of Oxone as an oxidant actually results in a lower atom economy than the traditional *m*CPBA version. In addition, the products are soluble under the reaction conditions and isolation requires extraction with ether.

Chromium-based oxidants such as pyridinium chlorochromate are often used in the undergraduate laboratory to oxidize primary and secondary alcohols. Environmental and safety concerns around these reagents have prompted investigation into more benign conditions (Section 6.2.1). Hulce and Marks have described a procedure for oxidizing alcohols that uses hydrogen peroxide as the oxidant and a sodium tungstate catalyst.[69] (Scheme 4.13). Since the reactants are immiscible with the aqueous phase, methyltrioctylammonium hydrogensulfate is added as a phase transfer catalyst. After heating in water for one to three hours, the products are

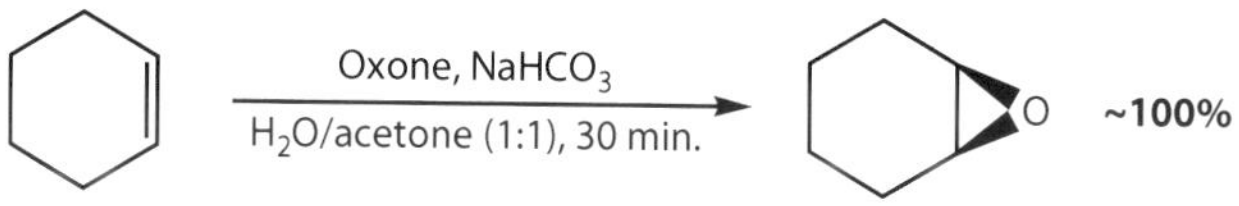

SCHEME 4.12 Epoxidation of cyclohexene with dimethyldioxirane generated in situ.

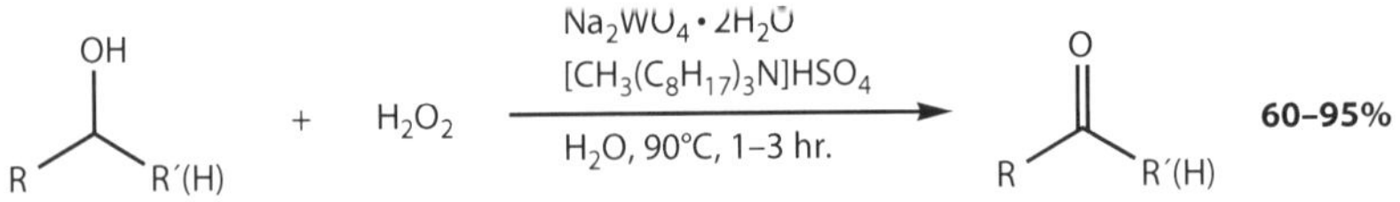

SCHEME 4.13 Oxidation of primary and secondary alcohols using hydrogen peroxide.

collected by either extraction or filtration, depending on the substrate used. In the case of solid ketones, purification is achieved by washing with cold water.

The haloform reaction is one of the oldest transformations known[70] and is typically performed in aqueous alkaline solutions with a molecular halogen. Less hazardous versions of this reaction have been reported for the undergraduate laboratory that use aqueous household bleach (~6% NaOCl) as the chlorine source.[71,72] Ballard has reported a discovery-oriented experiment where heating a solution of 4-methoxyacetophenone in household bleach for thirty minutes yields the corresponding carboxylic acid as a solid (Scheme 4.14).[72] Lowering the pH of the solution leads to formation of an electrophilic substitution product via aromatic ring chlorination in a bleach-acetic acid solvent mixture (Section 6.2.3.1). Students carry out both reactions and use physical and spectral data to identify the two products.

4.3.4 Nucleophilic Additions and Substitutions

Multicomponent reactions combine three or more reactants to form a single product in one step. They tend to be highly atom-economical reactions, making them attractive from a green chemistry perspective. Several examples of aqueous phase multicomponent reactions can be found in the pedagogical literature, including a Hantzsch dihydropyridine synthesis[73] and, more recently, a Passerini reaction.[74] The traditional Passerini reaction is usually performed in chlorinated or alcoholic solvents and requires long reaction times. However, in an undergraduate experiment reported by Hooper and DeBoef,[74] the three-component coupling between benzoic acid, benzaldehyde, and *t*-butyl isocyanide can be rapidly carried out at room temperature when pure water is used as the solvent (Scheme 4.15). The product of the reaction is collected by simple filtration and recrystallized from aqueous ethanol.

Creatine is a popular sports supplement used by athletes to increase muscle mass. In experiments amenable to both undergraduates[75] and high school students,[76] creatine preparation via a nucleophilic addition reaction has been devised. The procedure proposed by Smith and Tan[75a] combines cyanamide and sarcosine (*N*-methylglycine) in aqueous sodium chloride (Scheme 4.16). Concentrated ammonium hydroxide is

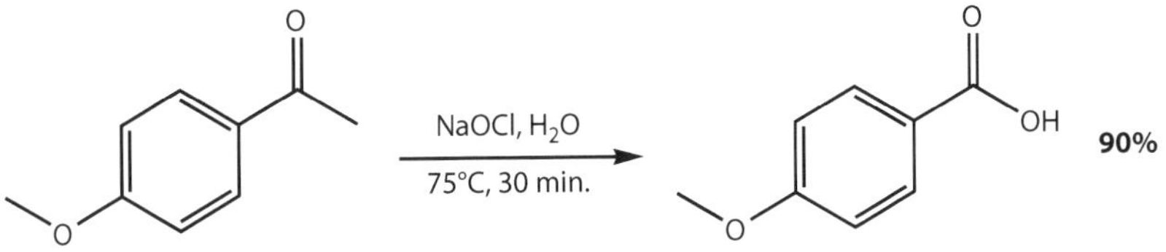

SCHEME 4.14 Haloform reaction using aqueous bleach as the chlorine source.

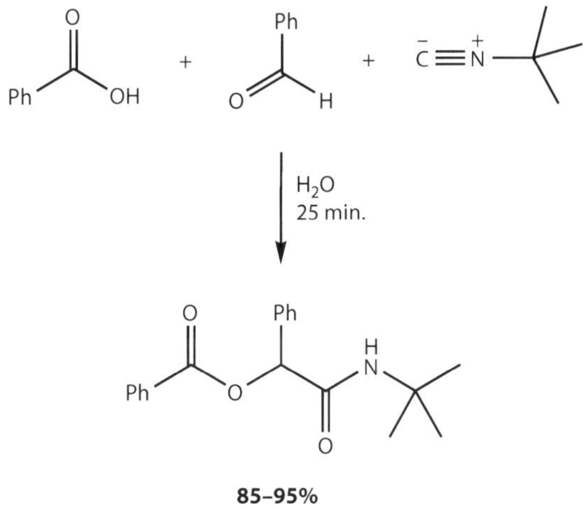

SCHEME 4.15 A multicomponent Passerini reaction.

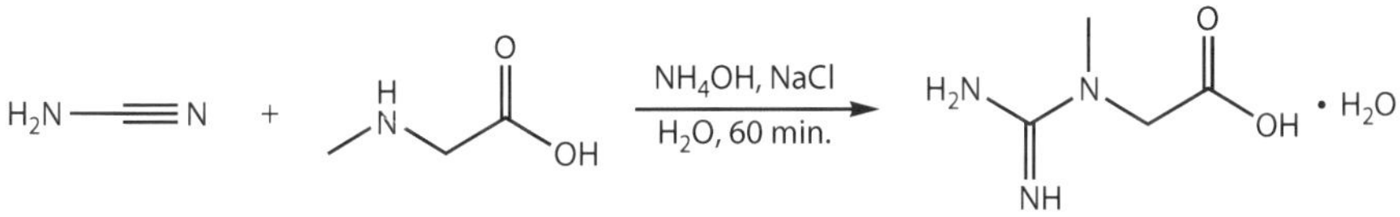

SCHEME 4.16 Synthesis of creatine.

added and the reaction is stirred for one hour. For complete precipitation of the product, the mixture is left to stand for one week, after which the product can be isolated by filtration and recrystallized from boiling water. Notably, this reaction proceeds with nearly 100% atom economy.

Wittig reactions have an inherently low atom economy due to use of triphenylphosphonium salts as a carbanion source (Section 3.5.4). Nevertheless, efforts have been made to develop greener alternatives for the undergraduate laboratory, including several in aqueous media.[77,78] In a variation of the traditional Wittig reaction, a Horner–Wadsworth–Emmons reaction has been exploited between trimethyl phosphonoacetate and various substituted benzaldehydes[78] (Scheme 4.17). The reactants are combined in water along with potassium carbonate and refluxed for fifteen minutes. The water-soluble phosphate anion by-product remains in solution at the end of the reaction, making product isolation

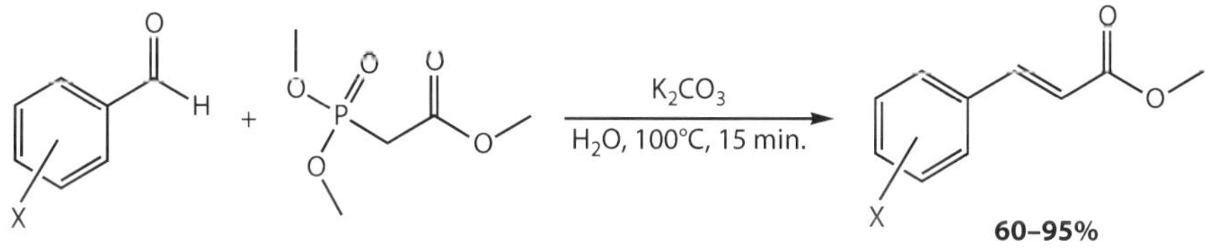

SCHEME 4.17 Horner–Wadsworth–Emmons preparation of unsaturated esters.

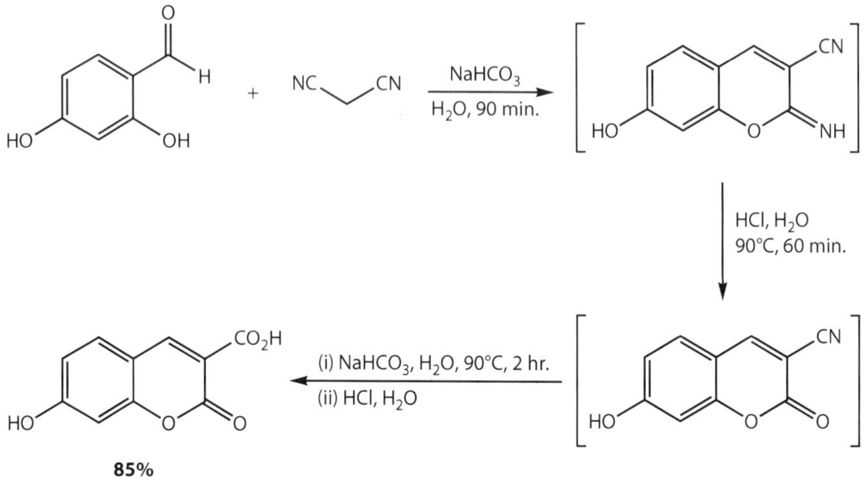

SCHEME 4.18 Four-step one-pot formation of a coumarin derivative.

possible by simple filtration. This offers a substantial improvement to the traditional Wittig reaction, where triphenylphosphine oxide removal is often difficult.[79] The reaction products are additionally of interest as sunscreen analogs.

Like multicomponent reactions, cascade reactions are often acclaimed for their high efficiency and operational simplicity. A four-step one-pot process has been developed for the aqueous synthesis of a coumarin derivative.[80] Beginning with a Knoevenagel condensation between malonitrile and an aromatic aldehyde, the product of this reaction then undergoes a Pinner reaction under the same conditions (Scheme 4.18). The resulting imine intermediate is then hydrolyzed by lowering the pH of the solution; subsequent acidification hydrolyzes the remaining cyano groups to give the target compound. All steps are carried out in water, and the reaction is controlled by varying the pH at different stages. Either of the hydrolysis steps may be omitted if isolation of the intermediates is desired or if the experiment is to be spread over multiple days. Both the final product and the intermediates precipitate out of solution, allowing for easy isolation by filtration.

As an alternative to the traditional Grignard reaction, an aqueous Barbier reaction has been reported for use in the teaching laboratory.[81] Development of these reactions has been an area of extensive research in the last two decades, making inclusion in the undergraduate curriculum very pertinent.[23] Elemental zinc is used to couple allyl bromide with benzaldehyde (Scheme 4.19). The reaction takes place in a 2:1 mixture of saturated ammonium chloride and tetrahydrofuran at an ambient temperature. After thirty minutes, the reaction is complete; however, isolation of the product requires extraction with ether.

Crouch et al.[82] recently published an aldol condensation under aqueous conditions (Scheme 4.20). Interestingly, the final product can be controlled by varying the reaction conditions. At ambient temperatures the crystalline aldol product is obtained with an average yield of 88%. If the reaction is heated to 50°C, the dehydration product is collected. Formation of the latter might seem unexpected given the reaction

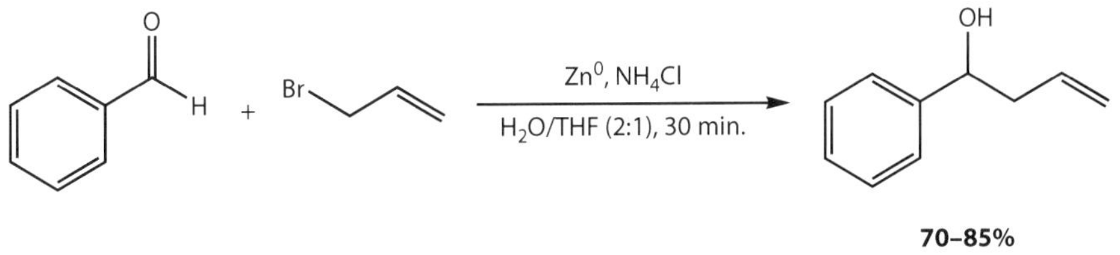

SCHEME 4.19 Aqueous Barbier reaction as a Grignard alternative.

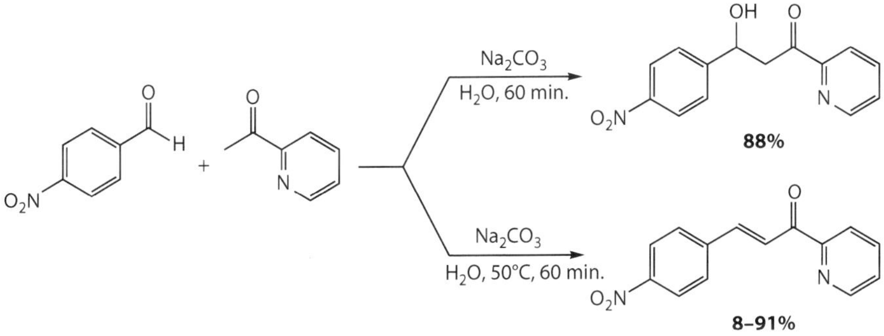

SCHEME 4.20 Temperature-controlled aldol condensation.

is taking place under aqueous conditions; however, the formation of an insoluble product drives the elimination of water.

Nucleophilic substitution mechanisms are generally slower in water than aprotic solvents due to increased nucleophile solvation.[83] However, there have been a few recent reports of aqueous S_N2 reactions. In one report by Dzyuba et al.,[84] a series of ionic liquids are formed by a two-step process. Nucleophilic substitution between 1-methylimidazole and 1-bromobutane is followed by ion exchange using salts such as KPF_6 (Scheme 4.21). The intermediate does not require isolation, and if the reaction scale is sufficient, the ionic liquid product can be collected by simple phase

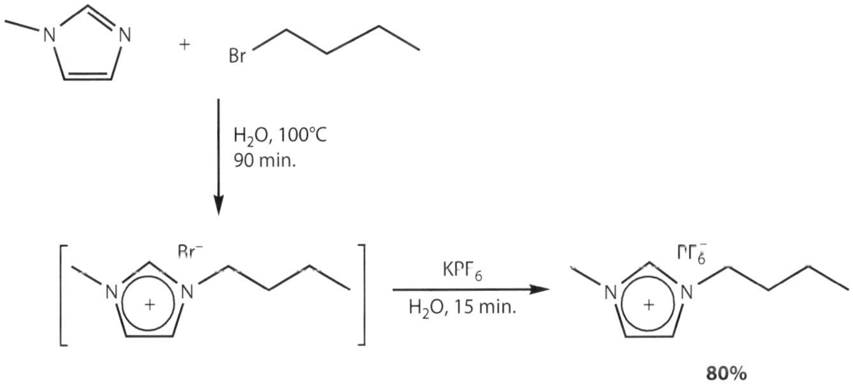

SCHEME 4.21 Ionic liquid formation by an aqueous S_N2 reaction.

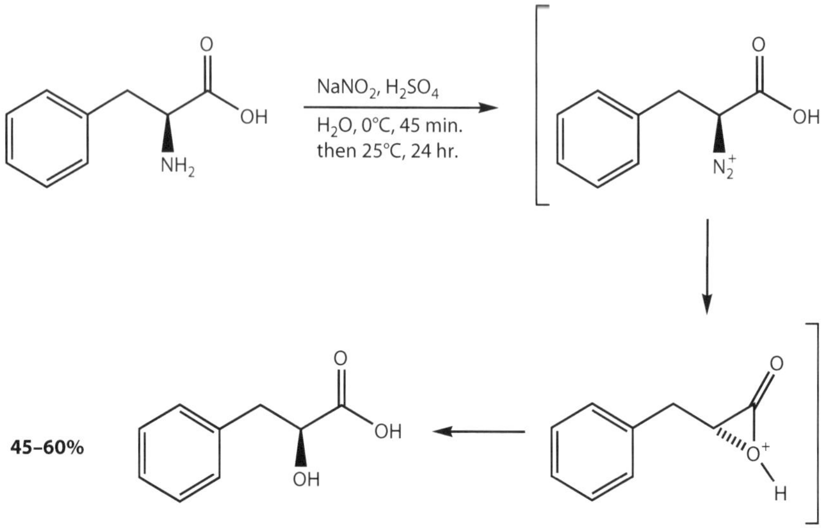

SCHEME 4.22 Aqueous S_N2 reactions on L-phenylalanine.

separation. Alternatively, extraction with dichloromethane can be performed for small-scale experiments. More details about the student preparation of ionic liquids are included in Section 5.5.2.1.

A second aqueous S_N2 experiment involves a double displacement on L-phenylalanine to produce (*S*)-2-hydroxy-3-phenylpropanoic acid, with overall retention of chirality.[85] Diazotization of L-phenylalanine forms a diazonium salt that undergoes a rapid intramolecular S_N2 displacement to form a strained lactone (Scheme 4.22). A second S_N2 displacement by water yields the solid α-hydroxy acid. Analysis by optical rotation allows students to confirm the retention of chirality in their final products.

4.3.5 Electrophilic Additions and Substitutions

Several experiments have been reported for electrophilic aromatic substitutions in water.[86,87] An aromatic nitration experiment designed by Jones-Wilson et al.[86] involves treatment of a water-soluble amino acid (L-tyrosine) with standard nitration conditions (HNO_3/H_2SO_4). The concentrated acids are premixed and added slowly to a cooled, aqueous solution of L-tyrosine (Scheme 4.23). After the addition, the

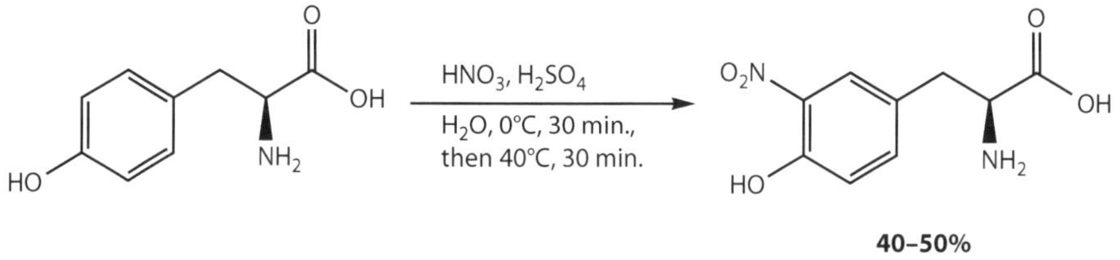

SCHEME 4.23 Aromatic nitration of L-tyrosine.

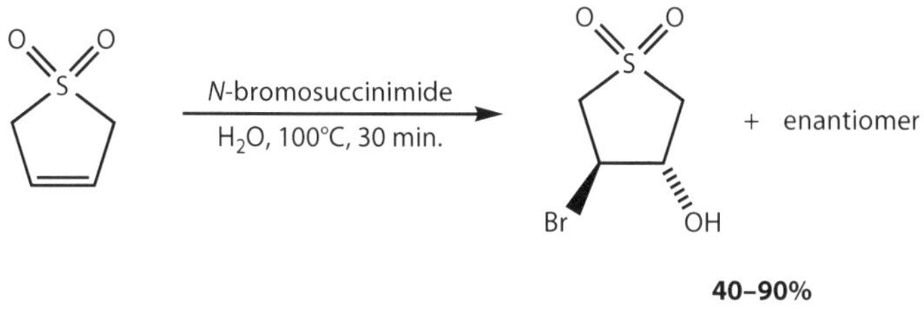

SCHEME 4.24 Bromohydrin formation using *N*-bromosuccinimide.

reaction is heated to 40°C for thirty minutes. The product is collected as a solid at the end of the reaction, and is purified by a wash with ethyl acetate, followed by recrystallization from water. A second example of an electrophilic aromatic substitution has been described as part of a highly efficient preparation of *meso*-diethyl-2,2′-dipyrromethane.[87] This reaction is discussed in Section 7.3.4.

Halohydration reactions lend themselves well to aqueous chemistry, given that water can serve both as a solvent and as a reagent. Greenberg has published an aqueous protocol for bromohydration of 3-sulfolene using *N*-bromosuccinimide as a safer alternative to molecular bromine[88] (Scheme 4.24). The reaction is complete after thirty minutes in refluxing water, and the product precipitates upon cooling. Recrystallization from hot water gives the desired product.

In a related reaction, Crouch et al. have reported an aqueous iodolactonization of 4-pentenoic acid[89] (Scheme 4.25). Molecular iodine is generated in situ from Oxone and potassium iodide, thus providing a safer alternative to the traditional reaction conditions. After one hour of stirring in water, the product is isolated by extraction with dichloromethane. The structure of the product is more difficult for students to predict compared to many halohydrin reactions, and requires them to critically analyze both infrared (IR) and proton nuclear magnetic resonance (NMR) spectral data.

4.3.6 Radical Reactions

Given the relatively high dissociation energy for an O-H bond,[90] water does not readily take part in radical reactions. Accordingly, many industrially important radical processes use water as a solvent, including the production of latex paints.[91] In the undergraduate laboratory, an aqueous radical dimerization was recently outlined for preparation of diapocynin, a biaryl with potential antioxidative and anti-inflammatory

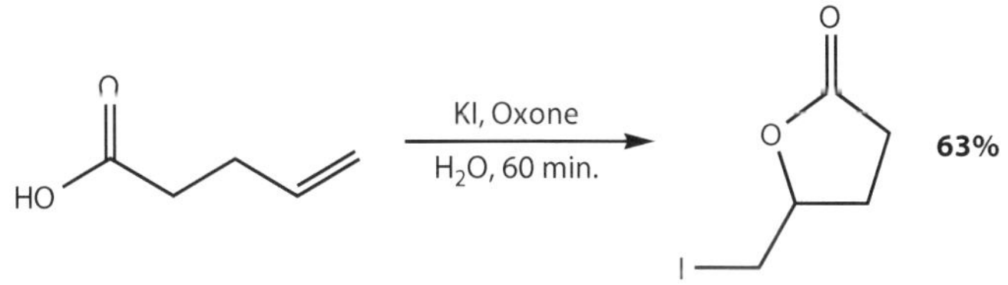

SCHEME 4.25 Iodolactonization of 4-pentenoic acid.

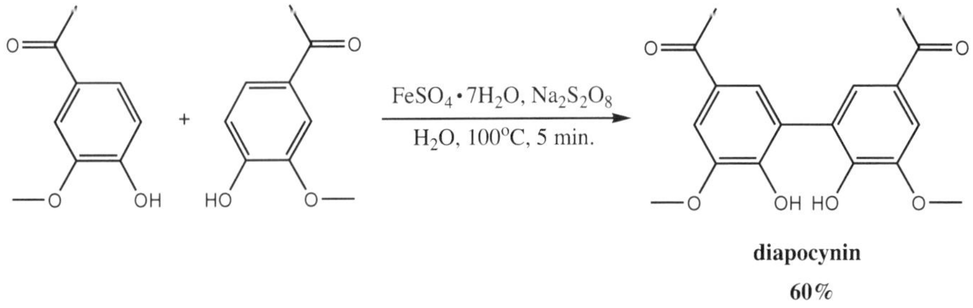

SCHEME 4.26 Diapocynin synthesis via radical dimerization.

properties (Scheme 4.26).[92,93] Iron(II) sulfate is used as the radical initiator and sodium persulfate acts as the oxidant. The reaction is complete after five minutes in refluxing water and the product collected by simple filtration. The enzyme horseradish peroxidase can also be used to effect this transformation (Section 6.2.7.2). In a similar experiment, Mak reported use of iron(III) chloride to dimerize 2-naphthol.[94] The reactants are heated for one to two hours in water, and the binaphthol product precipitates from solution (Scheme 4.27). If desired, the racemic product can be resolved by treating with (–)-*N*-benzylcinchonidinium chloride followed by acid hydrolysis. The chiral products, (*S*)-BINOL and (*R*)-BINOL, are important ligands commonly used in asymmetric catalysis.

4.4 LECTURE CASE STUDIES IN AQUEOUS CHEMISTRY

There are numerous examples of aqueous reactions that have been successfully implemented in industrial settings. These applications make excellent case studies for showcasing the viability of aqueous chemistry outside of an academic environment. The following sections highlight several such instances, which can be discussed in a stand-alone green chemistry course or infused into an existing organic offering.

4.4.1 Rhodium-Catalyzed Hydroformylation

Hydroformylation (also called the oxo process) is a catalytic reaction that converts alkenes into normal or iso-aldehydes (Scheme 4.28). More than 15 billion pounds

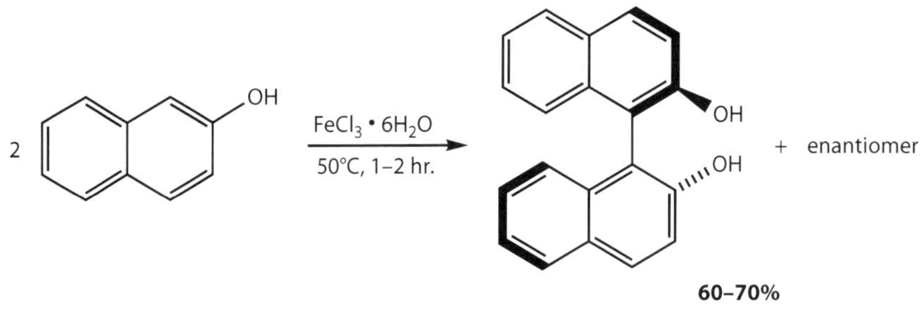

SCHEME 4.27 Radical dimerization of 2-naphthol.

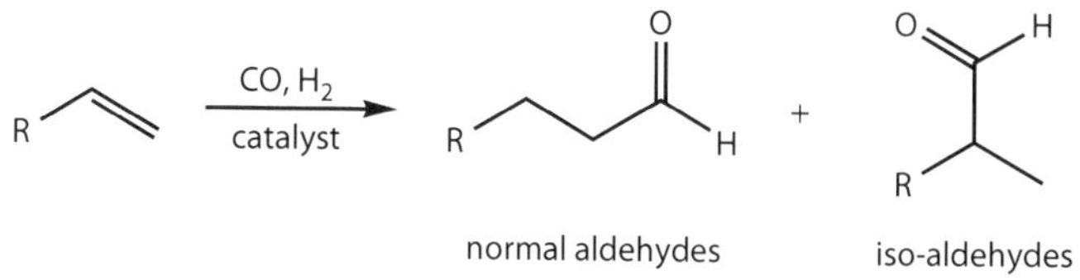

SCHEME 4.28 Hydroformylation reaction for production of normal and iso-aldehydes.

of aldehydes are produced from this reaction annually, making it one of the most important homogeneous catalytic reactions in use.[95] Despite its wide impact, difficulties with separating the metal catalyst from the reaction mixture have prompted chemists to seek alternatives to this traditionally homogeneous reaction.

In 1984, Ruhrchemie AG (RCH) and Rhône-Poulenc (RP) began production of *n*-butyraldehyde using the first commercial example of an aqueous biphasic reaction. Today, five plants are in operation worldwide,[96] and together they produce roughly 10% of the world's supply of C_4 and C_5 linear aldehydes.[97] The RCH/RP process involves hydroformylation of propene (or 1-butene) using syngas (CO, H_2) and a rhodium catalyst modified with the water-soluble phosphine ligand, TPPTS (Scheme 4.29).

The biphasic nature of the RCH/RP process means the rhodium catalyst is effectively immobilized in the aqueous phase by water-soluble TPPTS ligands. The product aldehyde, however, is only sparingly soluble in water, allowing it to be easily separated from the aqueous catalyst and gaseous reactants. Figure 4.1 shows a simplified version of the reaction mechanism,[98] highlighting the biphasic nature of the reaction. As the product aldehyde is formed, it is decanted from the reaction vessel, leaving behind the catalyst in aqueous solution. Maintenance of the propene and syngas pressure thus allows the reaction to operate in a continuous manner.

Implementation of the RCH/RP process has brought both economic and environmental benefits. It is estimated that the aqueous biphasic hydroformylation of propene is 10% less expensive to run than the traditional homogeneous system.[97] A significant portion of this saving is attributed to reduced energy costs for product separation that no longer requires distillation. In addition, loss of the precious metal catalyst is minimized by its immobilization in water. Improvements to its environmental impact can be seen by noting the reduced E-factor for the RCH/RP process: less than 0.1 compared to 0.6–0.9 for the traditional process.[99] Moreover, the total volume of wastewater produced is 70 times less for the biphasic reaction, which is a result of the continuous recycling of the catalyst-containing aqueous phase.

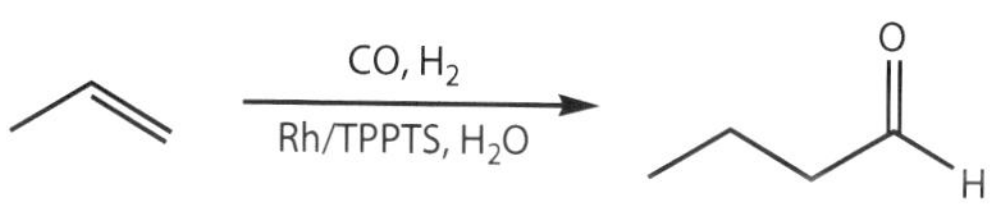

SCHEME 4.29 Hydroformylation of propene using the RCH/RP aqueous biphasic process.

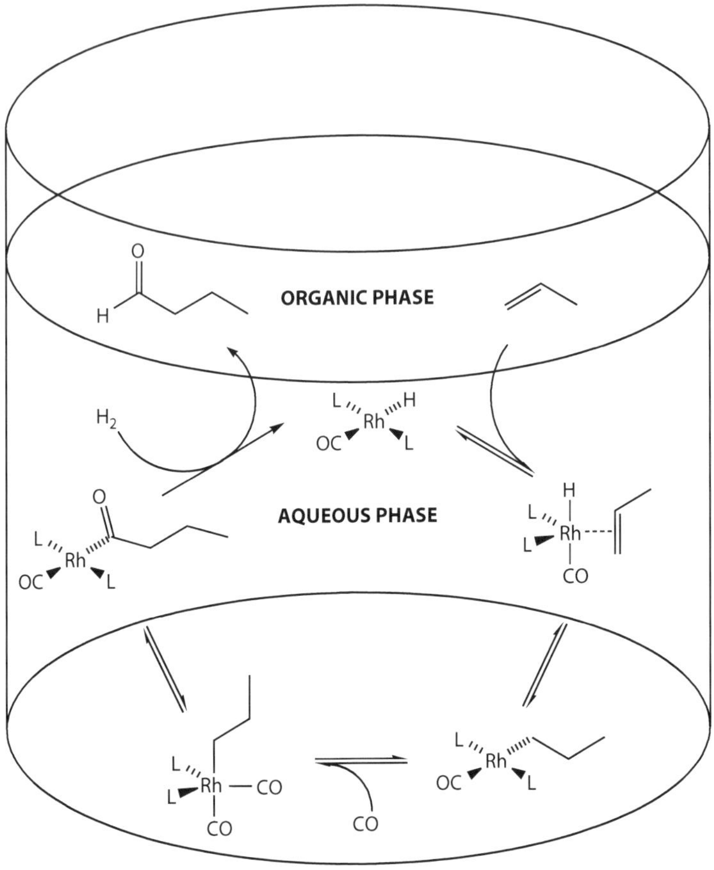

FIGURE 4.1 Simplified catalytic cycle for aqueous biphasic hydroformylation of propene.

4.4.2 Other Aqueous Biphasic Applications

Following the success of the RCH/RP system, several other industrial processes have come online that use aqueous biphasic conditions.[100] Unfortunately, the proprietary details of these reactions have not been fully disclosed, making an in-depth study of these applications more difficult. However, they are mentioned here to reinforce the potential for these biphasic reaction conditions to have widespread applicability beyond hydroformylation.

The aqueous biphasic hydrodimerization of butadiene and water (Scheme 4.30) was successfully commercialized by the Kuraray Corporation of Japan in 1991.[101] The immediate product from the reaction is 2,7-octadien-1-ol, which is further hydrogenated to 1-octanol, an important raw material for the plasticizer industry.

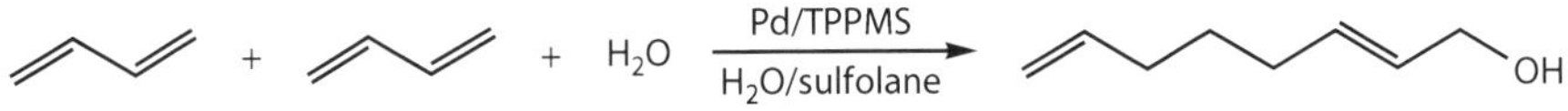

SCHEME 4.30 Aqueous biphasic hydrodimerization of butadiene and water.

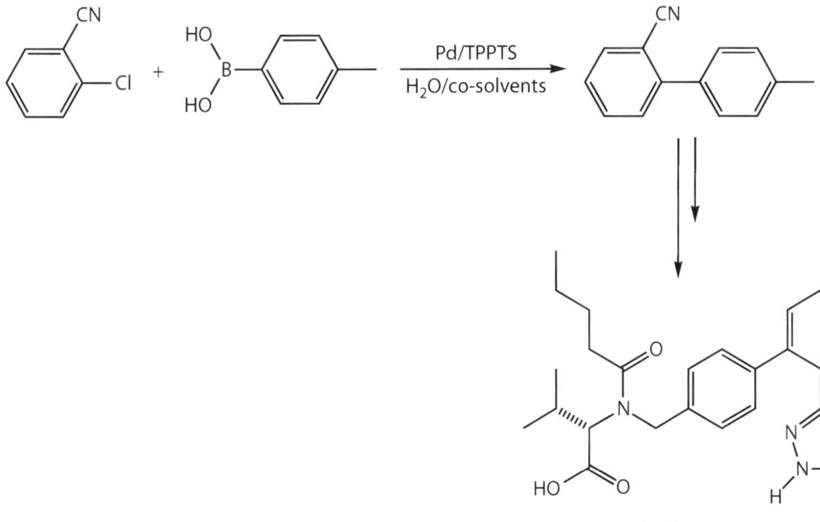

Valsartan

SCHEME 4.31 Aqueous biphasic Suzuki coupling.

The reaction takes place in a water-sulfolane solvent mixture containing a palladium catalyst rendered water soluble with monosulfonated triphenylphosphine (TPPMS).[101,102] The product is extracted into an organic solvent at the end of the reaction, while the catalyst-containing aqueous phase is returned to the reactor for reuse. The process is used to make 5,000 tonnes of 1-octanol per year.

Hoechst AG (now Clariant) has commercialized an aqueous biphasic version of the Suzuki reaction.[100] The process uses a TPPTS-based palladium catalyst to couple 4-tolylboronic acid with 2-chlorobenzonitrile (Scheme 4.31). The aqueous phase contains a polyhydridic alcohol (e.g., ethylene glycol) as a cosolvent, as well as small amounts of a sulfoxide or sulfone (e.g., dimethyl sulfoxide (DMSO)).[103] At the end of the reaction, the phases are separated, allowing for easy product isolation and recycling of the homogeneous catalyst. The biaryl product is generated in 98% yield on a 140 tonne per annum scale.[104] This compound is an important intermediate for the synthesis of "sartan" drugs such as valsartan (Diovan®), which is used in blood pressure regulation.

4.4.3 Synthesis of Rightfit™ Azo Pigments

In 2004, the Engelhard Corporation (now BASF) won a Presidential Green Chemistry Challenge Award for its development of Rightfit pigments.[105] These azo-based pigments produce colors in the yellow, orange, and red palette and, as such, replace many toxic chromium, lead, and cadmium-based compounds.[106] While other organic pigments are capable of producing these colors, they suffer from limitations such as high production costs, use of organic solvents for their preparation, and poor thermal stability.[105,106] In contrast, Rightfit pigments are inexpensive to make, require no organic solvents in their production, and exhibit high heat stability. These pigments

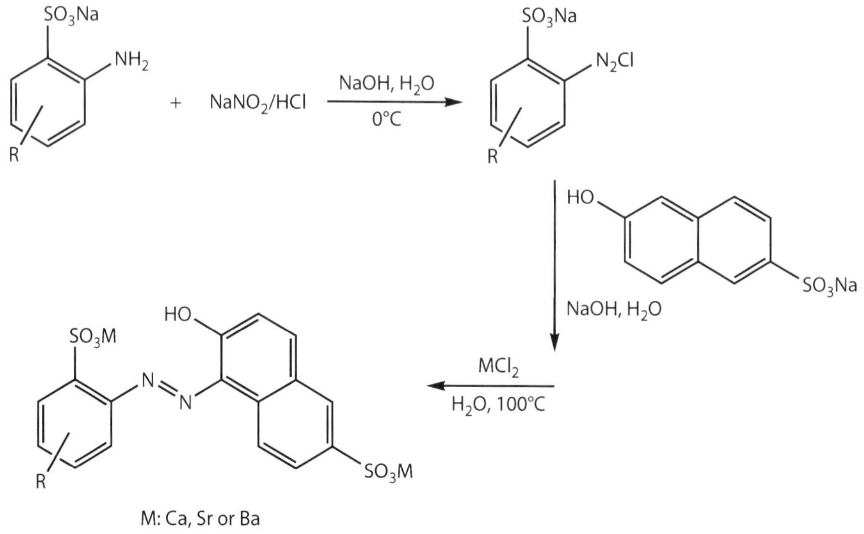

SCHEME 4.32 Synthesis of a Rightfit red pigment.

are additionally less harmful than their competitors. By incorporating anionic substituents into their design, Rightfit pigments have very low solubility in hydrophobic substances, and do not bioaccumulate in fatty tissues.[106] This property has led to FDA approval for their use in indirect food contact applications such as beverage packaging. In addition, it has also meant that their synthesis can be carried out entirely in water. Scheme 4.32 shows the synthesis for one of Engelhard's Rightfit red pigments.[107] The reactions take place in one pot, with water as the only solvent. Reagents are added sequentially and none of the intermediates require isolation. Variation of the substituents (R), protonation state, and cation (M) are used to tune the precise color of the pigment.[106] Students will be familiar with diazotizations and azo dye formations, making this an excellent case study in an introductory organic lecture class. Since the development of these pigments, Engelhard has been able to completely phase out its 6.5 million pound per annum production of heavy metal pigments and replace them with ones synthesized under aqueous conditions.[106]

4.5 CONCLUSION

Generations of synthetic chemists have been taught to regard water as something to be avoided in chemical reactions. This chapter aims to dispel this widespread belief, and to arm educators with the tools needed to ensure that future chemists place water at or near the top of their solvent selection list. Admittedly, the move toward aqueous phase chemistry is not without its challenges; however, the examples highlighted here serve as important reminders of what can be achieved with a little ingenuity. With any luck, the inclusion of aqueous phase chemistry in the undergraduate curriculum will inspire future minds to seek new ways of achieving sustainable chemistry with water as a solvent.

REFERENCES

1. (a) Wöhler, F. *Poggendorffs Ann.* 1828, 12, 253–256. (b) Wöhler, F. *Ann. Chim. Phys.* 1828, 37, 330–333.
2. Cohen, P. S., Cohen, S. M. *J. Chem. Educ.* 1996, 73, 883–886.
3. Baeyer, A., Villiger, V. *Ber. Dtsch. Chem. Ges.* 1899, 32, 3625–3633.
4. Buchner, E., Curtius, T. *Chem. Ber.* 1885, 18, 2371–2377.
5. Hofmann, A. W. *Chem. Ber.* 1881, 14, 2725–2736.
6. (a) Sandmeyer, T. *Ber. Dtsch. Chem. Ges.* 1884, 17, 1633–1635. (b) Sandmeyer, T. *Ber. Dtsch. Chem. Ges.* 1884, 17, 2650–2653.
7. (a) Kishner, N. *Russ. Phys. Chem. Soc.* 1911, 43, 582–595. (b) Wolff, L. *Liebigs Ann. Chem.* 1912, 394, 23–108.
8. Lindström, U. M. In *Organic Reactions in Water: Principles, Strategies and Applications*, Lindström, U. M., Ed. Blackwell, Oxford, 2007, xiii–xv.
9. Frankland, E. *Liebigs Ann. Chem.* 1849, 71, 171–213.
10. Grignard, V. C. R. *Hebd. Seances Acad. Sci.* 1900, 1322–1324.
11. Rideout, D. C., Breslow, R. *J. Am. Chem. Soc.* 1980, 102, 7816–7817.
12. For reviews of aqueous Diels-Alder Reactions, see (a) Garner, P. P. Diels-Alder Reactions in Aqueous Media. In *Organic Synthesis in Water*, Grieco, P. A., Ed. Blackie Academic & Professional, London, 1998, 1–46. (b) Otto, S., Engberts, J. B. F. N. *Pure Appl. Chem.* 2000, 72, 1365–1372.
13. Sheldon, R. A. *Chem. Ind.* 1992, 903–906. (b) Sheldon, R. A. *Green Chem.* 2007, 9, 1273–1283.
14. Constable, D. J. C., Jimenez-Gonzalez, C., Henderson, R. K. *Org. Process Res. Dev.* 2007, 11, 133–137.
15. Clark, J. H., Tavener, S. J. *Org. Process. Res. Dev.* 2007, 11, 149–155.
16. For example, the Diels-Alder reactions by Rideout and Breslow[11] were carried out at concentrations between 5 and 10 mM.
17. Myers, D. Solubilization and Micellar and Phase Transfer Catalysis. In *Surfactant Science and Technology*, 3rd ed. Wiley, Hoboken, NJ, 2005, 191–219.
18. Starks, C. M., Liotta, C. L., Halpern, M. *Phase-Transfer Catalysis: Fundamentals, Applications, and Industrial Perspectives.* Chapman & Hall, London, 1994.
19. Cornils, B., Herrmann, W. A., Eds. *Aqueous-Phase Organometallic Catalysis*, 2nd ed. Wiley-VCH, Weinheim, Germany, 2004.
20. Narayan, S., Muldoon, J., Finn, M. G., Fokin, V. V., Kolb, H. C., Sharpless, K. B. *Angew. Chem. Int. Ed.* 2005, 44, 3275–3279.
21. For further discussion on the terminology of aqueous reactions, see Hayashi, Y. *Angew. Chem. Int. Ed.* 2006, 45, 8103–8104.
22. Chanda, A., Fokin, V. V. *Chem. Rev.* 2009, 109, 725–748.
23. (a) Li, C. J. *Tetrahedron*, 1996, 52, 5643. (b) Li, C. J. *Chem. Rev.* 2005, 105, 3095–3165.
24. (a) Peters, W. *Chem. Ber.* 1905, 38, 2567–2570. (b) Sisido, K., Takeda, Y., Kinugawa, Z. *J. Am. Chem. Soc.* 1961, 83, 538–541. (c) Sisido, K., Kozima, S., Hanada, T. *J. Organomet. Chem.* 1967, 9, 99–107. (d) Sisido, K., Kozima, S. *J. Organomet. Chem.* 1968, 11, 503–513.
25. Blackmond, D. G., Armstrong, A., Coombe, V., Wells, A. *Angew. Chem. Int. Ed.* 2007, 46, 3798–3800.
26. For a recent review, see Butler, R. N., Coyne, A. G. *Chem. Rev.* 2010, 110, 6302–6337.
27. Wohlfarth, C. Permittivity (Dielectric Constant) of Liquids. In *Handbook of Chemistry and Physics*, 90th ed. [online], Lide, D. R., Haynes, W. M., Eds. Taylor and Francis, Boca Raton, FL, 2010, 6.148–6.169.
28. Reichardt, C. *Chem. Rev.* 1994, 94, 2319–2358.

29. (a) Sauer J., Sustmann, R. *Angew. Chem. Int. Ed.* 1980, 19, 779–807. (b) Huisgen, R. *Pure Appl. Chem.* 1980, 52, 2283–2302. (c) Reichardt, C. *Solvents and Solvent Effects in Organic Chemistry*, 3rd ed. Wiley-VCH, Weinheim, Germany, 2003.
30. (a) Blokzijl, W., Engberts, J. B. F. N. *Angew. Chem. Int. Ed.* 1993, 32, 1545–1579. (b) Otto, S., Engberts, J. B. F. N. *Org. Biomol. Chem.* 2003, 1, 2809–2820.
31. (a) Blokzijl, W., Blandamer, M. J., Engberts, J. B. F. N. *J. Am. Chem. Soc.* 1991, 113, 4241–4246. (b) Blokzijl, W., Engberts, J. B. F. N. *J. Am. Chem. Soc.* 1992, 114, 5440–5442.
32. (a) Breslow, R., Maitra, U., Rideout, D. *Tetrahedron Lett.* 1983, 24, 1901–1904. (b) Breslow, R., Maitra, U. *Tetrahedron Lett.* 1984, 25, 1239–1240.
33. (a) Grieco, P. A., Garner, P., He, Z. *Tetrahedron Lett.* 1983, 24, 1897–1900. (b) Grieco, P. A., Yoshida, K., Garner, P. J. *Org. Chem.* 1983, 48, 3137–3139.
34. (a) McDevit, W. F., Long, F. A. J. *Am. Chem. Soc.* 1952, 74, 1773–1777. (b) Dack, M. R. J. *Chem. Soc. Rev.* 1975, 4, 211–229.
35. Breslow, R., Guo, T. *Proc. Natl. Acad. Sci.* 1990, 87, 167–169.
36. (a) Blake, J. F., Jorgensen, W. L. *J. Am. Chem. Soc.* 1991, 113, 7430–7432. (b) Blake, J. F., Lim, D., Jorgensen, W. L. *J. Org. Chem.* 1994, 59, 803–805. (c) Jorgensen, W. L., Blake, J. F., Lim, D. C., Severance, D. L. *J. Chem. Soc., Faraday Trans.* 1994, 90, 1727–1732. (d) Chandrasekhar, J., Shariffskul, S., Jorgensen, W. L. *J. Phys. Chem. B* 2002, 106, 8078–8085. (e) Acevedo, O., Jorgensen, W. L. *J. Chem. Theory Comput.* 2007, 3, 1412–1419.
37. Furlani, T. R., Gao, J. L. *J. Org. Chem.* 1996, 61, 5492–5497.
38. Kong, S., Evanseck, J. D. *J. Am. Chem. Soc.* 2000, 122, 10418–10427.
39. Engberts, J. B. F. N. *Pure Appl. Chem.* 1995, 67, 823–828.
40. Butler, R. N., Cunningham, W. J., Coyne, A. G., Burke, L. A. *J. Am. Chem. Soc.* 2004, 126, 11923–11929.
41. Gajewski, J. J. *Acc. Chem. Res.* 1997, 30, 219–225 and references cited therein.
42. (a) Du, Q., Superfine, R., Freysz, E., Shen, Y. R. *Phys. Rev. Lett.* 1993, 70, 2313–2316. (b) Du, Q., Freysz, E., Shen, Y. R. *Science* 1994, 264, 826–828. (c) Shen, Y. R., Ostroverkhov, V. *Chem. Rev.* 2006, 106, 1140–1154.
43. Jung, Y., Marcus, R. A. *J. Am. Chem. Soc.* 2007, 129, 5492–5502.
44. (a) Zheng, Y., Zhang, J. *J. Phys. Chem. A* 2010, 114, 4325–4333. (b) Acevedo, O., Armacost, K. *J. Am. Chem. Soc.* 2010, 132, 1966–1975.
45. Dicks, A. P. *Green Chem. Lett. Rev.* 2009, 2, 9–21.
46. Pantess, D. A., Rich, C. V. *Chem. Educator* 2009, 14, 258–260.
47. Aktoudianakis, E., Chan, E., Edward, A. R., Jarosz, I., Lee, V., Mui, L., Thatipamala, S. S., Dicks, A. P. *J. Chem. Educ.* 2008, 85, 555–557.
48. Cheung, L. L. W., Aktoudianakis, E., Chan, E., Edward, A. R., Jarosz, I., Lee, V., Mui, L., Thatipamala, S. S., Dicks, A. P. *Chem. Educator* 2007, 12, 77–79.
49. Harper, B. A., Rainwater, J. C., Birdwhistell, K., Knight, D. A. *J. Chem. Educ.* 2002, 79, 729–731.
50. (a) Gilbertson, R., Doxsee, K., Succaw, G., Huffmann, L. M., Hutchison, J. E. Palladium-Catalyzed Alkyne Coupling/Intramolecular Alkyne Addition: Synthesis of a Benzofuran Product. In *Greener Approaches to Undergraduate Chemistry Experiments*, Kirchhoff, M., Ryan, M. A., Eds. American Chemical Society, Washington, DC, 2002, 4–7. (b) Doxsee, K. M., Hutchison, J. E. Experiment 13: Palladium-Catalyzed Alkyne Coupling/Intramolecular Alkyne Addition: Natural Product Synthesis. In *Green Organic Chemistry—Strategies, Tools, and Laboratory Experiments*, Brooks/Cole, Pacific Grove, CA, 2004, 189–196.
51. Viswanathan, T., Jethmalani, J. *J. Chem. Educ.* 1993, 70, 165–167.
52. Novak, B. M., Grubbs, R. H. *J. Am. Chem. Soc.* 1988, 110, 7542–7543.

53. (a) McKenzie, L. C., Huffman, L. M., Hutchison, J. E., Rogers, C. E., Goodwin, T. E., Spessard, G. O. *J. Chem. Educ.* 2009, 86, 488–493. (b) Huffmann, L. M., McKenzie, L. C., Hutchison, J. E. Diels-Alder Reaction in Water. http://greenchem.uoregon.edu/PDFs/GEMsID84.pdf (accessed December 23, 2010).
54. Winstead, A. J. *Chem. Educator* 2010, 15, 28–31.
55. Ballini, R., Barboni, L., Fringuelli, F., Palmieri, A., Pizzo, F., Vaccaro, L. *Green Chem.* 2007, 9, 823–838 and references cited therein.
56. Sauvage, X., Delaude, L. *J. Chem. Educ.* 2008, 85, 1538–1540.
57. Kolb, H. C., Finn, M. G., Sharpless, K. B. *Angew. Chem. Int. Ed.* 2001, 40, 2004–2021.
58. Huisgen, R., Szeimies, G., Möbius, L. *Chem. Ber.* 1967, 100, 2494–2507.
59. Sharpless, W. D., Wu, P., Hansen, T. V., Lindberg, J. G. *J. Chem. Educ.* 2005, 82, 1833–1836.
60. Dugger, R. W., Ragan, J. A., Ripin, D. H. B. *Org. Process Res. Dev.* 2005, 9, 253–258.
61. Hudak, N. J., Sholes, A. H. *J. Chem. Educ.* 1986, 63, 161.
62. Zaczek, N. M. *J. Chem. Educ.* 1986, 63, 909.
63. Fowler, R. G. *J. Chem. Educ.* 1992, 69, A43–A46.
64. Miles, W. H., Connell, K. B. *J. Chem. Educ.* 2006, 83, 285–286.
65. Pohl, N., Clague, A., Schwarz, K. *J. Chem. Educ.* 2002, 79, 727–728.
66. Ravía, S., Gamenara, D., Schapiro, V., Bellomo, A., Adum, J., Seoane, G., Gonzalez, D. *J. Chem. Educ.* 2006, 83, 1049–1051.
67. Williamson, K. L., Minard, R. D., Masters, K. M. Enzymatic Reactions: A Chiral Alcohol from a Ketone and Enzymatic Resolution of DL-Alanine. In *Microscale and Macroscale Organic Experiments*, 5th ed. Houghton Mifflin, Boston, 2007, pp. 785–791.
68. Broshears, W. C., Esteb, J. J., Richter, J., Wilson, A. M. *J. Chem. Educ.* 2004, 81, 1018–1019.
69. Hulce, M., Marks, D. W. *J. Chem. Educ.* 2001, 78, 66–67.
70. Serulas, M. *Ann. Chim.* 1822, 20, 165.
71. Lehman, J. W. Experiment 40: Haloform Oxidation of 4′-Methoxyacetophenone. In *Multiscale Operational Organic Chemistry: A Problem-Solving Approach to the Laboratory Course*. Prentice-Hall, Upper Saddle River, NJ, 2002, 322–328.
72. Ballard, C. E. *J. Chem. Educ.* 2010, 87, 190–193.
73. Norcross, B. E., Clement, G., Weinstein, M. *J. Chem. Educ.* 1969, 46, 694–695.
74. Hooper, M. M., DeBoef, B. *J. Chem. Educ.* 2009, 86, 1077–1079.
75. (a) Smith, A. L., Tan, P. *J. Chem. Educ.* 2006, 83, 1654–1657. (b) Lecher, C. S., Bernhardt, R. J. A Greener Synthesis of Creatine. http://greenchem.uoregon.edu/PDFs/GEMsID100.pdf (accessed December 23, 2010).
76. Lecher, C. S., Bernhardt, R. J. Synthesis of Creatine—A High School Procedure. http://greenchem.uoregon.edu/PDFs/GEMsID142.pdf (accessed December 23, 2010).
77. Broos, R., Tavernier, D., Anteunis, M. *J. Chem. Educ.* 1978, 55, 813.
78. Cheung, L. L. W., Lin, R. J., McIntee, J. W., Dicks, A. P. *Chem. Educator* 2005, 10, 300–302.
79. Boutagy, J., Thomas, R. *Chem. Rev.* 1974, 74, 87–99.
80. Fringuelli, F., Piermatti, O., Pizzo, F. *J. Chem. Educ.* 2004, 81, 874–876.
81. Breton, G. W., Hughey, C. A. *J. Chem. Educ.* 1998, 75, 85.
82. Crouch, R. D., Richardson, A., Howard, J. L., Harker, R. L., Barker, K. H. *J. Chem. Educ.* 2007, 84, 475–476.
83. Clayden, J., Greeves, N., Warren, S., Wothers, P. *Organic Chemistry*. Oxford University Press, New York, 2000, 332–334.
84. Dzyuba, S. V., Kollar, K. D., Sabnis, S. S. *J. Chem. Educ.* 2009, 86, 856–858.
85. Van Draanen, N. A., Hengst, S. *J. Chem. Educ.* 2010, 87, 623–624.
86. Jones-Wilson, T. M., Burtch, E. A. *J. Chem. Educ.* 2005, 82, 616–617.

87. Sobral, A. J. F. N. *J. Chem. Educ.* 2006, 83, 1665–1666.
88. Greenberg, F. H. *J. Chem. Educ.* 1985, 62, 638.
89. Crouch, R. D., Tucker-Schwartz, A., Barker, K. *J. Chem. Educ.* 2006, 83, 921–922.
90. Bond dissociation energy of HO-H is 497 kJ/mol. Taken from Luo, Y.-R. Bond Dissociation Energies. In *Handbook of Chemistry and Physics*, 90th ed. [online], Lide, D. R., Haynes, W. M., Eds. Taylor and Francis, Boca Raton, FL, 2010, 9.64–9.97. www.hbcpnetbase.com (accessed December 23, 2010).
91. (a) Lovell, P. A., El-Aasser, M. S. *Emulsion Polymerization and Emulsion Polymers.* Wiley, New York, 1997. (b) Chern, C.-S. *Principles and Applications of Emulsion Polymerization.* Wiley, Hoboken, NJ, 2008.
92. Dasari, M. S., Richards, K. M., Alt, M. L., Crawford, C. F. P., Schleiden, A., Ingram, J., Hamidou, A. A. A., Williams, A., Chernovitz, P. A., Luo, R., Sun, G. Y., Luchtefeld, R., Smith, R. E. *J. Chem. Educ.* 2008, 85, 411–412.
93. Klees, R. F., DeMarco, P. C., Salazsnyk, R. M., Ahuja, D., Hogg, M., Antoniotti, S., Kamath, L., Dordick, J. S., Plopper, G. E. *J. Biomed. Biotechnol.* 2006, 1–10.
94. Mak, K. K. W. *J. Chem. Educ.* 2004, 81, 1636–1640.
95. Van Leeuwen, P. W. N. M., Claver, C., Eds. Rhodium Catalyzed Hydroformylation [online]. Springer, 2000. http://lib.myilibrary.com?ID=20539 (accessed December 23, 2010).
96. Wiebus, E., Cornils, B. Water as a Reaction Solvent—An Industry Perspective. In *Organic Reactions in Water: Principles, Strategies and Applications*, Lindström, U. M., Ed. Blackwell, Oxford, 2007, 366–397.
97. Kohlpaintner, C. W., Fischer, R. W., Cornils, B. *Appl. Catal. A* 2001, 221, 219–225.
98. The aqueous reaction is presumed to go through the same mechanism as the Rh/TPP system. For references, see (a) Evans, D., Osborn, J. A., Wilkinson, G., *J. Chem. Soc. A* 1968, 3133–3142. (b) Evans, D., Yagupsky, G., Wilkinson, G. *J. Chem. Soc. A* 1968, 2660–2665. (c) Brown, C. K., Wilkinson, G. *J. Chem. Soc. A* 1970, 2753–2764.
99. Cornils, B., Wiebus, E. Environmental and Safety Aspects. In *Aqueous-Phase Organometallic Catalysis*, 2nd ed., Cornils, B., Herrmann, W. A., Eds. Wiley-VCH, Weinheim, Germany, 2004, 337–347.
100. Cornils, B. *Org. Process Res. Dev.* 1998, 2, 121–127.
101. Yoshimura, N. Hydrodimerization. In *Aqueous-Phase Organometallic Catalysis*, 2nd ed., Cornils, B., Herrmann, W. A., Eds. Wiley-VCH, Weinheim, Germany, 2004, 540–549.
102. (a) Yoshimura, N., Tamura, M. US 4417079. Kuraray Ind., 1983. (b) Yoshimura, N., Tamura, M. US 4510331. Kuraray Ind., 1985. (c) Tokitoh, Y., Yoshimura, N. US 5057631. Kuraray Ind., 1991. (d) Tokitoh, Y., Higashi, T., Hino, K., Murasawa, M., Yoshimura, N. US 5118885. Kuraray Ind., 1992.
103. (a) Haber, S., Kleiner, H.-J. US 5756804. Hoechst AG, 1998. (b) Haber, S., Egger, N. US 6140265. Clariant GmbH, 2000.
104. Haber, S. Fine Chemicals Synthesis. In *Aqueous-Phase Organometallic Catalysis*, 1st ed., Cornils, B., Herrmann, W. A., Eds. Wiley-VCH, Weinheim, Germany, 1998, 440–446.
105. The Presidential Green Chemistry Challenge Award Recipients 1996–2009, 2004 Designing Greener Chemicals Award. U.S. Environmental Protection Agency, Office of Pollution Prevention and Toxics, Washington, DC, 2009, 62–63.
106. Cann, M. C., Umile, T. P. Rightfit Pigments: Synthetic Azo Pigments to Replace Toxic Organic and Inorganic Pigments. In *Real-World Cases in Green Chemistry*, Vol. 2. American Chemical Society, Washington, DC, 2008, 53–61.
107. Bindra, A. P. US 6375733. Engelhard Corp., 2002.

5 Organic Chemistry in Greener Nonaqueous Media

Mr. Leo Mui

CONTENTS

5.1 INTRODUCTION

With the exception of some reactions that proceed efficiently without solvents, which are discussed in Chapter 3, the vast majority of organic reactions (and almost all extractions) occur in the solution phase. Solvents are used to mediate an even reaction temperature, control reactivity, and facilitate separation and transport of both reagents and products.[1] However, many traditional solvents used for organic reactions and extractions are flammable (for example, diethyl ether), toxic (*N,N*-dimethylformamide), carcinogenic (benzene), readily absorbed through skin (dimethylsulfoxide), environmentally deleterious (chloroform), or some combination thereof. It is understandable how elimination of these solvents from the undergraduate laboratory can improve safety, reduce negative environmental impact, and introduce future chemists to new, or "neoteric," solvents. This chapter discusses methods to evaluate solvent "greenness," introduces several greener nonaqueous alternative solvents that have been used in place of more traditional options in the undergraduate laboratory, and describes the comparative greenness of each one. Finally, an outlook on other green solvents that might potentially be implemented into future undergraduate organic experiments is provided.

5.2 MEASURES OF SOLVENT GREENNESS

When assessing the greenness of a solvent, its impact on health and safety inside the laboratory must be examined, along with its environmental impact. While it is relatively simple to evaluate solvent health and safety parameters, calculating "cradle to grave" environmental impacts can be very complex. Several different tools and methods to evaluate and select appropriate solvents have been published recently.[2–5] Some of these approaches, and particularly ones that are considered as most applicable to the undergraduate laboratory environment, are discussed in this section.

Capello et al. have compiled a quantitative examination of twenty-six organic solvents by combining the results of an environmental, health, and safety (EHS) evaluation with a life cycle assessment (LCA) for each solvent.[2] The EHS method evaluated nine potential hazards of a solvent: persistency, air hazard, water hazard, acute toxicity, chronic toxicity, irritation, release potential, fire/explosion hazards, and reactivity/decomposition hazards. Each hazard was rated with a score between zero and one and then summed to produce an EHS indicator score, which increases with hazard level. This simple EHS assessment can be displayed graphically as a segmented bar chart for quick and easy interpretation and comparison. The authors found that the lowest-scoring (safest) solvents included ethanol and ethyl acetate, whose largest hazard contribution was their high flammability. Acetonitrile and dioxane were among the higher-scoring solvents in the study.

After performing the EHS assessment, the authors proceeded to a LCA in order to quantify the emissions and resource use of each solvent from cradle to grave. Beginning with solvent production from petrochemicals through transportation, use, and disposal, the energy and resource inputs were calculated along with associated emissions and generated waste based on collected data. Two options for postdisposal treatment were analyzed for each solvent: one involved direct waste incineration, and the other included solvent recycling through distillation before eventual incineration. Through a complex method, the authors produced a cumulative energy demand (CED) score for each solvent by subtracting the energy gain from the incineration or distillation process from the total energy required for solvent use. A low CED score indicates a more environmentally favorable solvent. The LCA showed that simple solvents directly produced from petrochemicals (such as alkanes, diethyl

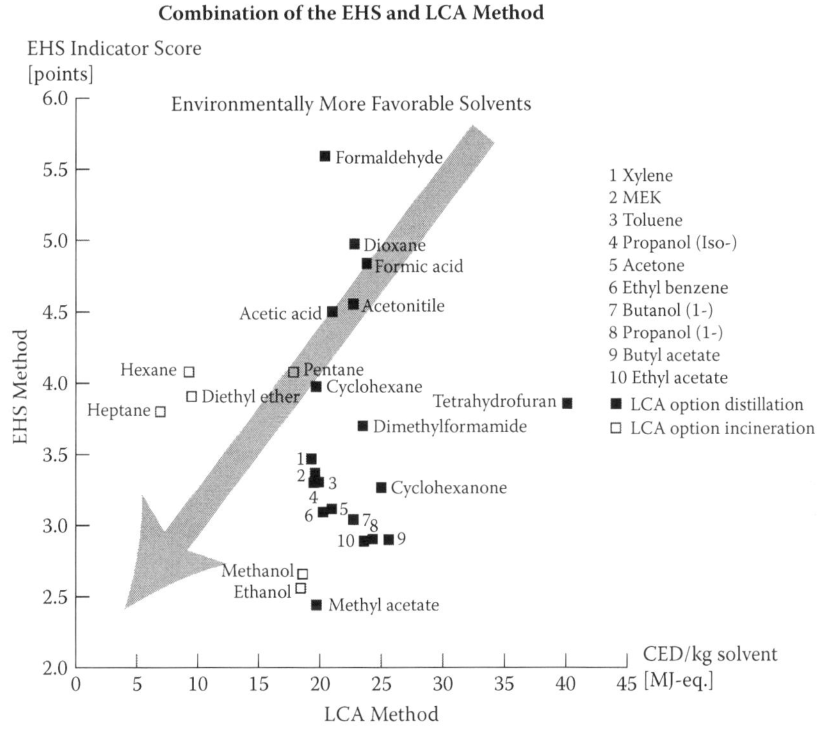

FIGURE 5.1 Environmental assessment of twenty-six organic solvents combining EHS and LCA methods. (From Capello, C., Fischer, U., Hungerbühler, K., *Green Chem.*, 9, 927–934, 2007. Reproduced with permission of the Royal Society of Chemistry.)

ether, and methanol) have the lowest CED. Solvents requiring more complex preparation, such as ethyl acetate and tetrahydrofuran, or solvents with low caloric value (leading to lower energy production during incineration, such as dioxane and *N,N*-dimethylformamide), predictably gave higher CED values.

Plotting the EHS results against the LCA results shows the relative greenness of each solvent analyzed (Figure 5.1). Solvents that appear in the lower left quadrant, such as methanol, ethanol, and methyl acetate, are more favorable than those that appear toward the upper right quadrant, such as formaldehyde (EHS hazard) and tetrahydrofuran (high CED). A drawback of this study is that no chlorinated solvents were included, nor any neoteric solvents (such as supercritical carbon dioxide [$scCO_2$], ionic liquids, or fluorous solvents) mostly due to lack of appropriate data. It is known, however, that the energy input required to produce ionic liquids at the laboratory scale is much greater than that for any of the solvents studied by the authors.[2,3]

The green alternative solvents that are absent from this analysis were reviewed by Clark and Tavener using a semiquantitative method.[3] The authors gave an arbitrary score to five solvent classes based on five criteria: usefulness, ease of separation and reuse, health and safety, economic cost, and cradle-to-grave environmental impact.

Their approach awarded relatively high scores to solvents from renewable sources (e.g., ethanol, glycerol, (*R*)-limonene, and ethyl L-lactate) and $scCO_2$. Ionic liquids and fluorous solvents received low grades even though they are generally described as green solvents in the chemical literature. More discussion on the specific advantages and disadvantages of $scCO_2$, ionic liquids, and fluorous solvents can be found in later sections of this chapter.

A team of researchers from Pfizer recently published a solvent selection tool to aid their company research and process chemists in choosing solvents that both improve safety and reduce environmental impact.[4] The authors aimed to create a very simple and easy-to-use guide for their scientists, so their evaluation broke down solvents into three straightforward categories: preferred, usable, and undesirable. Preferred solvents included water, acetone, ethanol, and ethyl acetate; usable solvents featured heptane, acetonitrile, and 2-methyltetrahydrofuran; while undesirable solvents contained hexane, diethyl ether, dichloromethane, and *N,N*-dimethylformamide. Each solvent was evaluated individually following three criteria: worker safety, process safety, and environmental and regulatory considerations. For worker safety, the chronic effects of the solvent on health (carcinogenicity, mutagenicity, and reprotoxicity) were weighted with highest concern, with skin absorption and toxicity following closely behind. Factors including flammability, emissions, and odor contributed to the process safety of each solvent. In evaluating for environmental concerns, the authors considered solvent ecotoxicity, water contamination properties, and persistence, among others.

Importantly, in addition to evaluating each solvent, the Pfizer chemists also created a *solvent replacement table* that guides researchers to a safer alternative solvent for the ones portrayed as undesirable. As examples, heptane is considered superior to pentane or hexane, triethylamine should replace pyridine as a basic solvent, and 2-methyltetrahydrofuran should find use instead of diethyl ether. The main advantage of Pfizer's qualitative approach to solvent evaluation is that it clearly shows employees which solvents are preferable without having to train them regarding how to interpret the advice. This simplicity is also highly desirable in the undergraduate laboratory, so students can quickly evaluate the relative greenness of a particular experimental procedure.

5.3 SUPERCRITICAL CARBON DIOXIDE

5.3.1 Introduction

Supercritical carbon dioxide is a nontoxic, nonflammable, inert, and easy-to-separate solvent that has minimal cradle-to-grave environmental impact.[1,3] The threshold limit value (TLV) of a chemical is the amount to which it is believed a person can be exposed on a daily basis for a working lifetime without detrimental health effects. $ScCO_2$ has a very high TLV at 5,000 ppm, making it safer than many traditional solvents to humans (for comparison, acetone has a TLV of 750 ppm and chloroform has a TLV of 10 ppm).[6] Its natural abundance in the atmosphere reduces the impact of its production emission to the environment, as most laboratory grade CO_2 comes from distillation of air, and the same CO_2 is returned to the atmosphere at the end of its

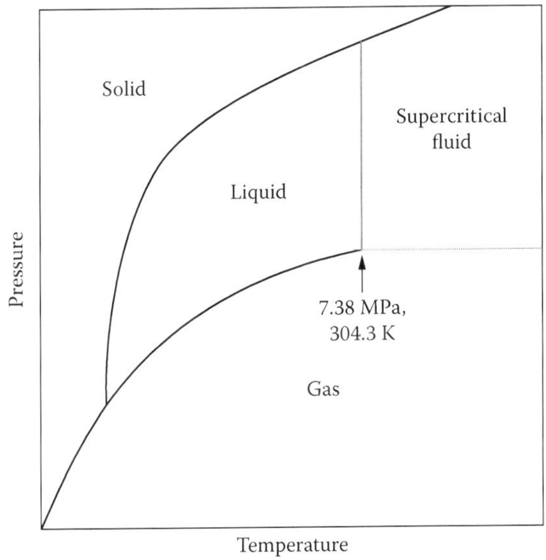

FIGURE 5.2 Phase diagram of carbon dioxide.

lifetime. Because of these safety reasons, it is currently used in industry as a solvent for extractions (as a replacement for dichloromethane in the decaffeination of coffee beans) and syntheses (as an alternative to chlorofluorocarbons in the polymerization of fluoromonomers), among other functions.[6] The main hindrances to use of $scCO_2$ are its low solubility toward many compounds, the energy cost of CO_2 purification and compression, and the capital cost of $scCO_2$ reactors. Clark and Tavener rated $scCO_2$ with a score of 18/25 on their green solvent scale, scoring just one point less than water.[3]

Supercritical fluids (SFs) are a state of matter achieved when a substance is heated above its critical temperature (T_c) and pressurized above its critical pressure (P_c).[7] Carbon dioxide has an easily accessible critical point at a T_c of 304.3 K and P_c of 7.38 MPa (Figure 5.2).[8] As a liquid-gas system approaches the critical point, the miniscus separating the two phases begins to disappear, until the entire system becomes homogeneous. At any temperature and pressure that exceeds the critical point, the liquid and gas phases cease to be distinguishable.[7] The physical properties of SFs are intermediate between a liquid and a gas. The solubility properties of SFs are similar to those of liquids,[9] while their zero surface tension, high diffusivity, and low viscosity are gas-like, allowing for better wettability, mass transfer, and faster extractions.[8] These properties make SFs good candidates for extracting substances from a solid matrix (for example, a food product), as the solvent can penetrate deeply into the matrix like a gas, yet still dissolve substances like a liquid. Use of $scCO_2$ in the processing of foods and food products is increasing in light of health concerns from residual and emitted solvents.

Carbon dioxide is a small linear molecule with no net dipole, so it is not suprising that, while it can act as a good solvent for small volatile substances, it is a poor solvent for large polar molecules.[6] However, CO_2 does have a large quadrupole, which somewhat increases the solubility of polar substances in it compared to alkanes.[10] The solubilizing power of $scCO_2$ correlates to its density, which increases with temperature and pressure.[11] An interesting application of this phenomenon is that selective fractional extraction can be achieved by simply adjusting the pressure of $scCO_2$.[11] However, the capital and energy costs increase nonlinearly with pressure, making the process of extracting more polar molecules with $scCO_2$ commercially unviable without using cosolvents or an auxiliary substance.[6] In 2007, the biotechnology company NovaSterilis earned the Presidential Green Chemistry Challenge Small Business Award for its application of $scCO_2$ in sterilization of biological products.[12] Historically used sterilants include gamma radiation (lethal to cells) and ethylene oxide (a volatile and carcinogenic gas). These are clearly unsustainable approaches that compromise the structures of sensitive biological materials. $ScCO_2$ is used under conditions of moderate pressure and low temperature in the presence of water/peracetic acid to sterilize items including biopolymers, vaccines, and graft tissue. This represents environmentally benign technology that inactivates microbes in a short time period, and facilitates sterilization of packaged products that directly reach hospital operating rooms.

5.3.2 Undergraduate Organic Chemistry Using Supercritical CO_2

The vast majority of published experiments for the undergraduate curriculum involving $scCO_2$ are either designed for the physical chemistry laboratory or involve extraction of natural compounds in the organic/analytic laboratory. Despite the publication of reviews regarding organic reactions[13] and synthesis of nanoparticles[14] in $scCO_2$, no procedure for an undergraduate laboratory involving a reaction in $scCO_2$ has been published.

The decaffeination of coffee using $scCO_2$ was introduced in Germany on an industrial scale in 1978, and supercritical fluid extraction (SFE) of nicotine from tobacco was implemented in Japan in 1984.[9] Mauldin et al. have developed an SFE undergraduate laboratory based on extraction of these two common drugs using $scCO_2$ and subsequent analysis by gas chromatography–mass spectrometry (GC-MS)[15] (Figure 5.3). The nicotine and caffeine products are collected in methanol, which is then diluted for quantification by GC-MS. Students compare the extraction with $scCO_2$ alone and the extraction with coffee beans and tobacco leaves saturated with methanol before $scCO_2$ was passed through, and discover that the cosolvent procedure

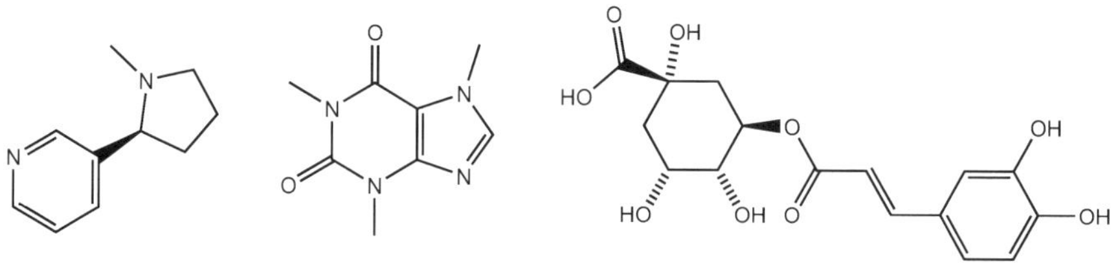

(S)-(−)-nicotine **caffeine** **chlorogenic acid**

FIGURE 5.3 Structures of (*S*)-(−)-nicotine, caffeine, and chlorogenic acid.

yields a greater amount of extract for both analytes. From these results, students conclude that both caffeine and nicotine are more soluble in the CO_2/CH_3OH cosolvent system, as hydrogen bonds formed between the products and methanol are stronger than the dispersion forces with CO_2. Additionally, as nicotine and caffeine are generally found hydrogen bonded to chlorogenic acid in the solid matrix, methanol acts as a hydrogen bond competitor to allow easier separation of the products from the matrix. The SFE of caffeine can be compared with more traditional procedures in terms of yield and greenness. Many procedures historically involve boiling of tea bags or coffee grounds in water followed by extraction using a chlorinated solvent (or sometimes a recrystallization from benzene).[16–18] An improved procedure that replaces chlorinated solvents with 1-propanol still requires a recrystallization from toluene,[19] and another involves a Soxhlet extraction in refluxing ethanol for two hours followed by workup involving MgO, H_2SO_4, CH_2Cl_2, and KOH.[20]

A potential problem with introduction of $scCO_2$ into the undergraduate laboratory is equipment cost, especially for larger classes. McKenzie et al. have developed an inexpensive and operationally simple experiment that uses liquid CO_2 (instead of $scCO_2$) as the extraction solvent.[21] The extraction of (*R*)-limonene from citrus rind is a good demonstration of natural product extraction in the organic laboratory, as this compound is present in very high concentrations, is nontoxic, and emits a very pleasant odor[22] (Figure 5.4). The procedure requires no specialized equipment, as the extraction vessel is simply a 15 mL polypropylene centrifuge tube with a sealed cap (Figure 5.5). After grated orange peel is placed into the tube, it is filled with crushed dry ice, tightly capped, and placed into a 40–50°C water bath. Within fifteen seconds, liquid CO_2 appears as the pressure builds up. At this moment all three subcritical phases of CO_2 are present, which the authors believe indicates that the system is near the triple point (–56.6°C, 0.52 MPa). As the liquid CO_2 forms, it extracts (*R*)-limonene oil from the solid rind, which collects on a filter paper trap. The tube is refilled with crushed dry ice two to three times to yield approximately 0.1 mL (1%–2% mass recovery) of an oil that is approximately 97% (*R*)-limonene by GC-MS analysis. Although the amount of (*R*)-limonene obtained from this liquid CO_2 extraction is slightly lower than that obtained by a previous method using hexanes,[23] this procedure avoids highly flammable solvents and also produces a cleaner crude product, as liquid CO_2 is more selective than hexanes. It is also a very visual and appealing demonstration of the CO_2 phase diagram, and is a cheaper alternative to $scCO_2$ when discussing green solvents. However, the authors found that this method was limited to extraction of (*R*)-limonene from orange peel. Other natural products that can be extracted with $scCO_2$ (e.g., caffeine from coffee beans) were not found in

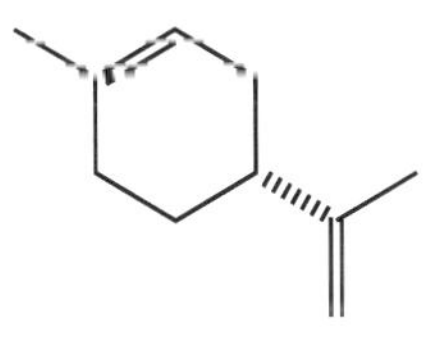

FIGURE 5.4 Structure of (*R*)-limonene.

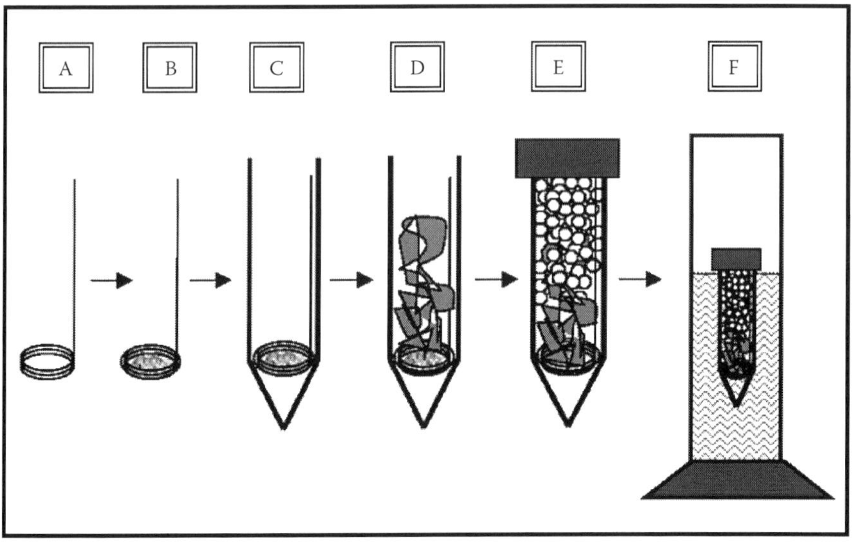

FIGURE 5.5 Extraction of (*R*)-limonene from citrus rind with liquid carbon dioxide. (From McKenzie, L. C., Thompson, J. E., Sullivan, R., Hutchison, J. E., *Green Chem.*, 6, 355–358, 2004. Reproduced with permission of the Royal Society of Chemistry.)

sufficiently high concentrations (or are too polar) to be extracted using liquid CO_2, which is less penetrating than a supercritical fluid.

In Figure 5.5, a solid trap is constructed by (A) bending copper wire into coils and a handle, (B) placing filter paper or metal screen between the wire coils, and (C) placing the solid trap in a centrifuge tube. For extraction, (D) grated orange peel is placed in the tube, and (E) the tube is filled with crushed dry ice and sealed with a cap. (F) The prepared centrifuge tube is placed in the water in the graduated cylinder, and the liquefaction and extraction occur over the following three minutes.

5.4 FLUOROUS SOLVENTS

5.4.1 Introduction

In organic chemistry, the separation of products from reactants and catalysts is regularly the most labor- and resource-intensive step of a reaction. Fluorous solvents are fluorinated compounds (usually alkanes or aromatics) that are miscible with other organic solvents at a higher temperature, but become separated in a biphasic system at a lower temperature.[24] Figure 5.6 illustrates several examples of typical fluorous solvents. In one general approach, reactants are dissolved in an organic solvent and combined with a fluorous catalyst dissolved in a fluorous solvent to form a biphasic system. When this mixture is heated to the consolute temperature, a single phase exists where the catalyst can mediate the transformation of reactants to products. After cooling down below the consolute temperature, the solvents separate back into two phases with the products dissolved in the organic phase. Only the catalyst

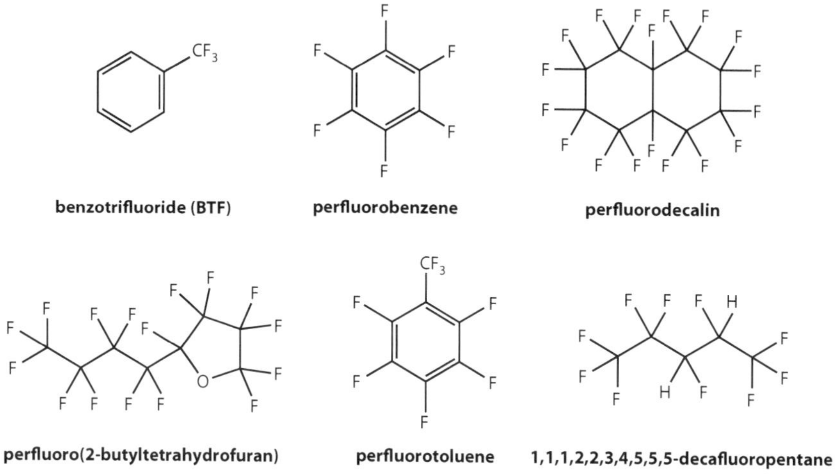

FIGURE 5.6 Some representative fluorous solvents.

remains in the fluorous phase, ready for reuse in the next reaction without having to change solvents. To ensure complete partitioning of the fluorous catalyst into the fluorous phase, the compound should have at least 60% fluorine by molecular weight. In other cases, a reaction can be held in a single-phase organic solvent that dissolves the reactants and products as well as the fluorous catalyst. After reaction is complete, the first solvent is removed and either a fluorous solvent is added to precipitate the product or a second organic solvent is added to precipitate the fluorous catalyst. This method requires more solvent than the liquid-liquid biphasic extraction method described previously, although it often ends up using less material than traditional purification methods. Other applications of fluorous compounds, such as fluorous reverse phase silica gel chromatography, fluorous reagents, and fluorous protecting groups, are outside the scope of this chapter, but are discussed in the pedagogical review article by Ubeda and Dembinski.[24]

It is important to note that while fluorous solvents facilitate solvent recycling and reduced use of traditional organic solvents, there are significant environmental downsides to using highly fluorinated substances. Although not known to be directly toxic to plants and animals, fluorinated compounds have high persistence in the environment, can bioaccumulate, and can be degraded to volatile shorter-chain perfluorocarbons, which are powerful greenhouse gases.[25] Their resource-demanding synthesis gives them a poor cradle-to-grave environmental impact factor, and also leads to high costs. Clark and Tavener assigned fluorous solvents a score of 12/25, the lowest score out of the five types of solvents evaluated, believing that the above concerns overwhelm any of the associated useful properties.[3] In addition, the greenness of certain organic solvents used in the biphasic systems is of some concern.

A cautionary tale about calling solvents "environmentally benign" without extensive investigations is worth telling here. An undergraduate laboratory textbook entitled *Zero Effluent Laboratory* was published by Morton et al. in 1974 as a prelude to the modern-day green chemistry movement.[26,27] In this text, an experiment was

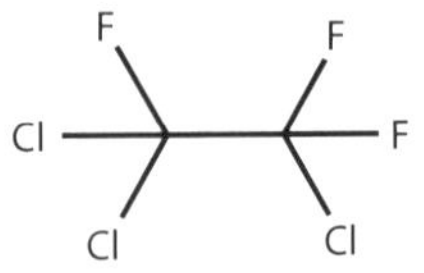

FIGURE 5.7 Structure of Freon 113.

described where diethyl ether was replaced by the "environmentally benign" Freon 113 for the extraction of trimyristin from nutmeg[27] (Figure 5.7). Freon 113 was chosen because of its excellent recyclability, stability, nonflammability, and relatively high threshold limit, but is now known to cause detrimental damage to the ozone layer.

5.4.2 Undergraduate Organic Experiments Using Fluorinated Solvents

There are not many undergraduate laboratory experiments or demonstrations performed using fluorous solvents, possibly due to their relative novelty and high costs. However, to demonstrate the special ability of fluorous solvents to change miscibility with organic solvents, Rábai used combinations of fluorophilic and organophilic dyes in a visually appealing manner.[28] Solvents with different colored dyes are placed in a sealed glass tube that is heated and cooled to change the number of liquid layers. As two layers mixed, its relative density changed with respect to the third layer, enabling layers to be moved up and down the tube. It should, however, be noted that the syntheses of the different dyes are difficult, hazardous, and not green.

One laboratory experiment uses the fluorinated solvent benzotrifluoride (BTF; also known as α,α,α-trifluorotoluene) as a replacement for carbon tetrachloride in the radical bromination of hydrocarbons, to demonstrate the relative radical reactivities of different types of C-H bonds.[29] BTF is an amphiphilic fluorinated compound, and unlike most of the fluorous solvents discussed previously, it forms a homogenous

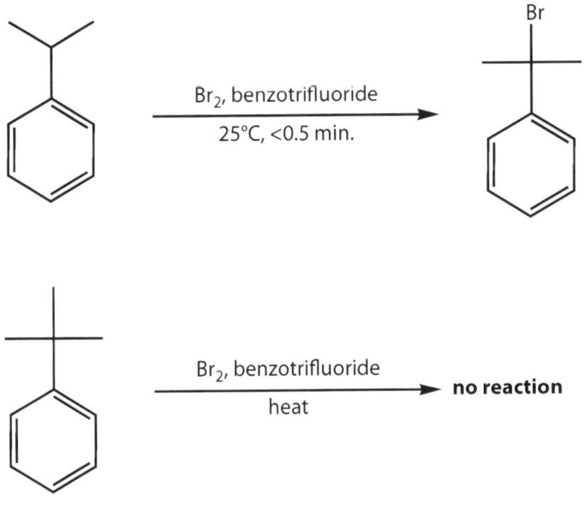

SCHEME 5.1 Benzotrifluoride as a solvent for radical bromination reactions.

mixture with typical organic solvents at all temperatures.[24] A brown solution of Br_2 in BTF is added to a series of hydrocarbons, and the time it takes for the solution to become colorless (after all the Br_2 has reacted) is recorded by students. The rate of decolorization is directly related to the relative stability of radical formation in the hydrocarbons, and students observe that the relative reactivity series follows the order 3° benzyl > 2° benzyl > 1° benzyl > 3° alkyl > 2° alkyl > 1° alkyl (Scheme 5.1). The results are comparable to a traditional demonstration that uses the now banned carbon tetrachloride as solvent,[30] with BTF being less toxic and comparatively more environmentally friendly. The 2010 cost of BTF was approximately $65/L.[31]

5.5 IONIC LIQUIDS

5.5.1 Introduction

Technically, any salt in a molten state can be termed an ionic liquid—above 801°C, even sodium chloride can be considered as such. However, the most common definition of the term today (and the definition to be used in this chapter) only considers salts that exist in the liquid state below 100°C, although the commonly used ionic liquids are in fact liquids at room temperature. A variety of organic, inorganic, organometallic, polymeric, and even gaseous compounds are very well solvated in ionic liquids, and the great variability of the solvent structure also allows for very specific "tuning" of the solvent's ability to solubilize specific compounds.[32] The main advantage of using these neoteric solvents from a green perspective is that they have near-zero vapor pressures, and therefore exhibit a reduced risk of exposure to people and the environment when compared to volatile organic solvents. This additionally allows for volatile products to be easily distilled from ionic liquids, saving the use of organic solvents in an extraction, and also opening up the possibility of solvent reuse with minimal purification.

Although they have been described as "environmentally benign,"[33] it is inappropriate to classify all ionic liquids, and all reactions undertaken in ionic liquids, as green. Some ionic liquids can emit volatile hazardous decomposition products such as hydrogen fluoride (HF) gas,[34] and others have even been shown to be flammable.[3] As much of the developments surrounding ionic liquids have been made in the past twenty years, the environmental fate of these substances is largely unknown.[3,32,35] One of the most commonly used ionic liquids (1-butyl-3-methylimidazolium hexafluorophosphate; Figure 5.8) has an LD_{50} value of 300–500 mg/kg (rat, oral).[32] This means the substance is more acutely toxic than solvents such as dichloromethane (1,600 mg/kg)[36] and pyridine (891 mg/kg),[36] although it does not have an inhalation route. In terms of life cycle impact, the heavy resource cost of synthesis and uncertain disposal strategies represent major disadvantages to the use of ionic liquids.[3] Indeed, researchers have remarked that ionic liquids can have a greater cradle-to-grave impact than traditional solvents for some reactions.[37] Considering that ionic liquid synthesis and reaction workup will likely require use of volatile organics, solvents become moved around the life cycle rather than eliminated. One of the most important green properties of ionic liquids is their recyclability, and the more times they are recycled, the greener their use becomes. However, if the recycling process requires heavy use of volatile solvents, the

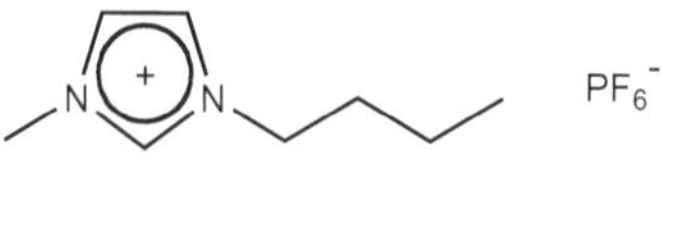

1-butyl-3-methylimidazolium hexafluorophosphate
[bmim][PF_6]

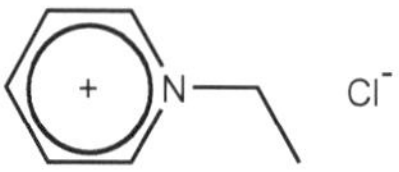

N-ethylpyridinium chloride
[epy][Cl]

FIGURE 5.8 Two commonly used ionic liquids.

green benefits again disappear.[38] It is interesting to note that Clark and Tavener gave ionic liquids a relatively low score of 13/25 in their solvent rating system.[3]

There are two general methods to abbreviate the systematic names of simple ionic liquids. For example, the ionic liquid 1-*b*utyl-3-*m*ethyl*im*idazolium hexafluorophosphate is often know as either as [bmim][PF_6] or [C_4mim][PF_6], with variations on whether square brackets are used for the anion (Figure 5.8). Although some have argued the benefits of the latter abbreviation,[39] the former appears most commonly in literature, and so this style is maintained throughout this chapter. *A*lkyl*py*ridinium-based cations are abbreviated as [apy]; for example, *N*-*e*thyl*py*ridinium chloride is shortened to [epy][Cl]. Ionic liquids based on phosphonium and other cations are less common and will not be discussed further here.

One immediately noticeable physical property of ionic liquids is their high viscosity compared to more common solvents. In organic chemistry, dimethylsulfoxide is regarded as one of the more viscous solvents; however, [bmim][PF_6] and [bmim][BF_4] are 100 times thicker at 25°C.[32] Ionic liquid viscosity can be decreased by addition of an organic cosolvent or by heating. At 50°C, the viscosity of [bmim][PF_6] drops to a quarter of its value at room temperature. Less viscous ionic liquids have been developed, such as [bmim][CH_3COO], [bmim][NTf_2], and [emim][NTf_2][32] (Figure 5.9).

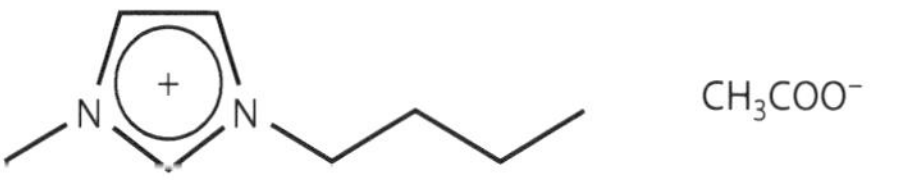

[bmim][CH_3COO]

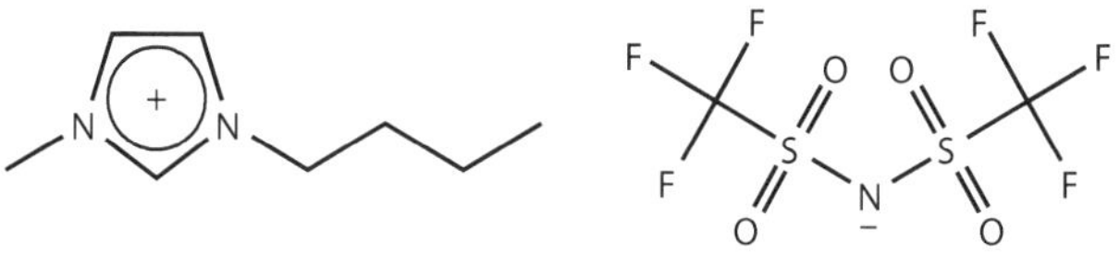

[bmim][NTf_2]

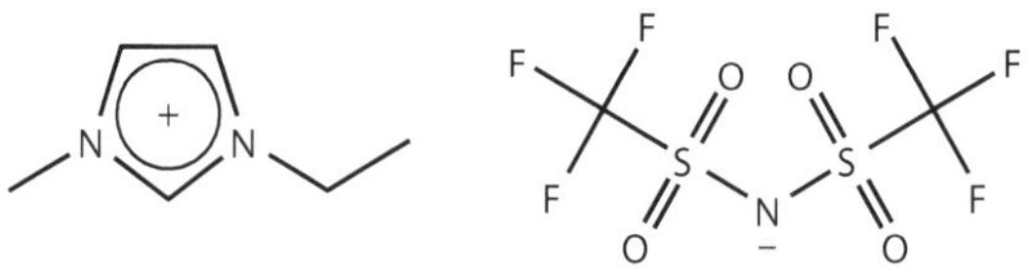

[emim][NTf_2]

FIGURE 5.9 Some less viscous ionic liquids.

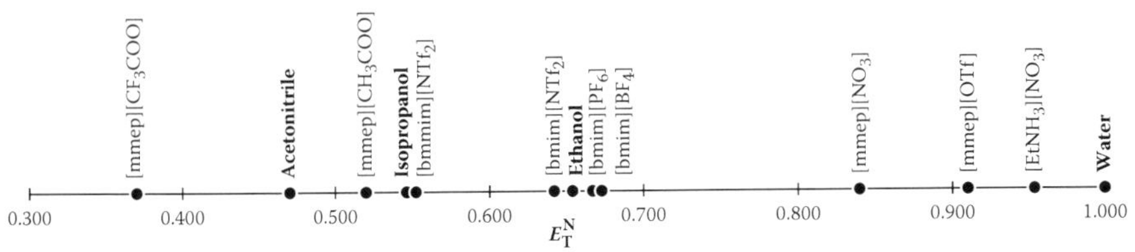

FIGURE 5.10 Ionic liquid polarities based on the solvatochromatic E_T^N scale.

Ionic liquids display a wide range of polarity, from those approaching water to some being less polar than acetonitrile based on the solvatochromatic E_T^N scale[32,40,41] (Figure 5.10). It is observed that [bmim] solvents have polarities similar to those of ethanol regardless of the counteranion, while 1-methyl-1-(2-methoxyethyl)pyrrolidinium ([mmep]) salts have a large range of polarities, which are highly dependent on the anion.[32,40] Despite having very similar E_T^N values, [bmim][BF_4] is miscible with water while [bmim][PF_6] is not,[42] so the structure of the anion controls the solubility of ionic liquids. The most common hydrophobic ionic liquids today have [PF_6] or [NTf_2] as the anion. Both suffer from disadvantages, with the former having the potential to release HF gas on hydrolysis, and the latter being expensive.[32] Hydrophilic ionic liquids are less problematic, and common examples include [emim]$_3$[PO_4], [emim][N(CN)$_2$], and [mmep][N(CN)$_2$][32] (Figure 5.11). Hydrophobic and hydrophilic ionic liquids are useful in liquid-liquid extractions to separate products and catalysts in different phases, for facile purification and potential reuse of the catalyst.

The variable polarity of ionic liquids makes it appropriate to discuss them in the context of solvent effects on S_N2 reactions in an introductory organic chemistry course. Bowman analyzed available experimental kinetic results of *n*-butylamine undergoing nucleophilic substitution with methyl *p*-nitrobenzene sulfonate in solvents with a range of polarity and hydrogen bonding behavior[33] (Scheme 5.2). S_N2 reactions occur more readily in polar aprotic solvents where the nucleophile is soluble and its deactivation via hydrogen bonding is prevented. The Kamlet-Taft parameters π^* (polarity), α (hydrogen bond donating ability), and β (hydrogen bond accepting ability) have been used to describe the properties of traditional organic solvents, but can be applied to ionic liquids as well. The relative rate constant for the substitution reaction with an uncharged nucleophile is much greater in an ionic liquid than a

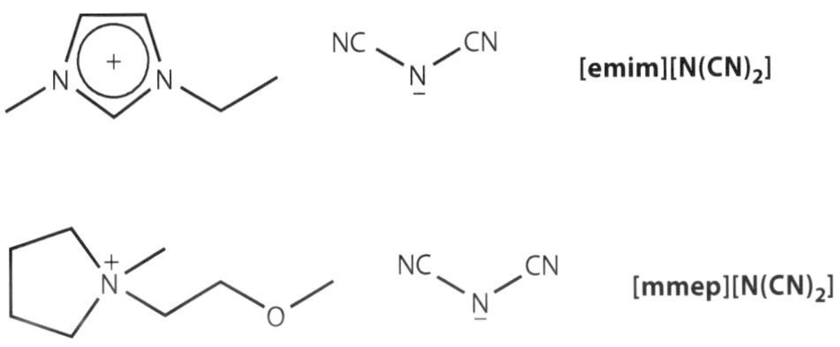

FIGURE 5.11 Structures of two hydrophilic ionic liquids.

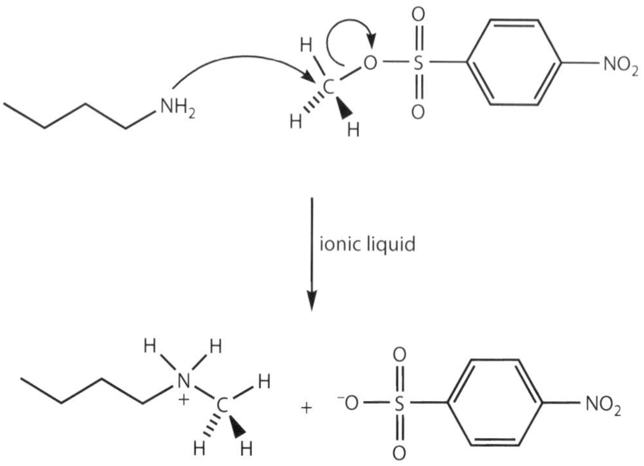

SCHEME 5.2 S_N2 reaction of *n*-butylamine with methyl *p*-nitrobenzene sulfonate.

molecular organic solvent. The polar nature of an ionic liquid (larger π*) helps it to stabilize the transition state (which involves a charge-separated activated complex) more than the reactants. A rate dependence on hydrogen bonding is also apparent. Solvents with larger α-values retard the substitution by reducing the nucleophilicity of *n*-butylamine, and solvents with larger β-values increase the reaction rate by interacting more with the amino hydrogens in the activated complex than the same hydrogens in the ground state, reducing the energy of activation.

Ionic liquids have been called "task-specific" solvents,[39] "designer solvents,"[3] and "tailor-made solvents"[33] because of the variety stemming from their modular design, as both the cation and anion can be altered relatively easily. Varying either ion results in solvent property changes (e.g., differences in melting point, viscosity, polarity, and solubility). This allows a chemist to pick and choose a solvent that best fits a particular reaction. Applications of ionic liquids are not limited to their use as a solvent, as they have been used as catalysts for a variety of transformations.[43,44]

Ionic liquids have found a wide range of applications in industry. BASF has used ionic liquids to replace reagents such as triethylamine and phosgene, and have incorporated them as solvents for cellulose.[45a,45b] Cellulose from almost any resource (cotton, filter paper, pulp, etc.) can be easily dissolved in [bmim][Cl] on microwave heating. This cellulose can be reconstituted in water in a controlled form (e.g., as fibers). If additives (e.g., dyes, clays, or enzymes) are combined with the mixture prior to reconstitution, it is possible to synthesize blended materials. Significantly, the [bmim][Cl] is recyclable by either cation exchange or "salting out," which are more energy-efficient approaches than distillation. The Central Glass Company currently uses an ionic liquid as the solvent for a Sonogashira coupling reaction in the synthesis of a pharmaceutical intermediate.[45a] Sonogashira chemistry usually requires an organic solvent such as toluene or tetrahydrofuran, and a stoichiometric amount of base (e.g., pyrrolidine or piperazine). It should be noted that an undergraduate Sonogashira coupling using ethanol as a greener solvent has been reported (Section 5.7.2). Central Glass discovered that reaction of an aryl bromide with a

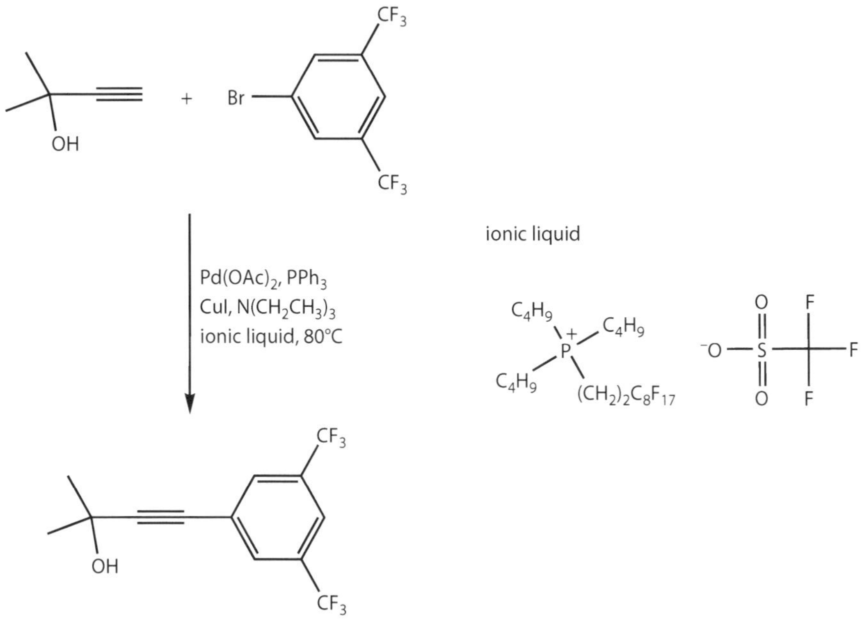

SCHEME 5.3 Commercial Sonogashira coupling in an ionic liquid.

terminal alkyne proceeded efficiently using a tetra-alkylphosphonium ionic liquid (Scheme 5.3). The desired internal alkyne product is extracted with hexane, and the ionic liquid-catalyst mixture (containing palladium and copper salts) can be recycled several times.

5.5.2 Undergraduate Organic Experiments Using Ionic Liquids

Undergraduate students are accustomed to using solvents from the shelf to synthesize products in the organic laboratory, but how often are they required to prepare their own solvent? The relative simplicity of the preparation of imidazolium-based ionic liquids introduces students to this new media, allowing them to use their homemade solvent in a subsequent reaction. Students can also evaluate the relative greenness of ionic liquid production and implementation. In addition, having students prepare ionic liquids is a good economic alternative to purchasing them from commercial suppliers.

5.5.2.1 Preparation of Ionic Liquids

It is possible to undertake a facile preparation of a low-melting-point ionic liquid without doing a chemical reaction at all. When urea (mp = 134°C) and choline chloride (mp = 303°C) (Figure 5.12) are mixed together in a 2:1 mole ratio in the absence of a solvent at 80°C, the two white solids eventually "melt" into a colorless ionic liquid that freezes at 12°C.[46,47] The mechanism behind this extreme example of freezing point depression is a strong hydrogen bonding interaction between the amino groups in urea and the quaternary ammonium group in choline chloride that disrupts the ionic bond between choline and chloride ion.[46] Both components are derived from

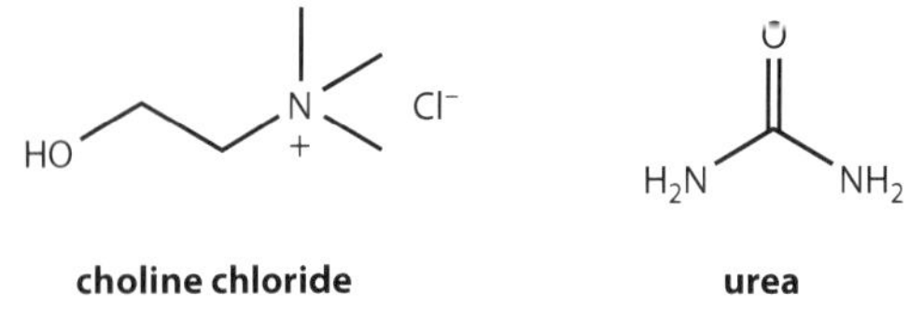

FIGURE 5.12 Structure of choline chloride and urea.

natural substances (choline chloride is vitamin B_4, and urea is a fertilizer) that are made in Mt quantities and readily available. Abbott et al. have described this special type of ionic liquid as a "deep eutectic solvent" (the lowest melting point of a mixture).[46,47] Deep eutectic solvents that are prepared from natural components like choline chloride–urea can be thought of as greener than the regular ionic liquids described in this section, as their toxicity and environmental fates are already well known, and their preparation requires fewer resources.

The synthesis of imidazolium-based ionic liquids is not as simple and straightforward as the example above, but it is appropriate enough to be incorporated into an intermediate undergraduate organic laboratory (Scheme 5.4). A wide variety of [bmim] ionic liquid syntheses have been published, but they generally involve the quaternization of commercially available 1-methylimidazole with an *n*-butyl halide, tosylate, mesylate, or triflate. The best synthetic strategy would be to carry out the quaternization under neat conditions, but an apparatus that can maintain accurate temperature is necessary to produce a clean product.[39,48] Generally, even if the reaction is performed under neat conditions, some sort of organic solvent is required in the workup during liquid-liquid extraction, and it is also helpful for mass transfer since, as previously discussed, many ionic liquids are very viscous.[48] Anion metathesis can be undertaken if an anion other than the leaving group is desired. For water-soluble ionic liquids, anions like BF_4^- and CF_3COO^- are introduced by the addition of the corresponding silver salt to an aqueous solution of the halide, precipitating out one equivalent of silver halide as a by-product.[32] Water-insoluble ionic liquids, with anions like PF_6^- and NTf_2^-, are prepared by adding the corresponding potassium salt to an aqueous solution of the halide, then collecting the nascent ionic liquid phase.[32] It is apparent that the synthesis of hydrophobic ionic liquids is much greener than hydrophilic ones.

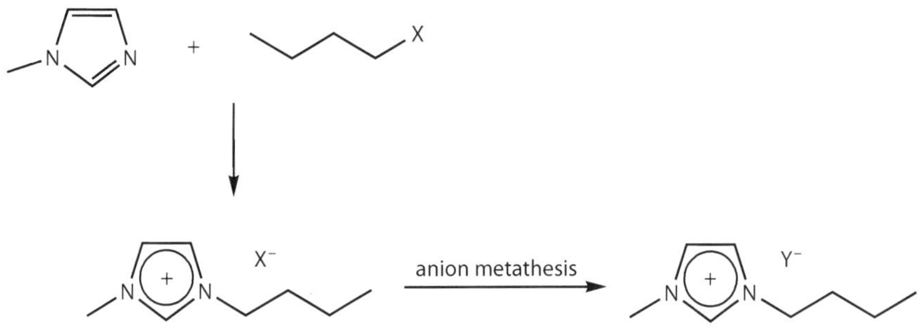

SCHEME 5.4 General synthesis of a [bmim] ionic liquid.

The published syntheses of [bmim] ionic liquids in the undergraduate laboratory cover a large range of greenness. One of the main purposes of making and using ionic liquids is because doing so reduces the need for volatile organic solvents. As such, procedures that demand a twenty-four-hour reflux in toluene[49] or a forty-eight-hour reflux in acetonitrile[50] should be considered carefully—if not for lack of greenness, then for the impracticality of running long refluxes in an undergraduate laboratory! A better synthesis of [bmim][Br] and [bmim][PF_6] was proposed by Dzyuba et al.[39] Equimolar quantities of 1-methylimidazole and 1-bromobutane are refluxed at high concentrations (12.5 M) in water for ninety minutes, or under neat conditions if precise temperature control is available (Section 4.3.4). The resulting [bmim][Br] is typically >99% pure by ^1H nuclear magnetic resonance (NMR), and anion metathesis is carried on in one pot by adding KPF_6 to the aqueous solution and stirring to produce two phases. The one disadvantage to this preparation is that high yields (80%) are only achieved if the hydrophobic [bmim][PF_6] is extracted from the water layer with the toxic and volatile dichloromethane. Significant yield loss is found by simple removal of the ionic liquid layer. Instructions for the synthesis of [bmim][NTf_2], [bmim][BPh_4], and [bmim][NO_3] can be found in the supplementary information of the original article.[39]

Stark et al. have described an undergraduate laboratory where students analyze the relative greenness of five different reaction conditions in the synthesis of [hmim][Cl] or [hmim][Br].[51] In this experiment, students consider five green metrics: yield, atom economy, reaction mass efficiency, E-factor, and energy efficiency of the different conditions (Scheme 5.5). The water-soluble ionic liquids are collected by extraction with diethyl ether, and the energy requirements of using and removing the volatile organic solvent are factored into the calculations. Unfortunately, this procedure does not form very pure ionic liquids. To reach >95% purity, the aqueous

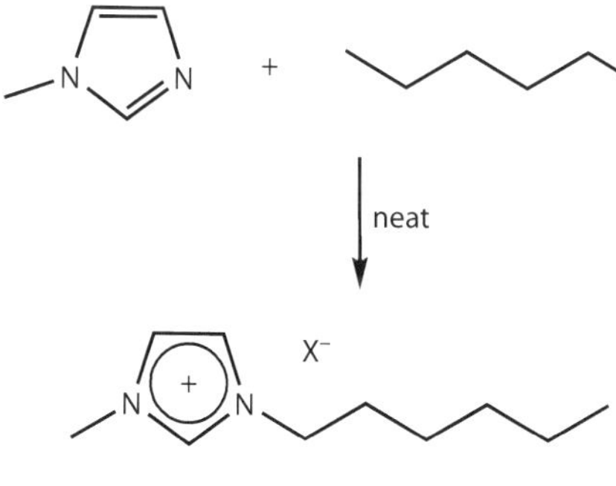

Conditions:

(i): X – Cl, oil bath, 100°C, t = 3 hr.; (ii): X = Cl, μW, 100°C, t = 3 hr.;

(iii): X = Cl, oil bath, 70°C, t = 6 hr.; (iv): X = Br, oil bath, 70°C, t = 6 hr.;

(v): X = Cl, oil bath, 100°C, t = 6 hr.

SCHEME 5.5 Analysis of ionic liquid preparations under various conditions.

layer must be extracted over ten times. By comparing results for conditions (i)-(v) (Scheme 5.5), students conclude that the best condition is (iv), as it receives the highest marks across the board for all five green metrics, and is also the lowest in terms of the chemical cost per gram of ionic liquid produced. However, investigation into the environmental and health effects shows that 1-bromohexane is more toxic than 1-chlorohexane. Students are expected to discuss the balance of efficiency and toxicity in the assessment of the greenness of this process. Interestingly, the microwave condition (ii) does not benefit from a "microwave effect" to enhance the reaction rate. With the microwave acting as a heating device, the reaction required three hours to afford an average yield of 55%. Previous reports have demonstrated that microwave-assisted syntheses of ionic liquids required less than one minute.[32] These results are also in complete contrast to those reported by Deetlefs and Seddon.[48] Imidazolium-based ionic liquids are said to undergo "ionic oscillations" in addition to standard dipole rotation under microwave irradiation. This causes especially efficient heating, making synthesis much more rapid and less energy-intensive. A conclusion is that, at bench scales (<2 kg), microwave-assisted ionic liquid synthesis is generally much greener than methods employing conventional heating.[48,52]

5.5.2.2 Undergraduate Organic Reactions in Ionic Liquids

After having synthesized an ionic liquid, students can use their "homemade" solvent for various organic reactions as a replacement for traditional volatile organic solvents. The Mannich reaction is a three-component carbon-carbon bond-forming process involving initial formation of an imine from an aldehyde and an amine, followed by nucleophilic attack at the imine by the enol form of a ketone. Mak et al. have published a double Mannich condensation between benzaldehyde, 3-pentanone, and ammonium acetate conducted in [bmim][BF_4] at slightly elevated temperatures[50] (Scheme 5.6). As part of the exercise, students use mass spectrometry and ^{1}H NMR spectroscopy along with the Karplus equation to determine the exact structure of the product. The average student yield using [bmim][BF_4] as solvent (60–65%) is comparable to performing the reaction in ethanol (52%). The authors noted that approximately 90% of the [bmim][BF_4] solvent is recoverable for reuse, and have discovered that no significant decrease in product yield is noticeable after three reaction cycles. However, a severe drawback to reuse of the ionic liquid is the requirement to wash the solvent three times with diethyl ether, followed by rotary evaporation and high-vacuum removal of water. Although this reaction represents one of the few published reactions in an ionic liquid for the undergraduate laboratory, students should question the greenness of using ionic liquids, especially if the recovery process requires use of an extremely volatile and flammable organic solvent.

An undergraduate experiment involving acetylation of glucose and cellulose in [bmim][Cl] has been proposed by Riisager and Bösmann[49] (Scheme 5.7). The ionic liquid-mediated acetylation of glucose uses a slight excess of acetic anhydride and sonication to produce the penta-acetylated product in high yield. If this experiment, currently described at the 10 mmol scale, can be scaled down to microscale levels, it may represent a good demonstration of reaction improvement using ionic liquids. A previous procedure requires use of iodine, which is highly toxic and volatile, as well

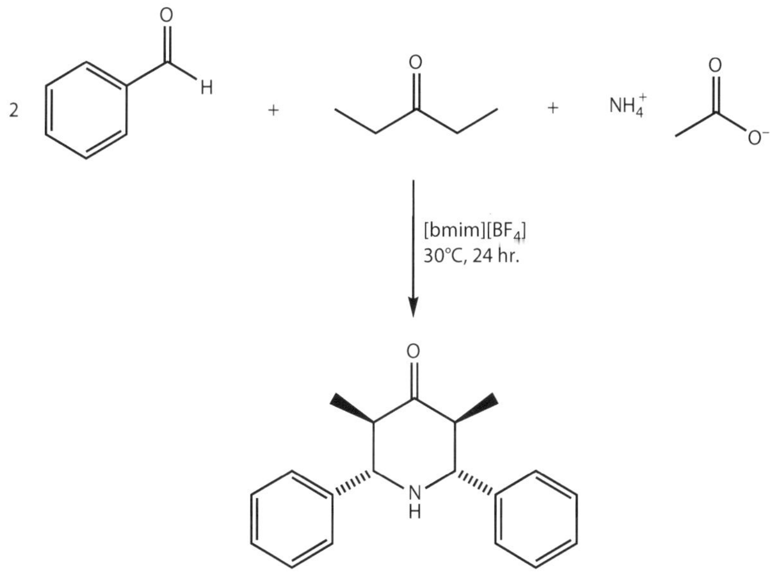

3,5-dimethyl-2,6-diphenyl-4-piperidinone
60–65%

SCHEME 5.6 Double Mannich reaction in [bmim][BF_4] solvent.

as a large excess of acetic anhydride.[53] The acetylation of cellulose (a high molecular weight polymer of D-glucose) can be performed in a similar fashion, except that acetyl chloride is used as the acetylating agent in place of acetic anhydride.[49,53] The high solubility of cellulose in [bmim][Cl] compared to most solvents (Section 5.5.1) is key to ensuring a high degree of acetylation, allowing the reaction to be conducted with no excess acetyl chloride. Since the beginning of the twentieth century, cellulose acetate has found widespread industrial application in the textile, membrane, and cigarette industries.[54]

A very important carbon-carbon bond-forming reaction in modern organic chemistry is the ruthenium-catalyzed metathesis of two olefins, which won the Nobel Prize in Chemistry in 2005.[55] The reaction showcases some important aspects of green chemistry as a single low molecular weight by-product (ethene) is formed and a highly selective catalyst is often used in very low loadings. A concern regarding this reaction

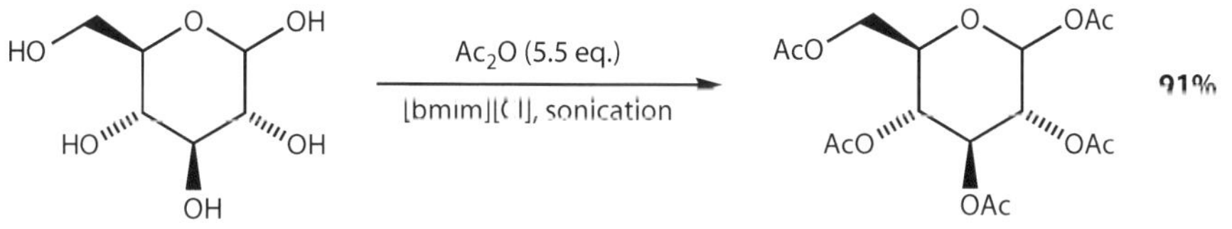

D-glucose

SCHEME 5.7 D-Glucose acetylation in [bmim][Cl] solvent.

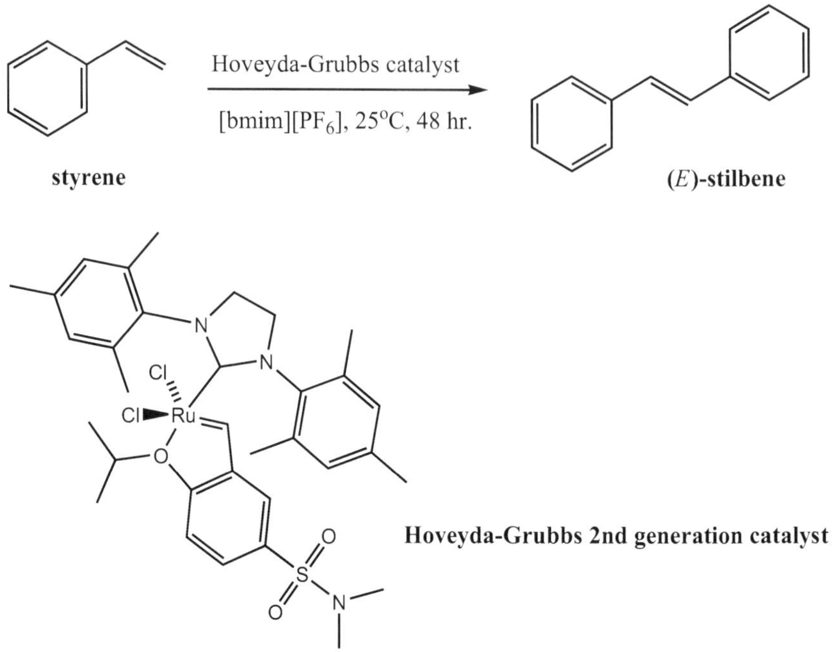

SCHEME 5.8 Ruthenium-catalyzed olefin metathesis in [bmim][PF_6] solvent.

on an industrial scale is that chromatography must be used to separate the catalyst from the product, making it essentially unrecyclable in traditional organic solvents.[56] As one of the major benefits of ionic liquids is their recyclability, it is appropriate for undergraduates to test out this feature for the recycling of a dissolved ruthenium-based catalyst. In Fujita's procedure, students carry out the self-cross metathesis of styrene to form (*E*)-stilbene using 2.5 mol% of the ruthenium-based Hoveyda-Grubbs catalyst in [bmim][PF_6].[56] The product is extracted with petroleum ether, which unfortunately is a volatile and flammable solvent, and can be purified by silica gel column chromatography if desired. The catalyst is not extracted and remains in the ionic liquid layer, which is dried under a vacuum before an additional equivalent of styrene is added for a second reaction cycle (Scheme 5.8).

While ionic liquids can be considered greener reaction media when compared to volatile organic solvents, they still carry the baggage of being expensive, toxic, resource-intensive to synthesize, and having an unknown fate in the environment. Use of volatile organic solvents during workup to extract products or purify the ionic liquid is also problematic. The coupling of ionic liquids as a reaction solvent and supercritical carbon dioxide as an extraction solvent may be one of the potential solutions to this problem. Blanchard et al. first reported a near-quantitative extraction of naphthalene from [bmim][PF_6] into scCO_2 in 1999.[57] Since the publication of this seminal work, ionic liquid-scCO_2 systems have been used in the asymmetric synthesis and isolation of ibuprofen,[58] as well as various biocatalytic processes.[59,60] Although many of the most commonly used ionic liquids have questionable environmental lifetimes, biodegradable ionic liquids that may have reduced toxicity and ecotoxicity have been synthesized.[61] Early examples involve addition of enzymatically hydrolyzable functional groups such as esters and amides to the alkyl chain, and more recent examples have seen use of lactate or octylsulfate as the anion to further increase biodegradability.[61] It will be interesting to see if these "designer" ionic liquids make an appearances in mainstream undergraduate curricula in the years to come.

5.6 LIQUID POLYMERS

5.6.1 Introduction

Like ionic liquids, liquid polymers are another group of neoteric solvents that may be considered green due to low volatility and nonflammability. However, these solvents have additional benefits, as they exhibit low ecotoxicity, rapid biodegradability, and lower costs than ionic liquids.[62] Polyethylene glycol (PEG) is widely used in academic green chemistry research as a phase transfer catalyst, as well as a solvent for a variety of organic transformations.[63] Other polymers, especially polydimethylsiloxane (PDMS), are now beginning to find applications in organic chemistry laboratories.[62] In industry, PEG has been utilized in a wide variety of roles in electroplating, cosmetics and personal care items, agricultural antidusting formulations, and as a solvent for dyes in paints and inks due to its relatively low cost.[64] In medicinal chemistry, short PEG segments are often covalently bound to make pharmaceuticals more soluble, bioavailable, and stable against a host's enzymes in a process aptly termed "PEGylation".[65,66] PEG is currently made from the polymerization of ethylene dioxide, a petrochemical product. However, new methods of manufacturing from renewable, nonpetroleum feedstocks are in development.[67]

A single type of liquid polymer can assume a range of properties, as its mean chain length (molecular weight) is variable. The general nomenclature for liquid polymers indicates the average molecular weight after the abbreviated polymer name. The average is reported since most liquid polymers used as solvents are not monodisperse.[63] As an example, PEG-400 is polyethylene glycol with an average weight of 400 Da. Different PEGs can have molecular weights from 200 Da to >10,000 Da. At room temperature, those weighing less than 600 Da are colorless viscous liquids that can act as solvents by themselves, while those weighing over 800 Da are white waxes that are often used as a cosolvent in water. All liquid PEGs are completely miscible with water, and solid PEGs are generally very soluble as well.[63] PEGs are less polar than water, ionic liquids, methanol, and dimethyl sulfoxide (DMSO), and more polar than dichloromethane. However, other polymers, such as polypropylene glycol, can have polarities similar to those of toluene.[62] While PEGs are insoluble in very nonpolar solvents, such as hexane and diethyl ether, they readily dissolve in dichloromethane, toluene, acetone, and ethanol. PEGs also have the benefit of being largely resistant to acids, bases, heat, oxygen, and oxidative species, as well as reducing agents.

In organic transformations, PEGs and other polymers have been used for a wide range of reactions with varying degrees of success.[1,68] Heldebrant et al. have discussed

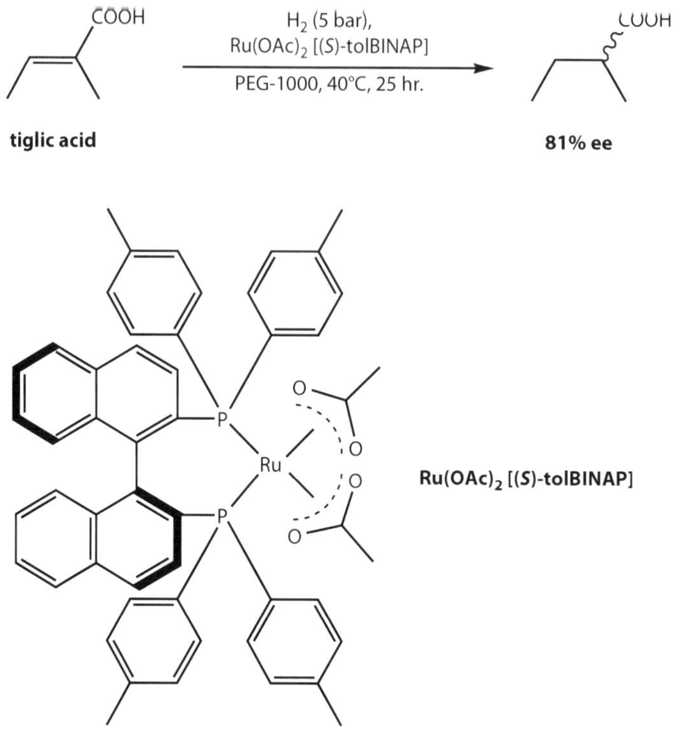

SCHEME 5.9 Ruthenium-catalyzed asymmetric hydrogenation in PEG-1000 solvent.

use of various liquid polymers as solvents for the asymmetric hydrogenation of tiglic acid with a commercially available chiral ruthenium acetate catalyst.[62] Although the enantiomeric excess (ee) of the hydrogenation in PEG-1000 (81%) is lower than that of methanol (91%) and [bmim][PF_6] (87%), the product can be fully extracted by $scCO_2$, and the catalyst and solvent can be reused several times with no reduction in yield or ee (Scheme 5.9).

Although PEGs are very soluble in water, when they are mixed with another water-soluble polymer or specific water-soluble salts above a critical concentration, the mixture will separate into two *aqueous* phases.[69] This phenomenon results in an aqueous biphasic system (ABS), and is a widely used extraction technique. This strategy has been employed since the mid-twentieth century, and thus well before the green chemistry movement.[63,69] An ABS is an important separation method in biological chemistry, as many biomolecules often break down when taken out of an aqueous environment. Shamery et al. have designed an undergraduate analytical and separation sciences laboratory using an ABS to extract a porphyrin-based dye between the PEG and salt phases.[70]

One of the most important green benefits of PEGs and other commonly used liquid polymers is that very comprehensive toxicological and environmental data

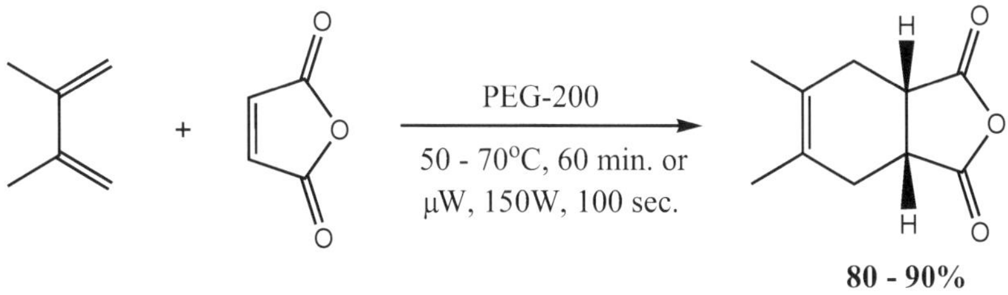

SCHEME 5.10 Diels-Alder reaction in PEG-200 solvent.

have already been compiled about them, unlike very new substances such as fluorous solvents and ionic liquids. Certain PEGs are edible and "generally recognized as safe" for human consumption by the U.S. Food and Drug Administration,[63] and polypropylene glycol (PPG) is approved for use in cleansers and shampoos. In the environment, these polymers display very low toxicity toward aquatic organisms, and are biodegraded by bacteria in soil and sewage.[62] However, some liquid polymers (especially polysiloxanes and others that have been modified to have capped ends) are more difficult to biodegrade, and may be considered persistent.[62]

5.6.2 Undergraduate Organic Reactions in Liquid Polymers

The Diels-Alder reaction is widely taught in undergraduate organic chemistry courses to demonstrate cycloadditions with 100% atom economy. Green versions of this reaction under solvent-free conditions and in aqueous solution are discussed in Sections 3.5.2.2 and 4.3.2, respectively. McKenzie et al. developed an example of a Diels-Alder reaction between 2,3-dimethyl-1,3-butadiene and maleic anhydride in PEG-200.[67] The reaction can be conducted under either conventional or microwave heating, with both giving similar yields (80–90%) and purities. After heating, the white crystalline product is isolated by pouring the PEG solution into water and vacuum filtering, thus avoiding use of volatile solvents (Scheme 5.10).

The Suzuki coupling is another reaction that is commonly encountered in the undergraduate curriculum, as an exemplar of palladium-catalyzed carbon-carbon bond formation. It is also stimulating for students as the biaryl products are topical nonsteroidal anti-inflammatory drug (NSAID) analogs. Some undergraduate Suzuki reactions under aqueous conditions are described in Section 4.3.1.[71] Alternatively, an equimolar ratio of an aryl boronic acid (X = $-CH_3$ or $-OCH_3$) and an aryl bromide (Y = H or CH_3), along with 0.00045 mol% of Pd^0, is refluxed in a 1:5 w/w PEG-400:dilute aqueous sodium carbonate solvent for twenty minutes (Scheme 5.11).[26] The product biaryl precipitates as a white solid, typically with a 70–80% yield upon collection by filtration. This reaction is convenient, as it can be conducted in air, uses a very low catalyst loading, and avoids unnecessary ligands and volatile solvents for both the reaction and the workup.

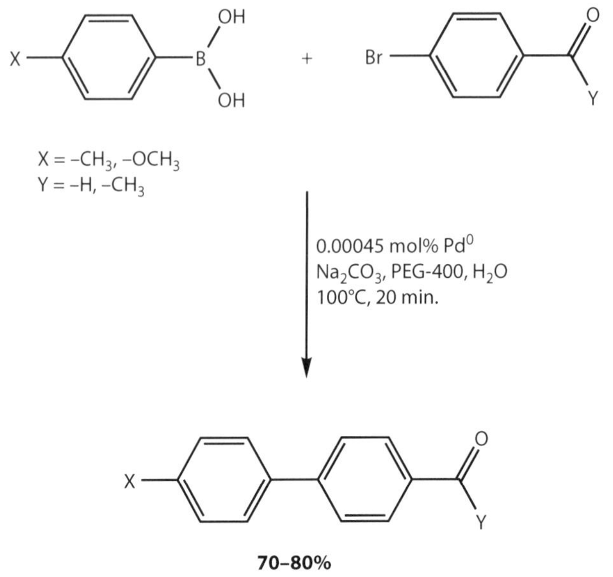

SCHEME 5.11 Aqueous Suzuki reaction with PEG-400 cosolvent.

5.7 ADDITIONAL GREENER SOLVENTS

5.7.1 Introduction

The previous sections of this chapter discussed new, greener alternatives to traditional volatile organic solvents, and each one has its own advantages and disadvantages that may not be compatible for all reactions and settings. There will be situations where a volatile solvent is required as part of a synthetic process, but for most of these cases, chemical educators can replace undesirable solvents like diethyl ether and dichloromethane with safer and greener choices, such as ethanol and ethyl acetate. Pfizer's solvent replacement table[4] is a good resource for some of the more commonly used organic solvents. This section describes several successful examples of "greening up" organic laboratory experiments by using more preferable solvents. Following this is an introduction to some new, greener solvents that have yet to be comprehensively incorporated into undergraduate laboratories.

5.7.2 Undergraduate Organic Experiments

Many introductory organic chemistry courses cover alkene halogenations and S_N2 reactions within their curricula, and as such it is beneficial to give students an opportunity to undertake these reactions firsthand. Most undergraduate textbooks declare molecular bromine in carbon tetrachloride or (nowadays) dichloromethane as the typical conditions for alkene bromination, without mention of the corrosive nature of Br_2 and the deleterious nature of chlorinated solvents. McKenzie et al. have developed an undergraduate laboratory that compares the greenness of the bromination of (*E*)-stilbene in three conditions.[72] The first protocol is a modified traditional

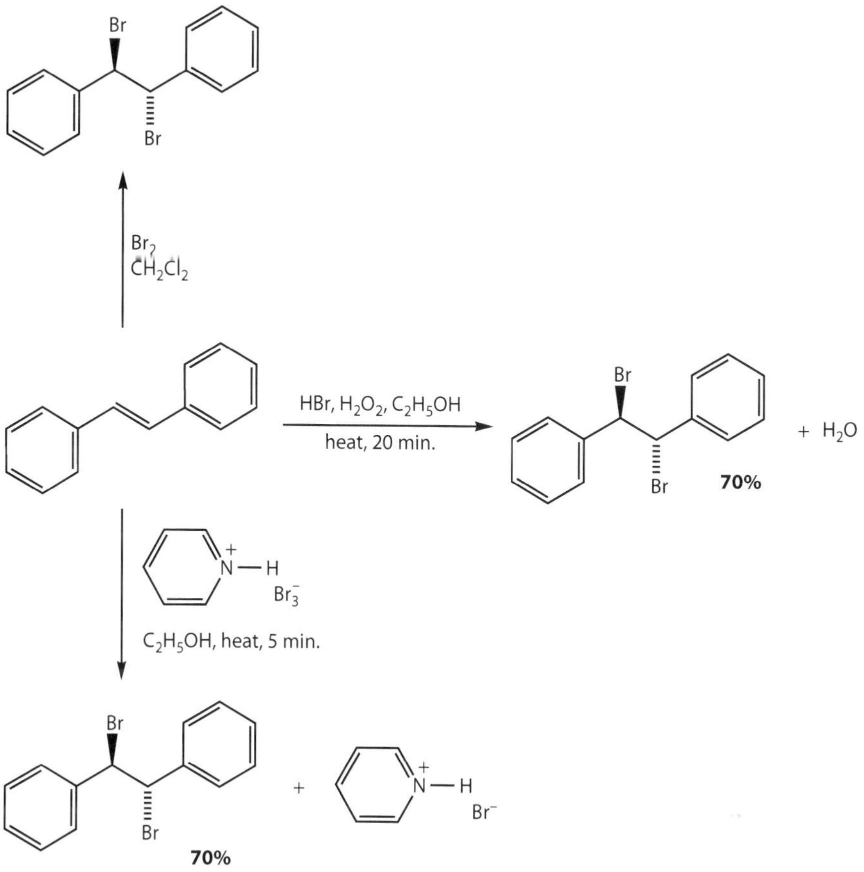

SCHEME 5.12 Bromination of (*E*)-stilbene under three conditions.

approach, using liquid bromine in dichloromethane (Scheme 5.12). The second condition uses pyridinium tribromide in ethanol to create bromine in situ, and the third uses H_2O_2 in ethanol to oxidize HBr to Br_2 in situ. The two latter conditions are greener than the traditional method, as they avoid the toxic dichloromethane by using safer ethanol, and they use less hazardous brominating materials. While pyridinium tribromide offers a greener bromination than the traditional method, it is corrosive and causes significant damage to metal balance trays. The third condition is the greener of the two, as it is more atom economical, producing only water as the by-product instead of an equivalent of pyridinium bromide. While HBr and hydrogen peroxide are corrosive, the amount of material used is much less than for the solid pyridinium tribromide, and exposure is more controlled, as these reagents can be delivered by automated pipettes. The brominated product is insoluble in ethanol and precipitates out of solution after reaction completion, allowing for facile product isolation by filtration, followed by purification by washing with cold ethanol. This compound can be "recycled" to (*E*)-stilbene in around 70% yield by heating with zinc powder in ethanol.[72]

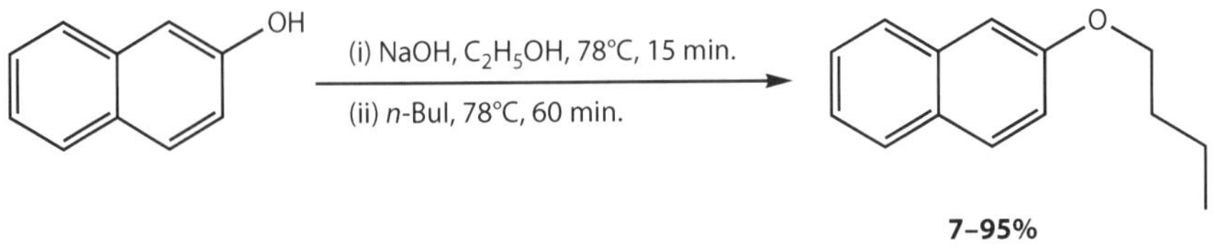

SCHEME 5.13 Williamson ether synthesis in ethanol solvent.

Esteb et al. showed that 2-butoxynaphthalene, an artificial raspberry and strawberry flavoring agent, can be synthesized by an S_N2 reaction from 2-naphthol in refluxing ethanol (Scheme 5.13).[73] This reaction is presented in Section 1.4 and discussed from an atom-economy perspective. The crystalline ether product can be isolated by pouring the ethanolic solution into a beaker of ice water and then performing vacuum filtration. The large range of student yields reflects the importance of keeping the water cold during the crystallization. Although S_N2 reactions are normally conducted under polar aprotic conditions, solvents such as dimethylformamide (DMF) and DMSO are not ideal, as they are toxic and highly skin penetrating, respectively. In theory, use of a polar aprotic solvent may cause the reaction to proceed more rapidly by enhancing naphthoxide nucleophilicity, so students can be asked to evaluate the advantages and disadvantages of switching to the more solvating ethanol in this reaction.

As outlined previously, palladium-catalyzed carbon-carbon bond-forming reactions are extremely important in organic synthesis. The Sonogashira reaction typically involves a terminal alkyne coupling with an aryl or vinyl halide under basic conditions in the presence of catalytic Cu^+ and Pd^0. A previous undergraduate experiment undertook this coupling under a nitrogen atmosphere in a 1:1 tetrahydrofuran:triethylamine solvent system.[74] Goodwin et al. employed alternative conditions using the more desirable ethanol as solvent and palladium(II) acetate in place of $(Ph_3P)_2PdCl_2$ as a catalyst. This reaction is performed in the open atmosphere with solid piperazine hexahydrate replacing the malodorous triethylamine.[26,75] The product of this reaction can be utilized further in a multistep organic synthesis (Scheme 5.14).

Use of *N*-*p*-methoxyphenyl (*N*-PMP) imino esters in the synthesis of unnatural amino acids and β-lactam drugs is well established.[76] However, traditional methods for preparation of these imino esters use dichloromethane as the reaction solvent. Bennett et al. have described imine formation via reaction of ethyl glyoxalate and *p*-anisidine in ethyl acetate that acts as the first step in a multistep asymmetric synthesis of an unnatural amino acid.[77] Ethyl acetate represents a less toxic and more environmentally friendly solvent, and reactants are added in higher concentrations than in previous reports to reduce overall solvent consumption. For this reaction, ethyl glyoxalate is purchased as a 50% solution in toluene. The two starting materials are simply dissolved in ethyl acetate along with 4 Å molecular sieves (used to remove the water generated from the condensation reaction) and left for a week before the dark red liquid product is collected by vacuum distillation (Scheme 5.15). The collected product is determined as pure enough to be carried on to the next step of the reaction.

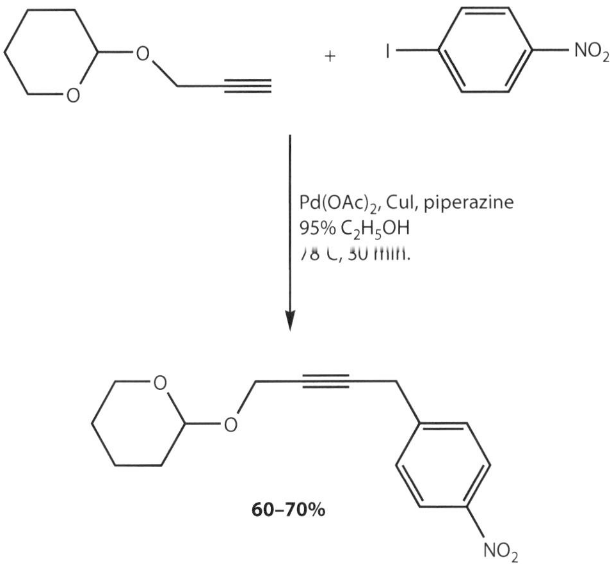

SCHEME 5.14 Sonogashira coupling in ethanol solvent.

Amiet and Urban have devised an iodochlorination of styrene that replaces carcinogenic carbon tetrachloride with petroleum spirits (40–60°C).[78] The synthesis is set up as a discovery-based exercise that has variable results that students must interpret. Iodine monochloride is added to styrene followed by sodium methoxide in methanol, to form a variety of elimination products (Scheme 5.16). Carbon tetrachloride is

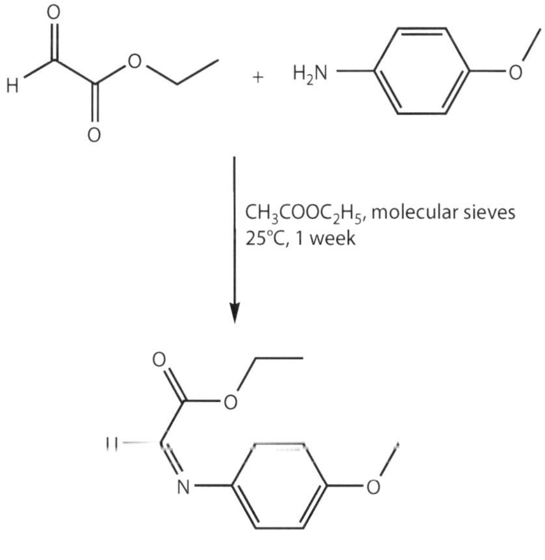

68–96%

SCHEME 5.15 *N*-PMP imino ester synthesis in ethyl acetate solvent.

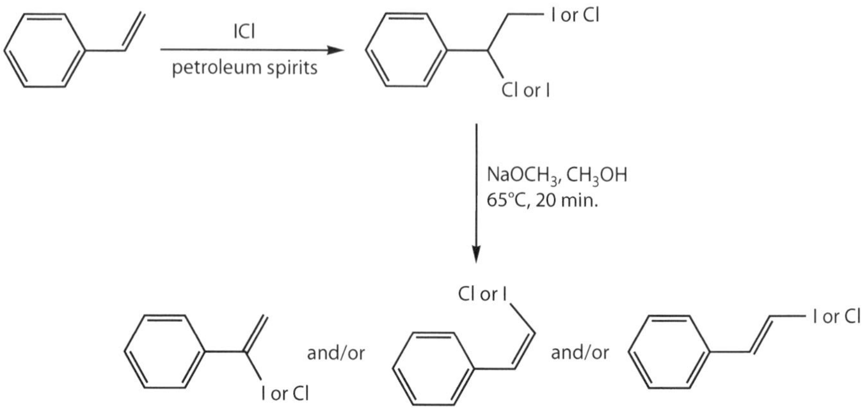

SCHEME 5.16 Iodochlorination of styrene in petroleum spirits.

additionally designated as a controlled substance under the Montreal Protocol[79] due to its ability to deplete stratospheric ozone. In this regard it is important to educate students about restrictions governing use of CCl_4 (and related substances, including bromomethane and 1,1,1-trichloroethane).

Very recently, Teixeira et al. proposed a "question-driven laboratory exercise" where students suggest green modifications to a standard Grignard reaction.[80] An obvious major concern is use of diethyl ether or tetrahydrofuran as the standard solvent. Students consider whether a pure hydrocarbon such as cyclohexane is an appropriate alternative, or a potential combination of cyclohexane and diethyl ether.

5.8 FUTURE OUTLOOK

Green solvents derived from renewable sources represent another category of possible replacements for traditional organic solvents derived from petroleum products, but have thus far not been used in many undergraduate experiments. This class of solvents earned a very respectable score of 19/25 from Clark and Tavener, equaling the rating given to water.[3] The most widely used (and probably oldest) solvent from renewable sources is ethanol, and two other important examples are ethyl L-lactate and 2-methyltetrahydrofuran (Figure 5.13). One of the sources of ethyl L-lactate is from processing carbohydrates from corn, as discussed in Section 1.4.[81] Ethyl L-lactate is a polar solvent that is slightly acidic, so it is not compatible for any acid-sensitive substances.[3] Because it has a chiral moiety, interesting chiral induction is a possibility, although only modest enantioselectivity is observed in the asymmetric reduction of acetophenone in this solvent.[82] Ethyl L-lactate is edible, non-ozone depleting, noncorrosive, nonecotoxic, and rapidly biodegrades in the environment into carbon dioxide and water.[83,84] As a solvent it has shown great promise for preparation of aldimines from aromatic aldehydes and aniline derivatives.[85] These reactions usually require extensive heating, yet occur within minutes at room temperature using ethyl L-lactate. However, it is subject to degradation to ethyl pyruvate, water,

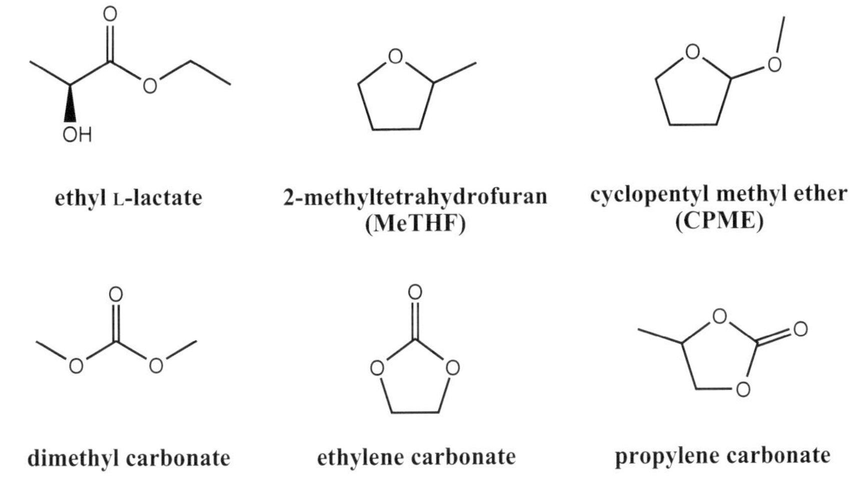

FIGURE 5.13 Structures of some alternative greener solvents.

lactic acid, ethanol, and small quantities of potentially explosive peroxides if not stored under an inert atmosphere.[83]

2-Methyltetrahydrofuran (MeTHF) has been reported as a viable green replacement solvent for tetrahydrofuran and dichloromethane in a variety of organic processes, including organometallic reactions and biotransformations.[86–89] Unlike tetrahydrofuran, which is a petrochemical with an elaborate and resource-intensive preparation,[2] MeTHF is derived from reduction of furfural, which can be obtained from corn or sugarcane[87] (Scheme 5.17). In this instance, the polysaccharide hemicellulose is hydrolyzed on heating with acid to form a mixture of five-carbon sugars, including xylose. Sugar dehydration and elimination of three water molecules generates furfural. MeTHF falls between diethyl ether and tetrahydrofuran in terms of polarity and Lewis base strength. It boils at 80°C and freezes at –76°C, allowing higher reaction temperatures than other ethers.[89] One of its most attractive properties

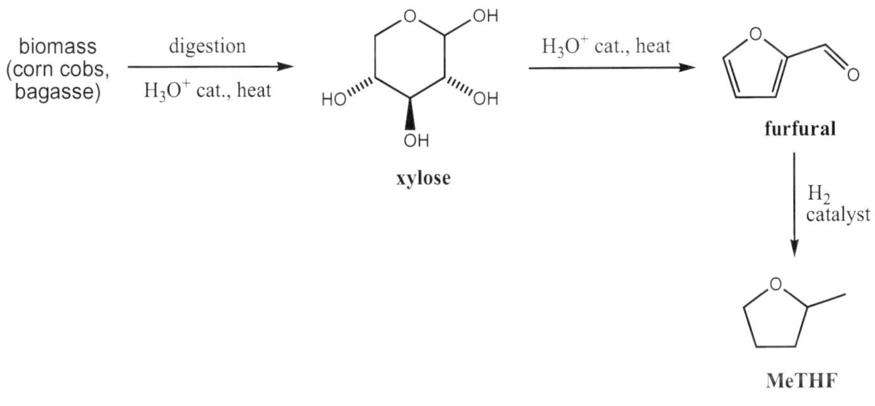

SCHEME 5.17 Synthesis of 2-methyltetrahydrofuran from biomass.

is that it cleanly forms two phases with water, so unlike tetrahydrofuran, it can act as a green solvent for aqueous extractions as a direct replacement for dichloromethane.[86] A possible alternative solvent to MeTHF that forms peroxides less readily is cyclopentyl methyl ether (CPME)[90] (Figure 5.13). Although not derived from renewable resources, it can be considered green because of the ease of its recovery and reuse. CPME has a higher boiling point (106°C) than either tetrahydrofuran or MeTHF, leading to a reduction in many reaction times, and a low heat of vaporization so less solvent is lost during reflux. In addition, CPME forms an azeotrope rich with water, and is therefore more easily dried than other ethereal solvents.

Other potential solvents derived from renewable resources include nonpolar terpenes such as (*R*)-limonene and pinene,[1] polar solvents like glycerol,[3,91] as well as fatty acid esters including methyl soyate.[92,93] There is also current interest in dimethyl carbonate (DMC) as a nontoxic, biodegradable solvent (Figure 5.13).[94] Catalytic oxidation of various functional groups has been reported under mild conditions utilizing DMC as the solvent and H_2O_2/CH_3ReO_3 as the oxidizing agent.[95] Cyclic carbonates (for example, ethylene and propylene carbonate) (Figure 5.13) are additionally garnering attention.[96] These solvents exhibit low flammability and toxicity, and can be synthesized by reaction of ethene or propene with CO_2, representing 100% atom economy. Ethylene carbonate can act as a green alternative to the polar aprotic solvents *N*,*N*-dimethylformamide and dimethylsulfoxide. One issue with these cyclic carbonates is their comparatively high viscosities, which may prove problematic at an industrial scale.

A new class of green solvents exhibiting "switchable" polarity and phasic properties has been developed by Jessop et al.[97–99] A recent development showed that a hydrophobic solvent that forms a biphasic system with water can be switched into a hydrophilic solvent that is miscible with water simply by bubbling in CO_2.[98] Jessop's solvent is an amidine with four long alkyl chains that is hydrophobic under neutral conditions. When CO_2 is bubbled through in the presence of water, the imine nitrogen is protonated to form a charged polar solvent that forms a single phase with water (Scheme 5.18). The reaction can be reversed by heating for 60 minutes at 80°C or by bubbling air slowly through the solution at room temperature over several hours. In this regard, these solvents are complementary to fluorous solvents, which have switchable phasic behavior with organic solvents.

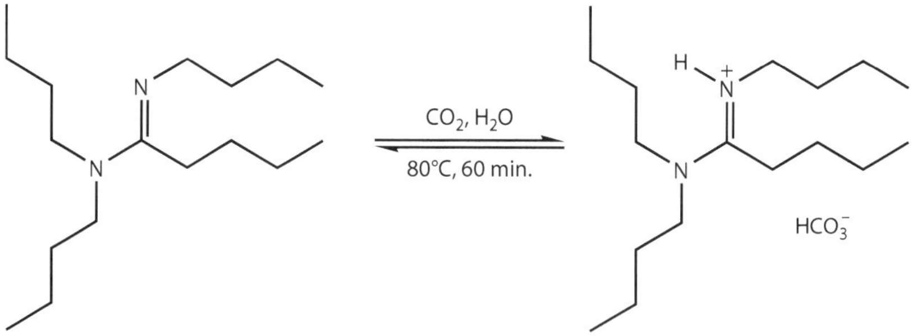

SCHEME 5.18 Switchable amidine solvents.

This solvent was used in a green extraction of soybean oil from soybean flakes that avoids volatile organics and the distillation step of traditional methods. The authors believe that, unlike with fluorous alternatives, persistence and bioaccumulation are not significant concerns. The solvent is easily hydrolyzed in water, although it should be noted that no extensive environmental and toxicological studies have been performed.[98] This class of solvent is derived from petrochemicals, and is rather complex, so their production cannot be considered green; however, they are good examples of novel, greener reusable extraction solvents that potentially fit into an undergraduate laboratory.

5.9 CONCLUSION

The main lesson to be learned, and to be taught, about the different greener alternative solvents described in this chapter is that no solvent can be considered to be green just because it has been described as such in the green chemistry literature. Whether the use of a new solvent actually makes a reaction or an extraction safer and more environmentally friendly should be carefully considered on a case-by-case basis. Showcasing green reaction media in undergraduate lectures and laboratories clearly introduces students to the possibility of solvent replacement in their future careers as responsible chemists. Importantly, it also gives them a valuable opportunity to engage in critical thinking and evaluation when weighing the pros and cons of alternative reaction conditions.

REFERENCES

1. Jessop, P. G. *Can. Chem. News* 2007, 59 (2), 16–18.
2. Capello, C., Fischer, U., Hungerbühler, K. *Green Chem.* 2007, 9, 927–934.
3. Clark, J. H., Tavener, S. J. *Org. Process Res. Dev.* 2007, 11, 149–155.
4. Alfonsi, K., Colberg, J., Dunn, P. J., Fevig, T., Jennings, S., Johnson, T. A., Kleine, H. P., Knight, C., Nagy, M. A., Perry, D. A., Stefaniak, M. *Green Chem.* 2008, 10, 31–36.
5. Curzons, A. D., Constable, D. C., Cunningham, V. L. *Clean Technol. Environ. Policy* 1999, 1, 82–90.
6. Beckman, E. *J. Environ. Sci. Technol.* 2002, 36, 347A–353A.
7. McMurry, J., Fay, R. C. *Chemistry*, 4th ed. Prentice-Hall, Upper Saddle River, NJ, 2004, pp. 414–415.
8. Clark, J. H., Deswarte, F. *Educ. Chem.* 2008, 45 (3), 76–79.
9. Phelps, C. L., Smart, N. G., Wai, C. M. *J. Chem. Educ.* 1996, 73, 1163–1168.
10. Rayner, C. M., Oakes, R. S. Supercritical Carbon Dioxide. In *Green Reaction Media in Organic Synthesis*, Mikami, K., Ed. Blackwell Publishing, Oxford, 2005.
11. Luque de Castro, M. D., Valcarcel, M., Tena, M. T. *Analytical Supercritical Fluid Extraction*. Springer-Verlag, Berlin, 1994.
12. The Presidential Green Chemistry Challenge Award Recipients 1996–2009, 2007 Small Business Award. U.S. Environmental Protection Agency, Office of Pollution Prevention and Toxics, Washington, DC, 2009, 24–25.
13. Wai, C. M., Hunt, F., Ji, M., Chen, X. *J. Chem. Educ.* 1998, 75, 1641–1645.
14. Ye, X., Wai, C. M. *J. Chem. Educ.* 2003, 80, 198–204.
15. Mauldin, R. F., Burns, D. J., Keller, I. K., Koehn, K. K., Johnson, M. J., Gray, S. L. *Chem. Educator* 1999, 4, 183–185.
16. Moye, A. L. *J. Chem. Educ.* 1972, 49, 194.
17. Mitchell, R. H., Scott, W. A., West, P. R. *J. Chem. Educ.* 1974, 51, 69.

18. Taber, D. F., Hoemer, R. S. *J. Chem. Educ.* 1991, 68, 73.
19. Murray, S. D., Hansen, P. J. *J. Chem. Educ.* 1995, 72, 851–852.
20. Adam, D. J., Mainwaring, J., Quigley, M. N. *J. Chem. Educ.* 1996, 73, 1171.
21. McKenzie, L. C., Thompson, J. E., Sullivan, R., Hutchison, J. E. *Green Chem.* 2004, 6, 355–358.
22. Billington, S., Smith, R. B., Karousos, N. G., Cowham, E., Davis, J. *J. Chem. Educ.* 2008, 85, 379–380.
23. Garner, C. M., Garibaldi, C. *J. Chem. Educ.* 1994, 71, A146–A147.
24. Ubeda, M. A., Dembinski, R. *J. Chem. Educ.* 2006, 83, 84–92.
25. Tavener, S. J., Clark, J. H. Fluorine: Friend or Foe? A Green Chemist's Perspective. In *Advances in Fluorine Science—Fluorine and the Environment—Agrochemicals, Archaeology, Green Chemistry & Water*, Vol. 2, Tressaud, A., Ed. Elsevier, Amsterdam, 2006.
26. Goodwin, T. E. The Garden of Green Organic Chemistry at Hendrix College. In *Green Chemistry Education: Changing the Course of Chemistry*, Anastas, P. T., Levy, I. J., Parent, K. E., Eds. American Chemical Society Symposium Series 1011. American Chemical Society, Washington, DC, 2009.
27. *Zero Effluent Laboratory: An Educational Experiment, A Chemistry Professor's Viewpoint.* www.hendrix.edu/uploadedFiles/Departments_and_Programs/Chemistry/Green_Chemistry/ACS talk and lab.pdf (accessed December 23, 2010).
28. Rábai, J. Fun and Games with Fluorous Chemistry. In *Handbook of Fluorous Chemistry*, Gladysz, J. A., Curran, D. P., Horváth, I. T., Eds. Wiley, Weinheim, Germany, 2004.
29. Daley, J. M., Landolt, R. G. *J. Chem. Educ.* 2005, 82, 120–121.
30. Doheny, A. J., Loudon, G. M. *J. Chem. Educ.* 1980, 57, 507–508.
31. Sigma Aldrich Online Catalog 547948. α,α,α-Trifluorotoluene, ≥99%. www.sigmaaldrich.com (accessed December 23, 2010).
32. Jodry, J. J., Mikami, K. Ionic Liquids. In *Green Reaction Media in Organic Synthesis*, Mikami, K., Ed. Blackwell Publishing, Oxford, 2005.
33. Bowman, D. C. *Chem. Educator* 2006, 11, 64–66.
34. Swatloski, R. P., Holbrey, J. D., Rogers, R. D. *Green Chem.* 2003, 5, 361–363.
35. Pham, T. P. T., Cho, C., Yun, Y. *Water Res.* 2010, 44, 352–372.
36. J. T. Baker Material Safety Data Sheet (MSDS). www.jtbaker.com/europe/msds/ (accessed December 23, 2010).
37. Zhang, Y., Bakshi, B. R., Demessie, E. S. *Environ. Sci. Technol.* 2008, 42, 1724–1730.
38. Renner, R. *Environ. Sci. Technol.* 2001, 35, 411A–413A.
39. Dzyuba, S. V., Kollar, K. D., Sabnis, S. S. *J. Chem. Educ.* 2009, 86, 856–858.
40. Kaar, J. L., Jesionowski, A. M., Berberich, J. A., Moulton, R., Russell, A. J. *J. Am. Chem. Soc.* 2003, 125, 4126–4131.
41. Anslyn, E. V., Dougherty, D. A. *Modern Physical Organic Chemistry*. University Science Books, Sausalito, CA, 2006, pp. 148–149.
42. Zhao, D., Wu, M., Kou, Y., Min, E. *Catal. Today* 2002, 74, 157–189.
43. Lee, J. W., Shin, J. Y., Chun, Y. S., Jang, H. B., Song, C. E., Lee, S. *Acc. Chem. Res.* 2010, 43, 985–994.
44. Gordon, C. M. *Appl. Catal. A* 2001, 222, 101–118.
45. (a) Plechkova, N. V., Seddon, K. R. *Chem. Soc. Rev.* 2008, 37, 123–150. (b) The Presidential Green Chemistry Challenge Award Recipients 1996–2009, 2005 Academic Award. U.S. Environmental Protection Agency, Office of Pollution Prevention and Toxics, Washington, DC, 2009, 42–43.
46. Abbott, A., Davies, D. L. *Educ. Chem.* 2005, 42 (1), 12–15.
47. Abbott, A., Capper, G., Davies, D. L., Munro, H. L., Rasheed, R. K., Tambyrajah, V. *Chem. Commun.* 2001, 2010–2011.
48. Deetlefs, M., Seddon, K. R. *Green Chem.* 2010, 12, 17–30.

49. Riisager, A., Bösmann, A. Ionic Liquids as Benign Solvents for Sustainable Chemistry. In *Experiments in Green and Sustainable Chemistry*, Roesky, H. W., Kennepohl, D. K., Eds. Wiley VCH, Weinheim, Germany, 2009, 108–113.
50. Mak, K. K.W., Siu, J., Lai, Y. M., Chan, P. *J. Chem. Educ.* 2006, 83, 943–946.
51. Stark, A., Ott, D., Kralisch, D., Kreisel, G., Ondruschka, B. *J. Chem. Educ.* 2010, 87, 196–201.
52. Gorke, J., Srienc, F., Kazlauskas, R. J. *Chem. Commun.* 2008, 1235–1237.
53. Schatz, P. F. *J. Chem. Educ.* 2001, 78, 1378.
54. Cao, Y., Wu, J., Zhang, J., Li, H., Zhang, Y., He, J. *Chem. Eng. J.* 2009, 147, 13–21.
55. Casey, C. P. *J. Chem. Educ.* 2006, 83, 192–195.
56. (a). Fujita, M. Olefin Self-Cross Metathesis in an Ionic Liquid. In *Experiments in Green and Sustainable Chemistry*, Roesky, H. W., Kennepohl, D. K., Eds. Wiley VCH, Weinheim, Germany, 2009, pp. 114–117. (b). Fujita, M. *Chem. Educator* 2010, 15, 376–380.
57. Blanchard, L. A., Hancu, D., Beckman, E. J., Brennecke, J. F. *Nature* 1999, 399, 28–29.
58. Brown, R. A., Pollet, P., McKoon, E., Eckert, C. A., Liotta, C. L., Jessop, P. G. *J. Am. Chem. Soc.* 2001, 123, 1254–1255.
59. Dzyuba, S. V., Bartsch, R. A. *Angew. Chem. Int. Ed.* 2003, 42, 148–150.
60. Lozano, P. *Green Chem.* 2010, 12, 555–569.
61. Coleman, D., Gathergood, N. *Chem. Soc. Rev.* 2010, 39, 600–637.
62. Heldebrant, D. J., Witt, H. N., Walsh, S. M., Ellis, T., Rauscher, J., Jessop, P. G. *Green Chem.* 2006, 8, 807–815.
63. Chen, J., Spear, S. K., Huddleson, J. G., Rogers, R. D. *Green Chem.* 2005, 7, 64–82.
64. Singh, P., Pandey, S. *Green Chem.* 2007, 9, 254–261.
65. Knop, K., Hoogenboom, R., Fischer, D., Shubert, U. S. *Angew. Chem. Int. Ed.* 2010, 49, 6288–6308.
66. Wonganan, P., Croyle, M. A. *Viruses* 2010, 2, 468–502.
67. McKenzie, L. C., Huffman, L. M., Hutchison, J. E., Rogers, C. E., Goodwin, T. E., Spessard, G. O. *J. Chem. Educ.* 2009, 86, 488–493.
68. Ahluwalia, V. K., Varma, R. S. *Green Solvents for Organic Synthesis*. Alpha Science, Oxford, 2009.
69. Huddleson, J. G., Willauer, H. D., Griffin, S. T., Rogers, R. D. *Ind. Eng. Chem. Res.* 1999, 38, 2523–2539.
70. Shamery, T. L., Huddleson, J. G., Chen, J., Spear, S. K., Rogers, R. D. Aqueous Biphasic Systems for Liquid-Liquid Separations. In *Experiments in Green and Sustainable Chemistry*, Roesky, H. W., Kennepohl, D. K., Eds. Wiley VCH, Weinheim, Germany, 2009, 92–96.
71. (a) Aktoudianakis, E., Chan, E., Edward, A. R., Jarosz, I., Lee, V., Mui, L., Thatipamala, S. S., Dicks, A. P. *J. Chem. Educ.* 2008, 85, 555–557. (b) Pantess, D. A., Rich, C. V. *Chem. Educator* 2009, 14, 258–260. (c) Novak, M., Wang, Y.-T., Ambrogio, M. W., Chan, C. A., Davis, H. E., Goodwin, K. S., Hadley, M. A., Hall, C. M., Herrick, A. M., Ivanov, A. S., Mueller, C. M., Oh, J. J., Soukup, R. J., Sullivan, T. J., Todd, A. M. *Chem. Educator* 2007, 12, 414–418.
72. McKenzie, L. C., Huffman, L. M., Hutchison, J. E. *J. Chem. Educ.* 2005, 82, 306–310.
73. Esteb, J. J., Magers, J. R., McNulty, L., Morgan, P., Wilson, A. M. *J. Chem. Educ.* 2009, 86, 850–852.
74. Brisbois, R. G., Batterman, W. G., Kragerud, S. R. *J. Chem. Educ.* 1997, 74, 832–833.
75. Goodwin, T. E., Hurst, E. M., Ross, A. S. *J. Chem. Educ.* 1999, 76, 74–75.
76. Taggi, A. E., Hafez, A. M., Lectka, T. *Acc. Chem. Res.* 2003, 36, 10–19.
77. Bennett, J., Meldi, K., Kimmell II, C. *J. Chem. Educ.* 2006, 83, 1221–1224.

78. Amlet, R. G., Urban, S. *J. Chem. Educ.* 2008, 85, 962–964.
79. Rae, I. D. *J. Chem. Educ.* 2009, 86, 689.
80. Teixeira, J. M., Nedrow Byers, J., Perez, M. G., Holman, R. W. *J. Chem. Educ.* 2010, 87, 714–716.
81. Warner, J. C., Cannon, A. S., Dye, K. M. *Environ. Impact Assess. Rev.* 2004, 24, 775–799.
82. Hüttenhain, S. H. *Synth. Commun.* 2006, 36, 175–180.
83. McConville, J. T., Carvalho, T. C., Kucera, S. A., Garza, E. *Pharm. Technol.* 2009, 33, 74–84.
84. Bowmer, C. T., Hooftman, R. N., Hanstveit, A. O., Venderbosch, P. W. M., van der Hoeven, N. *Chemosphere* 1998, 37, 1317–1333.
85. Bennett, J. S., Charles, K. L., Miner, M. R., Heuberger, C. F., Spina, E. J., Bartels, M. F., Foreman, T. *Green Chem.* 2009, 11, 166–168.
86. Ripin, D. H. B., Vetelino, M. *Synlett* 2003, 2353.
87. Pace, V., Hoyos, P., Fernández, M., Sinisterra, J. V., Alcántara, A. R. *Green Chem.* 2010, 12, 1380–1382.
88. Simeó, Y., Sinisterra, J. V., Alcántara, A. R. *Green Chem.* 2009, 11, 855–862.
89. Aycock, D. F. *Org. Process Res. Dev.* 2007, 11, 156–159.
90. Wantanabe, K., Yamagiwa, N., Torisawa, Y. *Org. Process Res. Dev.* 2007, 11, 251–258.
91. Gu, Y., Jérôme, F. *Green Chem.* 2010, 12, 1127–1138.
92. Spear, S. K., Griffin, S. T., Granger, K. S., Huddleson, J. G., Rogers, R. D. *Green Chem.* 2007, 9, 1008–1015.
93. Srinivas, K., Potts, T. M., King, J. W. *Green Chem.* 2009, 11, 1581–1588.
94. Tundo, P., Selva, M. *Acc. Chem. Res.* 2002, 35, 706–716.
95. Bernini, R., Mincione, E., Barontini, M., Crisante, F., Fabrizi, G., Gambacorta, A. *Tetrahedron* 2007, 63, 6895–6900.
96. Andrews, I., Cui, J., Dudin, L., Dunn, P., Hayler, J., Hinkley, B., Hughes, D., Kaptein, B., Lorenz, K., Mathew, S., Rammeloo, T., Wang, L., Wells, A., White, T. D. *Org. Process Res. Dev.* 2010, 14, 770–780.
97. Phan, L., Andreatta, J. R., Horvey, L. K., Edie, C. F., Luco, A., Mirchandani, A., Darensbourg, D. J., Jessop, P. G. *J. Org. Chem.* 2008, 73, 127–132.
98. Jessop, P. G., Phan, L., Carrier, A., Robinson, S., Dürr, C. J., Harjani, J. R. *Green Chem.* 2010, 12, 809–814.
99. Lougheed, T. *Can. Chem. News* 2010, 62 (9), 22–23, 25.

6 Environmentally Friendly Organic Reagents

Dr. Loyd D. Bastin

CONTENTS

6.1 INTRODUCTION

As noted in the other chapters of this book, a number of organic reactions have been modified for use in the undergraduate teaching laboratory in order to incorporate green chemistry principles. Many of these transformations involve the replacement of hazardous reagents with more environmentally friendly substances. These updated experiments can serve as excellent examples to demonstrate how organic reactivity can be made greener. It is important to stress to undergraduates that "greening" a chemical reaction is very much a step-wise process, and that the resulting synthetic methodology will not be perfectly green. In order to incorporate green chemistry into the everyday vocabulary of organic synthesis, chemists must expand the definition of *green* to include *any* change in a process that results in a more environmentally friendly reaction.

The primary focus of this chapter is to profile the plethora of greener reagents available for adoption in the undergraduate organic laboratory. This is broadly organized by reaction type, with an emphasis on oxidations, reductions, halogenations, greener nonmetal catalysis, organocatalysis, metal catalysis, biocatalysis, and solid-supported reagents. The coverage of these topics is not intended to be all-encompassing. Rather, examples are chosen to highlight a wide range of reactivity pertinent for both introductory and advanced laboratory courses. Many new approaches have been developed during the last ten years that build on more classical transformations. Use of greener reagents is often coupled with other features, including more environmentally friendly solvents and energy-efficient reactivity. Where appropriate, the reader is directed to other chapter sections, which provide related exemplars in these areas.

6.2 GREENER REAGENTS IN THE UNDERGRADUATE ORGANIC LABORATORY

6.2.1 Carbonyl-Forming Oxidations

A number of green modifications have been reported for the oxidation of alcohols into carbonyl-containing products. These reactions conveniently fall into two general categories: those utilizing bleach (sodium hypochlorite) as the oxidant, and those that do not involve bleach. Examples falling into each division are outlined here.

6.2.1.1 Oxidations Using Bleach

Use of bleach to oxidize secondary alcohols into ketones was first reported by Stevens et al. in 1980, where 85%–96% yields were obtained for a variety of substrates.[1]

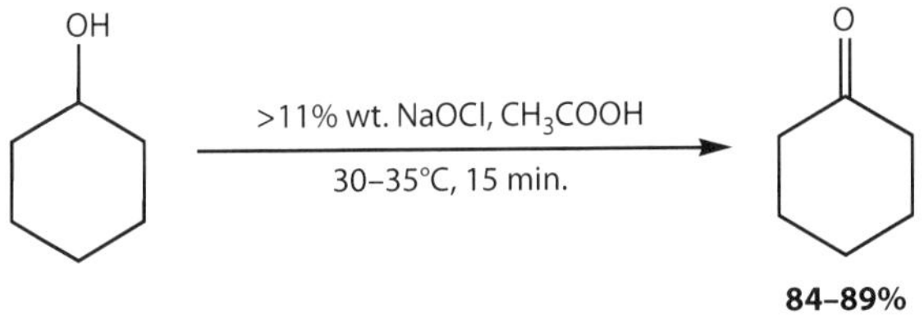

SCHEME 6.1 Oxidation of cyclohexanol using sodium hypochlorite.

Within a year, two adaptations for the undergraduate organic chemistry laboratory were reported in the *Journal of Chemical Education*. First, Zuczek and Furth modified the oxidation of cyclohexanol to cyclohexanone for use in the undergraduate laboratory.[2] This procedure utilizes >11% aqueous sodium hypochlorite and mild heat, resulting in an excellent yield (Scheme 6.1). Similarly, the room temperature conversion of isopropanol to acetone via reaction with NaOCl was revised for student use.[3] Bleach oxidation of secondary alcohols to ketones addresses a couple of green principles. Use of sodium hypochlorite rather than sodium or potassium dichromate eliminates the need for carcinogenic reagents, and the generation of chromium salt waste. Sodium hypochlorite additionally breaks down in the environment to form sodium chloride, water, and oxygen.

In 1982, alteration of previously reported methods led to the synthesis of a range of ketones from secondary alcohols.[4] This strategy features household bleach (5%–6% sodium hypochlorite) as the oxidizing agent and dichloromethane (rather than diethyl ether) as the extraction solvent. Student yields of 70%–87% for a variety of straight-chain and cyclic ketones (e.g., 2-hexanol and cycloheptanol) have been reported. Use of a less concentrated sodium hypochlorite solution reduces the hazards associated with the oxidizing agent. The switch from diethyl ether to dichloromethane was perpetuated by the lower flammability of the latter solvent. However, it is now widely accepted that dichloromethane is a potential carcinogen. Mohrig et al. have also described the successful use of household bleach for the oxidation of secondary alcohols to ketones[5] (Scheme 6.2). In addition to this modification, the oxidation procedure has been altered in two other important ways. The amount of acetic acid was reduced from 7.5 to 0.5 mL per gram of alcohol reactant, and each ketone product purified by distillation rather than extraction. The change of purification method eliminates the need for hazardous solvents and reduces the amount of waste generated in the experiment. Conversely, liquid distillation uses more energy than liquid-liquid extraction. A similar procedure for the oxidation of borneol to camphor

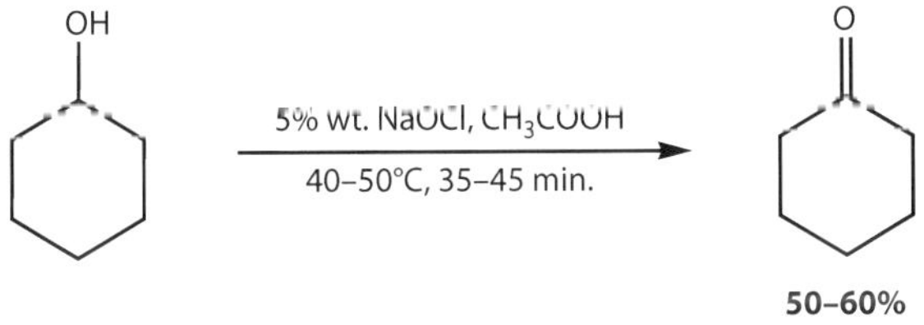

SCHEME 6.2 Modification of the bleach oxidation of cyclohexanol.

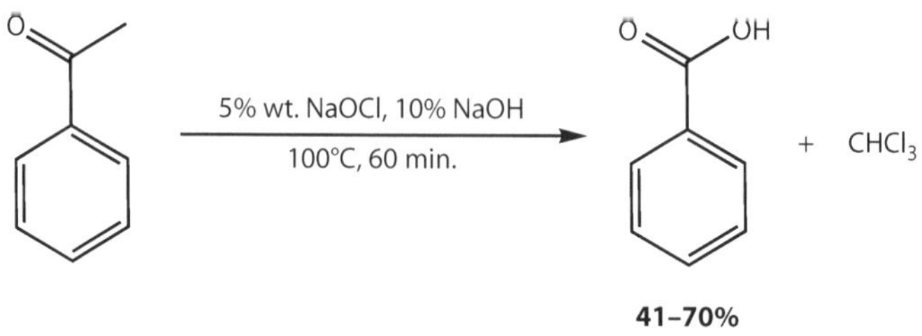

SCHEME 6.3 Bleach oxidation of acetophenone to benzoic acid.

has recently been published.[6] These experiments serve as excellent examples that a process can be modernized and made greener, but may not become totally green.

While the oxidation of secondary alcohols to ketones using household bleach in acetic acid is versatile and robust, Straub has reported that the system does not oxidize benzoin to benzil.[7] However, this transformation can be effected using household bleach, 2,2,6,6-tetramethyl-piperidine-1-oxyl (TEMPO), sodium bicarbonate, and potassium bromide in dichloromethane.[7]

In a related example, sodium hypochlorite has been used to oxidize acetophenone to benzoic acid[8] (Scheme 6.3). The benzoic acid does not typically require purification; however, the product can be recrystallized from hot water if necessary. As with the aforementioned alcohol reactions, oxidation of ketones to carboxylic acids is traditionally accomplished using hot, basic potassium permanganate or acidic dichromate solutions. This modification removes the generation of chromium waste and replaces a harsh oxidizer with a milder oxidizing reagent. However, it is important to note that this reaction does generate a chlorinated organic compound (chloroform) as a by-product.

In a similar study, Ballard recently designed a discovery-oriented experiment probing the pH dependence of an aromatic ketone bleach oxidation.[9] The procedures described do not require additional sodium hydroxide, thus eliminating the use of a corrosive reagent. This is discussed in more detail in Section 4.3.3.

6.2.1.2 Oxidations Using Nonbleach Reagents

As mentioned previously, chromium(VI) reagents (sodium dichromate, Jones reagent, pyridinium chlorochromate) are the classical ones used to transform a secondary alcohol into a ketone. Therefore, it is unsurprising that some greener oxidation methods are based on a solid-supported chromium redox cycle. These reactions are outlined in Section 6.2.8.1.

Greener strategies for oxidizing the side chain of an arene have been investigated.[10,11] These approaches use molecular oxygen to oxidize fluorene to 9-fluorenone (Scheme 6.4) rather than acidic sodium dichromate solutions. In addition to the elimination of chromium waste, the use of oxygen results in a substantially higher percentage yield. It can also serve as an example of when additional improvements are needed. Sodium hydroxide is toxic and corrosive, while the phase transfer catalyst Aliquat® 336 (Section 3.3.1) is also toxic and a severe irritant. The two reported procedures utilize different purification methods (recrystallization[10] and column chromatography[11]). Recrystallization is generally considered to be a

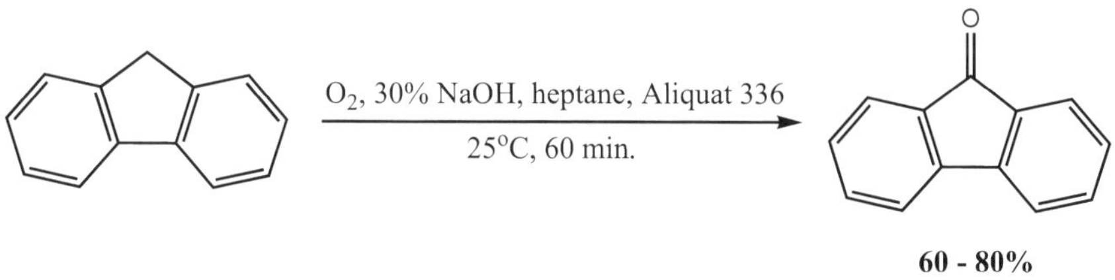

SCHEME 6.4 Oxidation of fluorene to 9-fluorenone with molecular oxygen.

greener purification method (due to the amount of solvent waste generated in column chromatography), but clearly the nature of the solvent(s) required is an additional consideration.

Gandhari et al. have developed a greener oxidation of aromatic aldehydes to carboxylic acids.[12] This procedure uses Oxone® (a 2:1:1 molar mixture of $KHSO_5$, $KHSO_4$, and K_2SO_4; Section 4.3.3) in water to oxidize benzaldehyde to benzoic acid (Scheme 6.5). Several derivatives of benzaldehyde can also be oxidized in this manner by using a 4:1 water:ethanol mixture as the solvent. Average yields of 65%–90% have been obtained by students for oxidation of 2-chlorobenzaldehyde, 4-chlorobenzaldehyde, 4-nitrobenzaldehyde, 4-bromobenzaldehyde, and 3-methoxybenzaldehyde. While all of the reagents employed in this strategy are irritants, use of Oxone provides a much safer and environmentally conscious alternative to $KMnO_4$ or $K_2Cr_2O_7$. These reactions avoid the use of an extraction solvent, since the desired product precipitates from the reaction solution and is easily isolated by filtration. Further purification is achieved by recrystallization from hot water. It must be emphasized, however, that these transformations are somewhat atom inefficient due to the high molecular weight of Oxone.

6.2.2 Carbonyl Reductions

Carbonyl functional group reductions have been the focus of significant reworking from a green perspective. The published modifications are simply classified by use of either biological or nonbiological reagents, and are described as such in Sections 6.2.2.1 and 6.2.2.2.

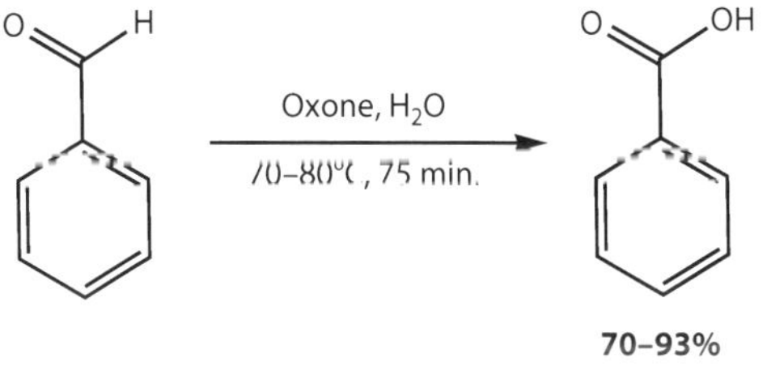

SCHEME 6.5 Oxidation of benzaldehyde to benzoic acid with Oxone.

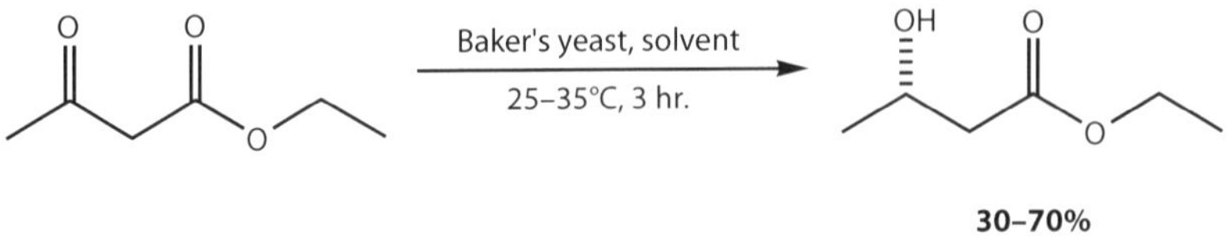

SCHEME 6.6 Baker's yeast enzymatic reduction of ethyl acetoacetate.

6.2.2.1 Reductions Using Biological Reagents

The reduction of β-ketoesters by Baker's yeast is a common undergraduate experiment that was originally designed to illustrate the stereoselective nature of many biochemical processes.[13–15] Several variations of this experiment have been published in the chemical education literature (Scheme 6.6).

The reaction conditions and workup methods are the major difference in the three experiments highlighted here.[13–15] Since Baker's yeast is a renewable resource and its reducing agent is an enzyme, this classical undergraduate experiment is also an excellent example of a green carbonyl reduction. However, the three reported experiments are not equal in terms of greenness. The procedure published by North uses nonfermenting yeast in petroleum ether and water for the reduction.[13] Petroleum ether, as its name implies, is derived from petroleum rather than a renewable source, and thus is not a particularly green solvent. The second experiment uses fermenting yeast in water, with a small amount of hexanes present.[14] This method is preferable, as it uses less petroleum-based solvent. In comparison, Pohl et al. have implemented this reduction using fermenting yeast and sucrose in an aqueous solvent system.[15] This approach is clearly superior from a green chemistry perspective, as it lacks usage of any organic solvent. Unfortunately, the enzymatic nature of these reactions creates a barrier to their widespread use as reducing agents. Baker's yeast can only be employed for the reduction of the ketone functionality in β-ketoesters. While the reaction conditions for a range of β-ketoesters have been reported in research journals,[16–18] the reduction of ethyl acetoacetate is the only documented example in the chemical education literature.

Carrot pieces are another biological source of an enzymatic reducing agent adapted to the undergraduate organic chemistry laboratory (Scheme 6.7). Ravía et al. have outlined the use of carrot fragments for the small-scale reduction of carbonyl compounds.[19] Similar to the Baker's yeast method, this reaction occurs in an aqueous solvent. The carrot method has the supplemental advantage in that it does not require any heating, therefore illustrating an additional green chemistry principle. While Baker's yeast reductions have only been reported for β-ketoesters, use of carrot parts

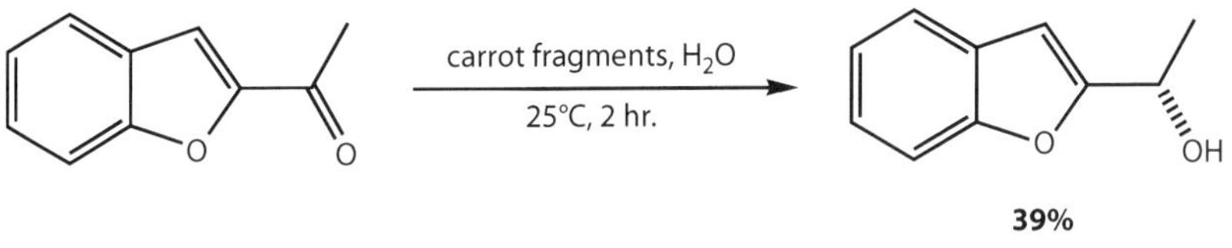

SCHEME 6.7 Carrot-based enzymatic reduction of benzofuran-2-yl methyl ketone.

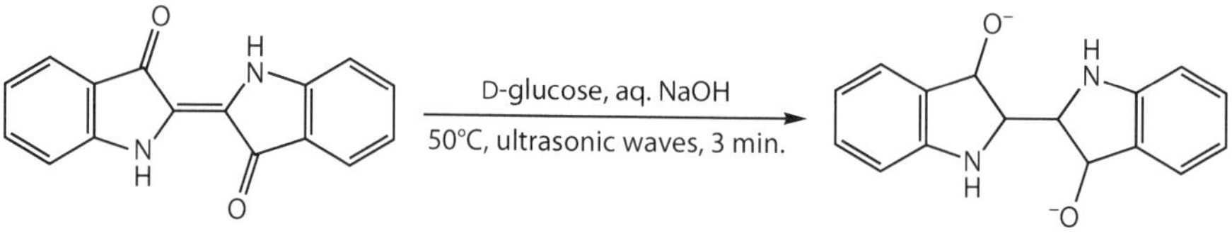

indigo **leuco-indigo (yellow)**

SCHEME 6.8 A greener reduction of indigo.

and cell cultures has been published in research articles to work with a variety of substrates.[20–24] However, benzofuran-2-yl methyl ketone is the only substrate to date that has been adapted for incorporation in the undergraduate laboratory. There are a number of reports in the research literature using other plant components[25–28] and biocatalysts[29] to reduce prochiral ketones.

6.2.2.2 Reductions Using Nonbiological Reagents

There are several nonbiological reagents that can accomplish the reduction of a carbonyl moiety under greener conditions compared to the traditional hydride transfer compounds (lithium aluminum hydride and sodium borohydride). The reduction of indigo to form leuco-indigo can easily be performed using glucose, sodium hydroxide, mild heating, and ultrasonic waves (Scheme 6.8). This method of reducing indigo[30] has several advantages over the classical aqueous sulfurous reducing agent reported by Boykin.[31] The latter reducing agent is replaced by a reducing sugar, and the required reaction temperature lowered from 75°C to 55°C by use of the ultrasonic waves.

O'Brien and Wicht have proposed an alternative reduction protocol that employs polymethylhydrosiloxane (PMHS) and catalytic tetra-*n*-butylammonium fluoride (TBAF) to reduce citronellal to citronellol[32] (Scheme 6.9). Citronellal is the major component of citronella oil, and has a characteristic lemon odor.

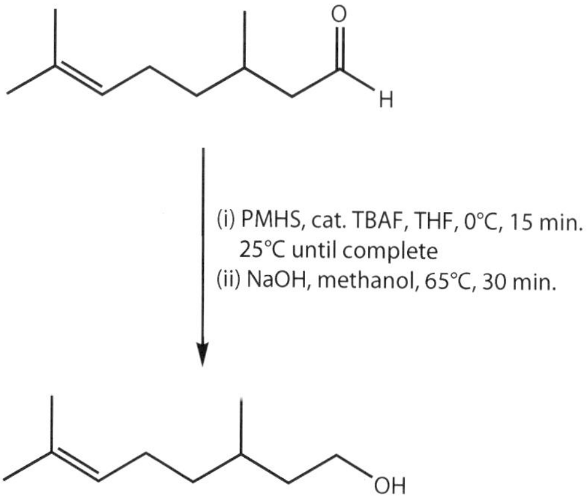

60%

SCHEME 6.9 Polymethylhydrosiloxane reduction of citronellal.

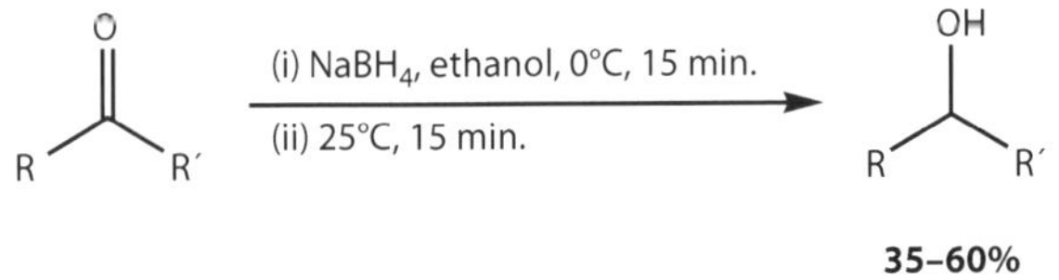

SCHEME 6.10 Sodium borohydride reduction of ketones.

It also exhibits effective antifungal and insect repellent properties. This procedure replaces the commonly used sodium borohydride with PMHS. PMHS is a nonviscous silicone-based polymer that has been shown to be a versatile reducing agent.[33] The polymer is stable to air and moisture, thus eliminating issues associated with sodium borohydride, and particularly lithium aluminum hydride. This laboratory experiment also introduces the concept of catalysis as a means of controlling hazards (catalytic release of hydride), in addition to chemical waste reduction. The approach also demonstrates use of a natural product as the starting material, in order to illustrate the green chemistry principle of obtaining materials from renewable sources.

The easiest and most robust carbonyl reduction method for the undergraduate laboratory does, however, remain the use of sodium borohydride (Scheme 6.10). A number of other undergraduate procedures have been designed to introduce students to sodium borohydride reductions. Baru and Mohan reported a chemoselective discovery-oriented approach that reduces vanillin acetate and 4-formylbenzoate to their corresponding alcohols in the presence of an ester.[38] While this experiment is fundamentally an effective introduction to carbonyl reduction chemistry, it also demonstrates several green chemistry principles. Sodium borohydride is a much safer reducing agent than lithium aluminum hydride. It exhibits short reaction times, is extremely versatile, and has limited moisture and air sensitivity. $NaBH_4$, therefore, allows the introduction of alcohols as alternative solvents. Ethanol is particularly green due to its potential production from renewable sources. Lecher has designed a low-solvent version of the $NaBH_4$ reduction of vanillin,[39] and Sections 4.3.3 and 7.3.3 detail similar transformations.

Pohl et al. have also described the use of sodium borohydride as a chemoselective reducing agent for ethyl acetoacetate.[15] In addition to undertaking the standard $NaBH_4$ reaction, students are exposed to stereoselective enzymatic reductions employing Baker's yeast and a stereoselective sodium borohydride reduction using L-tartaric acid (Scheme 6.11). This experiment allows for the discussion of a number of green chemistry principles during a single laboratory experiment. On writing their laboratory reports, students consider the following green methodologies less hazardous reagents and solvents, biological reagents, catalysis, atom economy, and energy reduction.

6.2.3 Halogenations

Along with oxidations and reductions, halogenation reactions are also commonly taught in an introductory organic chemistry course. Such reactions (particularly

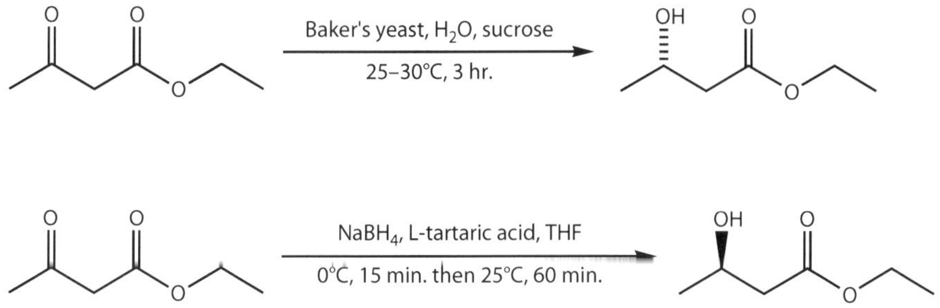

SCHEME 6.11 Stereoselective reductions of ethyl acetoacetate.

alkene brominations) are routinely showcased in both lecture and laboratory portions of such an offering, and thus a number of greener halogenations have been published in the pedagogical literature. Some of these modifications utilize bleach as a more environmentally friendly reagent (similar to those previously discussed for greener oxidation reactions). However, other reagents are known to act just as effectively in an undergraduate setting.

6.2.3.1 Halogenations Using Bleach

Most of the greener halogenation reactions involving bleach are electrophilic aromatic substitutions. Gilbertson et al. have proposed the use of sodium iodide and 6% sodium hypochlorite as a successful iodination reagent[40] (Scheme 6.12). Bleach is a less hazardous oxidizing agent than the traditional HNO_3, and sodium iodide is a much less hazardous reagent than molecular iodine. This method also uses ethanol, a fairly benign and renewable solvent. Doxsee and Hutchison have reported that this reaction proceeds very rapidly (twenty minutes) for the iodination of vanillin and results in a high percentage yield.[41] The same authors use the products of such iodination reactions as starting materials for an alkyne coupling reaction, which is run later in their course at the University of Oregon.[42] Eby and Deal have described a guided-inquiry version of the vanillin iodination, using salicylamide as the starting material.[43] Iodination of salicylamide occurs smoothly (five minutes), and provides an excellent opportunity for students to probe directing effects of the amide and hydroxyl groups toward further electrophilic substitution (Scheme 6.13).

As mentioned in Sections 6.2.1.1 and 4.3.3, Ballard has designed an experiment that probes the pH dependence of aromatic ketone oxidation with bleach. Under

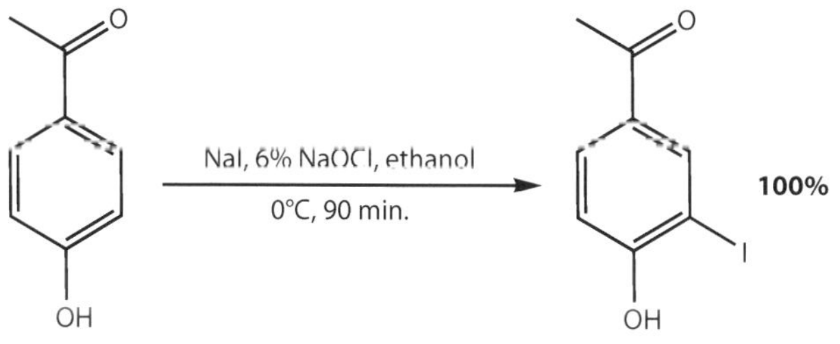

SCHEME 6.12 Electrophilic aromatic iodination of 4-hydroxyacetophenone.

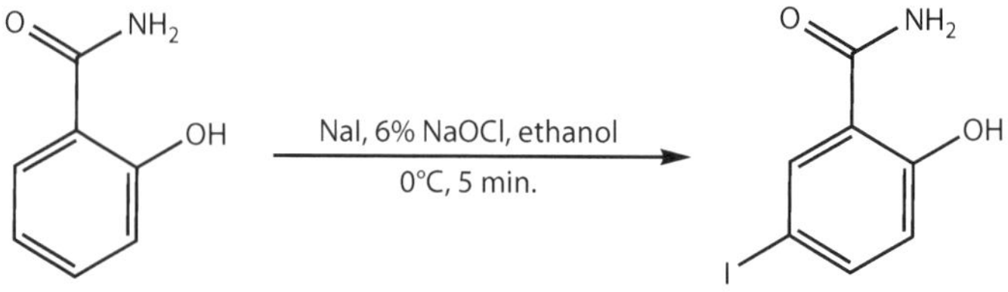

SCHEME 6.13 Electrophilic aromatic iodination of salicylamide.

acidic conditions, the aromatic ring undergoes an electrophilic chlorination reaction.[9] Bleach can also be used as an alkene chlorinating reagent. Moroz et al. used a combination of aqueous 1 M hydrochloric acid and bleach in dichloromethane to produce 1,2-dichloro-1,1,2,2-tetraphenylethane in nearly quantitative yield[44] (Scheme 6.14). While the reagents used in this procedure are not ideal from a green chemistry point of view, they are preferable to the usual choice of chlorine dissolved in dichloromethane.

Thus far, bleach has been described as a useful green reagent for a variety of organic transformations in the undergraduate organic laboratory. These have included oxidation of secondary alcohols, oxidation of alkyl aryl ketones, electrophilic aromatic halogenations, and chlorination of an alkene. In addition to these reactions, household bleach has also been combined with sodium hydroxide to initiate a Hofmann rearrangement of 3-nitrobenzamide[45] (Scheme 6.15). Molecular bromine and NaOH have been historically used to effect a Hofmann rearrangement. This greener procedure eradicates the use of hazardous bromine and utilizes dilute sodium hydroxide. These reagents also eliminate the competing hydrolysis to a carboxylic acid that is common during the traditional Hofmann rearrangement. Therefore, this version of the Hofmann rearrangement exemplifies a variety of green chemistry principles: preventing waste, maximizing experimental atom economy, and reducing solvent.

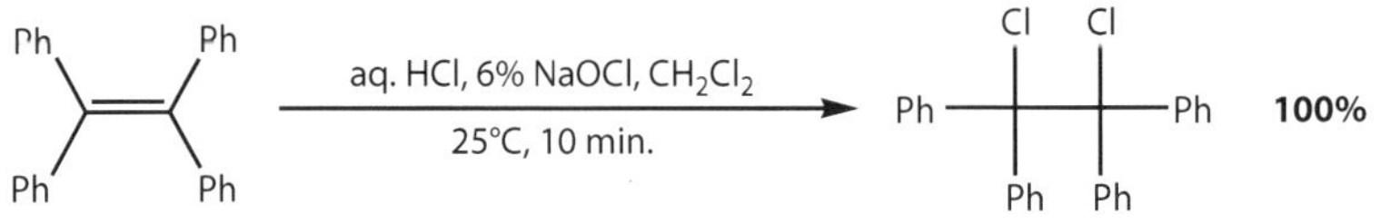

SCHEME 6.14 Chlorination of an alkene with bleach.

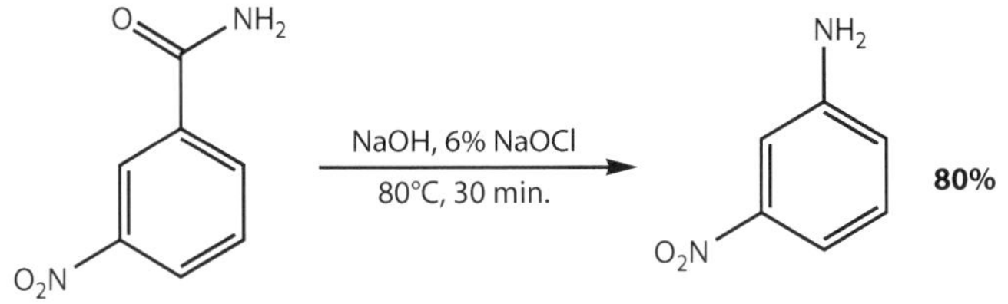

SCHEME 6.15 Hofmann rearrangement using bleach.

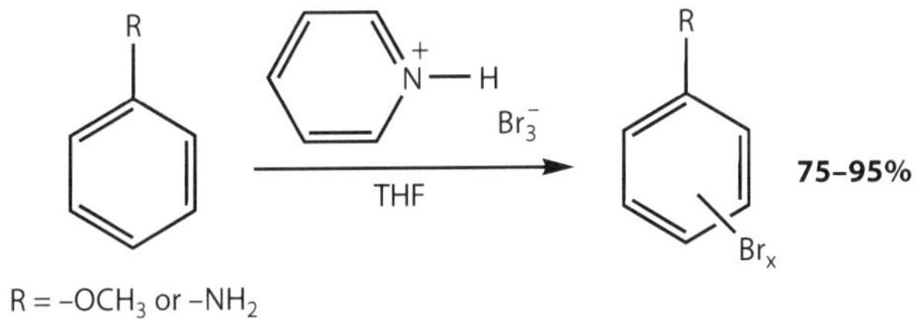

SCHEME 6.16 Electrophilic aromatic bromination using pyridinium tribromide.

6.2.3.2 Halogenations Using Nonbleach Reagents

A number of greener reagents have been developed and implemented in the undergraduate organic laboratory to accomplish the bromination of aromatic rings and alkenes. The versatility of pyridinium tribromide to effect the electrophilic aromatic bromination of a variety of amines and ethers has been reported[46] (Scheme 6.16). This procedure is selective for monosubstitution with aniline and fourteen aniline derivatives. The methodology is also applicable for the monobromination of anisole and six anisole derivatives. The aniline and aniline derivative reactions are fairly rapid (forty-five-minute reaction times). However, as expected, the reaction times for anisole and its derivatives are longer (one to twenty-five hours). Therefore, the aniline compounds are much more reasonable starting materials in an undergraduate laboratory environment. As described by Djerassi and Scholz,[47] pyridinium tribromide provides an in situ preparation of molecular bromine. The equilibrium created between pyridinium tribromide and pyridinium bromide/bromine controls the amount of Br_2 produced during the course of the reaction, and greatly decreases the hazards associated with this halogenation. In addition, aromatic brominations using pyridinium tribromide generally employ a less hazardous solvent (tetrahydrofuran or ethanol compared with the more commonly used dichloromethane). McKenzie et al. have reported the use of pyridinium tribromide dissolved in ethanol as an effective brominating agent for alkenes.[48] This approach and a related one using HBr and hydrogen peroxide to form molecular bromine in situ are discussed in more detail in Section 5.7.2.

Chandrasekhar and Dragojlovic have developed a related strategy toward bromination with HBr/H_2O_2.[49] Their modification replaces hydrobromic acid with sodium bromide and dilute hydrochloric acid. This bromination is additionally performed in an aqueous solvent rather than ethanol. However, more energy is needed in order to effect this transformation. Despite the increased energy requirements, using NaBr and dilute HCl replaces a corrosive reagent, and also eliminates ethanol as a flammable solvent. Use of ethanol would, on the other hand, be necessary if one wanted to extend these reaction conditions to other starting materials. A very interesting result of this work is that the bromination reaction is followed by an elimination to yield an aromatic product, a result that students are most likely not expecting (Scheme 6.17). Thus, this experiment can be used to develop an inquiry-based approach involving product deduction via proton nuclear magnetic resonance (NMR) spectroscopy.

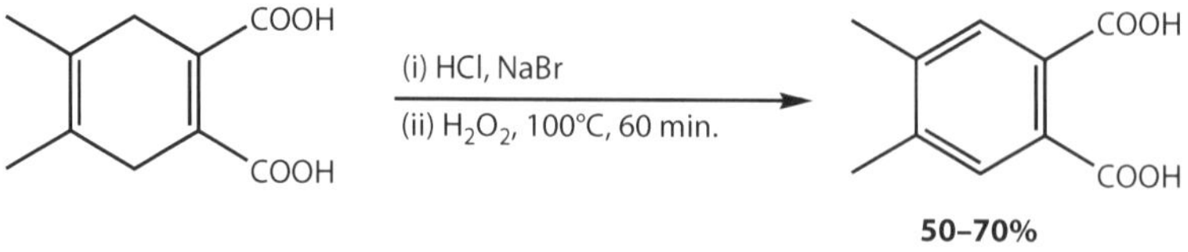

SCHEME 6.17 Alkene bromination and elimination to form an aromatic product.

6.2.4 Alternative, Greener Nonmetal Catalysts

6.2.4.1 Alcohol Dehydrations

One substitute for the strong acid catalyst usually employed in alcohol dehydration reactions is a weaker, nonnucleophilic acid such as phosphoric acid. The dehydration of various alcohols has appeared in a number of organic laboratory textbooks, utilizing either sulfuric or phosphoric acid.[50–53] However, the use of phosphoric acid as a greener reagent compared to sulfuric acid was first highlighted by Doxsee and Hutchison.[54] Their experiment employs phosphoric acid to eliminate water from cyclohexanol, yielding cyclohexene (Scheme 6.18). Using phosphoric acid as the catalyst in an alcohol dehydration exemplifies two green chemistry principles. First, phosphoric acid is a weaker acid than sulfuric acid, and therefore a less corrosive reagent. Second, the aqueous phosphoric acid also serves as the reaction solvent, thus eliminating use of organic solvents. As a practical extension to these principles, this experiment can be used to introduce both simple and fractional distillations. Cyclohexanol can alternatively be readily dehydrated using Nafion® NR50, a recyclable solid phase acidic resin[55] (Section 7.5). Finally, the cyclohexene product can act as the starting compound in the environmentally benign synthesis of adipic acid, thus reducing waste (Section 7.5).

6.2.4.2 Aromatic Alkylation

Use of graphite as an alternative catalyst to aluminum trichloride for the effective alkylation of aromatic compounds has been described in the research literature.[56] Lengthy reaction times and toxic alkylating agents hinder the general procedure from being useful in an undergraduate teaching laboratory. However, Sereda and Rajpara have recently adapted one of the graphite-catalyzed aromatic alkylation reactions for use as a teaching tool[57] (Scheme 6.19). While *p*-xylene and 2-bromobutane are flammable, neither is particularly toxic. The reaction effectively illustrates use of a renewable, safe catalyst with a reasonable reaction time and good percentage yields. The experiment can also be integrated with one- or two-dimensional proton NMR product

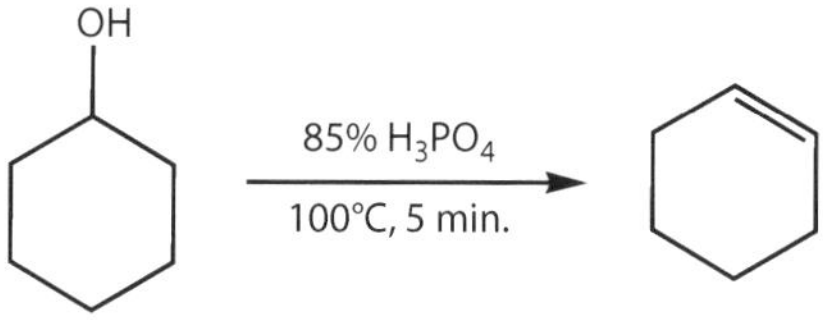

SCHEME 6.18 Dehydration of cyclohexanol.

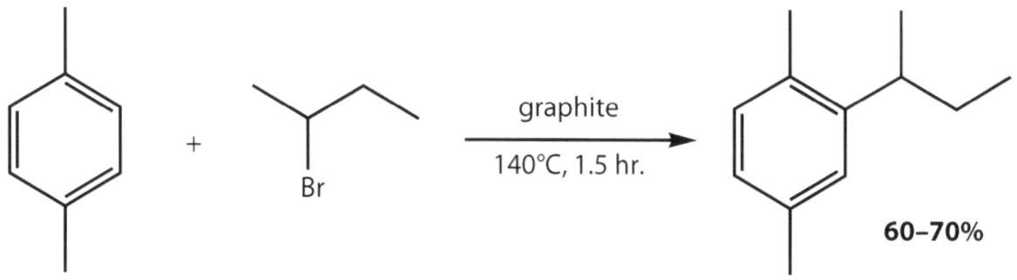

SCHEME 6.19 Graphite-catalyzed xylene alkylation.

analysis. The product correlation spectroscopy (COSY) spectrum serves as a simple example to introduce the use of proton coupling to determine organic structures.

6.2.4.3 Zeolite Multicomponent Reaction Catalysis

Zeolites are microporous, aluminosilicate materials and are known to be excellent catalysts for a range of organic functional group transformations.[58] Wetter et al. have designed an undergraduate experiment using H-β zeolite to catalyze a multicomponent reaction (MCR)[59] (Scheme 6.20). Amazingly, four reagents are combined into a single product in this instance. Multicomponent reactions are ideal from a green chemistry point of view, as they drastically reduce the amount of waste generated from solvents, purification methods, and side reactions, as well as exhibiting impressive atom efficiencies (Section 3.5.3.4).

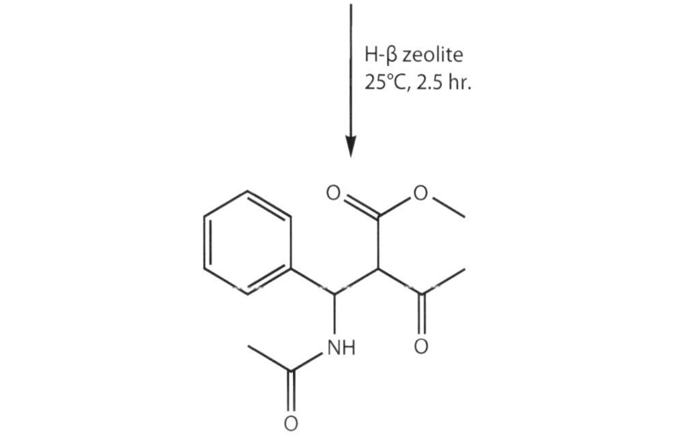

SCHEME 6.20 A zeolite-catalyzed multicomponent reaction.

6.2.5 Organocatalysis

Certain organic molecules, especially amines, are known to act as effective catalysts for a variety of transformations.[60] Two specific green chemistry experiments that showcase amine organocatalysis are covered in this chapter section.

6.2.5.1 Asymmetric Aldol Condensation

Due to its versatility, biological relevance, and significance as a carbon-carbon bond-forming process, the aldol condensation is one of the most important reactions taught in an introductory organic class. The impact of this reaction is evident by its prominence in contemporary organic laboratory textbooks.[61] The aldol condensation has also received much attention from the green chemistry community, and a number of solventless condensation procedures have been developed (Sections 3.5.2.1 and 7.3.4). Significantly, most of these involve the use of aromatic aldehydes and aromatic ketones as reactants. In these cases, the initially formed β-hydroxycarbonyl products undergo ready elimination to yield conjugated α,β-unsaturated carbonyls. Synthesis of a β-hydroxycarbonyl aldol product will often generate a new stereocenter, and facilitates discussion of potential enantioselective reactivity in the undergraduate curriculum. To that end, Bennett has developed an organocatalyzed aldol condensation between acetone and isobutyraldehyde, using L-proline as the catalyst[62] (Scheme 6.21). The reaction proceeds with reasonable yields and student enantiomeric excess (ee) values of ~70%. This enantioselective aldol condensation demonstrates a number of green values that include use of a reagent (acetone) as the solvent, use of a natural product as a catalyst, a lack of required product purification, and ambient temperature and pressure as reaction conditions. While the reaction has these green advantages, it also demonstrates the continual need for improvement. The procedure uses a large amount of acetone in order to suppress side reactions, but the excess acetone results in a low overall experimental atom economy. Additionally, acetone and ether (used to extract the product) are volatile and flammable liquids, and the reaction proceeds slowly (at least two days are needed to isolate the product).

6.2.5.2 Warfarin Synthesis

A second example of a green, enantioselective synthesis appropriate for the undergraduate laboratory concerns the one-step synthesis of the blood anticoagulant known as Warfarin® (Scheme 6.22). Wong et al.[63] have recently designed an experiment based on a simple and green procedure published in the research literature.[64] Students

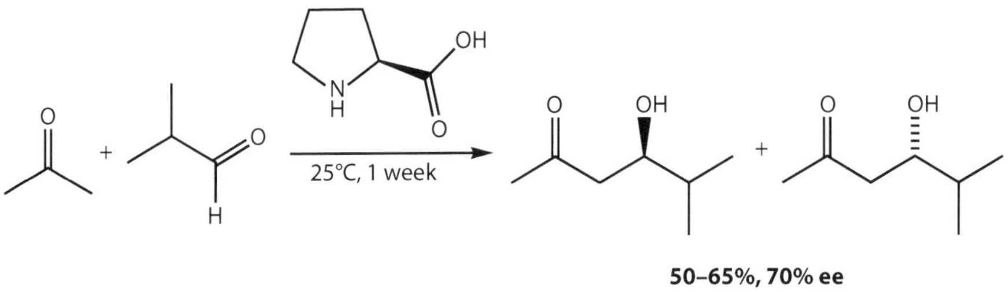

SCHEME 6.21 Asymmetric aldol condensation catalyzed by L-proline.

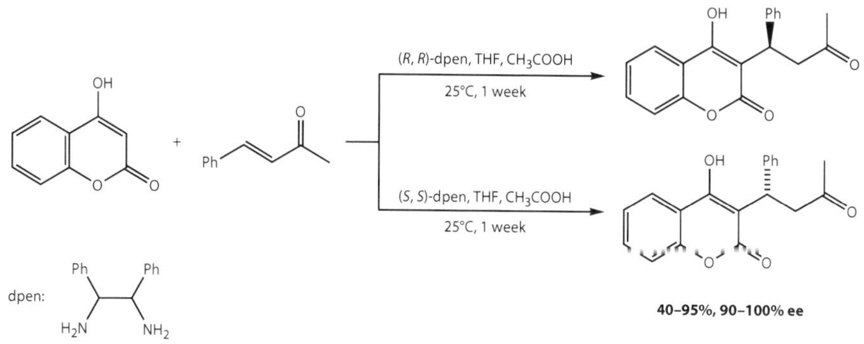

SCHEME 6.22 Warfarin synthesis via organocatalysis.

become introduced to enantioselective, organocatalyzed reactions, and the nature of optically active drugs. Reaction yields are very respectable, and the measured ee values are 90–100% after recrystallization. Although all of the reagents are irritants, and acetic acid and tetrahydrofuran (THF) are flammable, the reaction demonstrates a number of green principles. The reaction is performed at ambient temperature and pressure, and the chiral auxiliary ((*R*,*R*)- or (*S*,*S*)-1,2-diphenylethylenediamine) is used in a catalytic amount and can be recovered. If the catalyst is retrieved, the intrinsic and experimental atom economies are very high, at 100 and 92%, respectively. However, the reaction does require a solvent, and thus generates some organic waste.

6.2.6 Metal Catalysis

One of the richest sources of catalytic reactions involves the use of metals as catalysts. Metal catalysis has historically been downplayed in the undergraduate organic curriculum, often due to the cost of required reagents. However, many experiments have been designed in recent years that use relatively inexpensive, recyclable metals or metallic compounds. Therefore, curricular inclusion of metal-catalyzed reactions serves the dual purpose of demonstrating green chemistry principles as well as introducing mechanistic aspects of the observed reactivity. Together with the examples discussed here, the reader is referred to Sections 4.3.1 and 7.5, where several instances of metal catalyst recycling are outlined. These include recycling of sodium tungstate (alkene oxidative cleavage) and palladium on carbon (Suzuki reactions).

6.2.6.1 Alkyne Coupling Reactions

Some early examples of green, pedagogical metal-catalyzed reactions involved the coupling of alkynes using transition metal catalysts. Doxsee and Hutchison include one such approach in their laboratory textbook *Green Organic Chemistry—Strategies, Tools, and Laboratory Experiments*.[65] Here, an adaptation of the Glaser-Eglinton-Hay coupling is presented, which is commonly used to produce fungal antibiotics.[66] The customary Glaser-Eglinton-Hay coupling is accomplished by combining a terminal alkyne with copper(I) chloride in pyridine solvent. In the greener experiment, students oxidatively couple 1-ethynylcyclohexanol in the presence of CuCl,

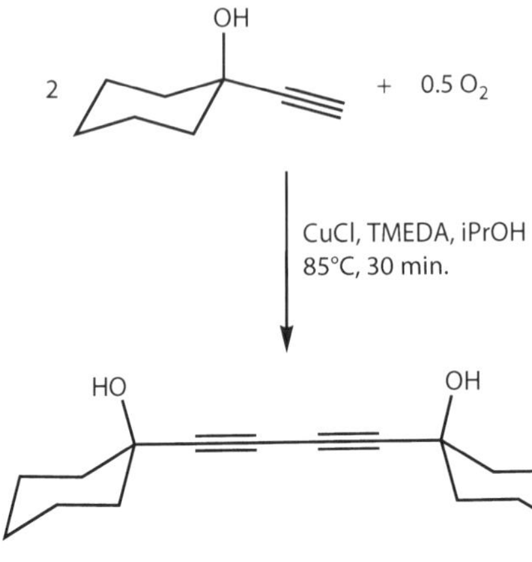

SCHEME 6.23 A Glaser-Eglinton-Hay coupling.

tetramethylethylenediamine (TMEDA), and isopropanol under aerobic conditions (Scheme 6.23). Pyridine from the original procedure is replaced with isopropanol and TMEDA. Isopropanol (which is significantly less hazardous than pyridine) serves as the reaction solvent. In comparison, TMEDA acts as a ligand that facilitates formation of the copper complex believed to be essential to the mechanism of the reaction. TMEDA is also much safer than pyridine and, because it is not serving as a solvent, can be used in a favorable catalytic amount.

As a second example of alkyne coupling, the use of a palladium catalyst to accomplish the reaction of an alkyne and an aromatic iodide has been disclosed.[67] The coupling is rapidly followed by an intramolecular alkyne addition to yield the observed benzofuran product. This reaction is an excellent example of an aqueous organometallic catalytic process that occurs under ambient conditions, and is covered in more detail in Section 4.3.1.

6.2.6.2 Metathesis Reactions

Metathesis reactions are very important organometallic conversions that have also not routinely appeared in the introductory organic laboratory. In 1999, France and Uffelman published a ring-opening metathesis polymerization (ROMP) reaction for the undergraduate inorganic or polymer laboratory.[68] A further ROMP reaction utilizing K_2RuCl_5 as a catalyst under aqueous conditions is discussed in Section 4.3.1. After Chauvin, Grubbs, and Schrock were jointly awarded the 2005 Chemistry Nobel Prize,[69] several new organic experiments were published that introduce students to metathesis reactions.[70–73] Taber and Frankowski developed an experiment that uses the commercially available Grubbs second-generation catalyst to cross-metathesize eugenol and *cis*-2-butene-1,4-diol[70] (Scheme 6.24). In addition to exemplifying catalysis, this reaction showcases some trade-offs that often exist when trying to make a process greener. As an example, the catalyst in this reaction does not require an inert atmosphere or distilled solvents, but petroleum ether is not an ideal solvent since it is flammable and obtained from petroleum. This experiment can also be used to demonstrate the synthesis of a pharmaceutically important product.[74]

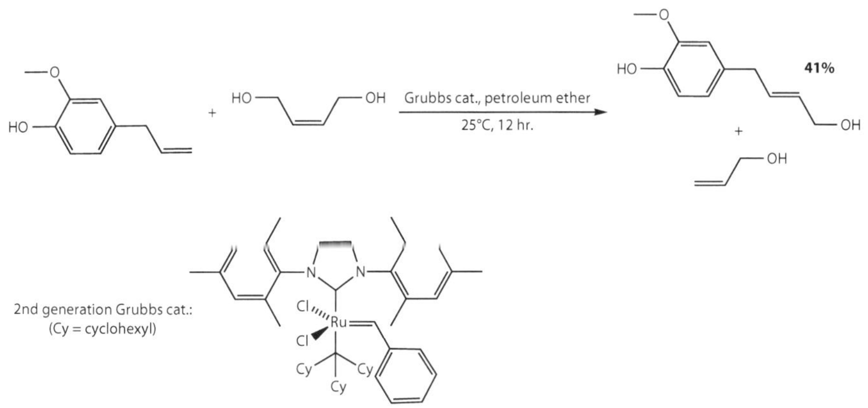

SCHEME 6.24 Cross-metathesis of eugenol and *cis*-2-butene-1,4-diol.

A related reaction that has been designed for the undergraduate laboratory is the ring-closing metathesis of diethyl diallylmalonate[71] (Scheme 6.25). In this experiment, students synthesize the ruthenium catalyst themselves, in addition to performing the metathesis reaction. There are options to synthesize one or two ruthenium catalysts (Scheme 6.26). Both ruthenium catalysts can be easily prepared in high yields (75 and 60%), but catalyst 1a takes over three hours to generate, and 1b takes an additional two or more hours. These timeframes may limit the utility of this experiment in an undergraduate laboratory venue. However, the catalyst could be prepared by the instructor for the whole class if time is an issue. While many of the solvents used in this metathesis are not particularly green (e.g., dichloromethane and THF), the metathesis reaction itself has some green features of merit, other than being catalytic. Like some previous examples, only ambient temperatures and pressures are required. The reaction is also high yielding and has a fairly high atom economy, if the catalyst is recovered.

6.2.6.3 Epoxide Ring Openings

While epoxides are among the most reactive and useful organic functional groups,[75,76] students are rarely exposed to them in the undergraduate organic laboratory. One of the many synthetically important reactions of epoxides is their rearrangement to carbonyl compounds using Lewis acids.[77–82] Due to the corrosive and toxic nature of typical Lewis acids employed in the reaction, this type of rearrangement has not

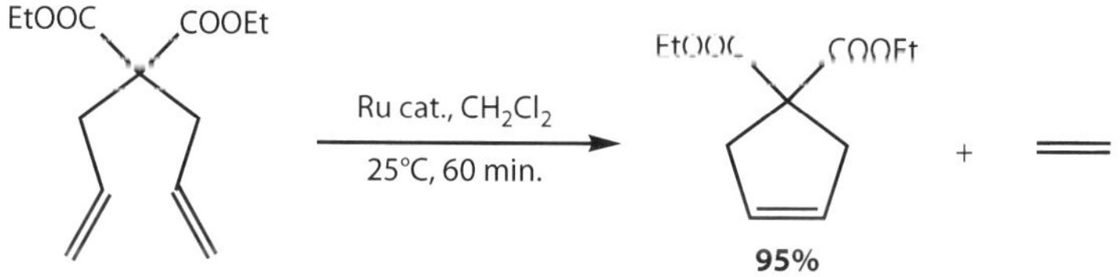

SCHEME 6.25 Ruthenium-catalyzed ring-closing metathesis.

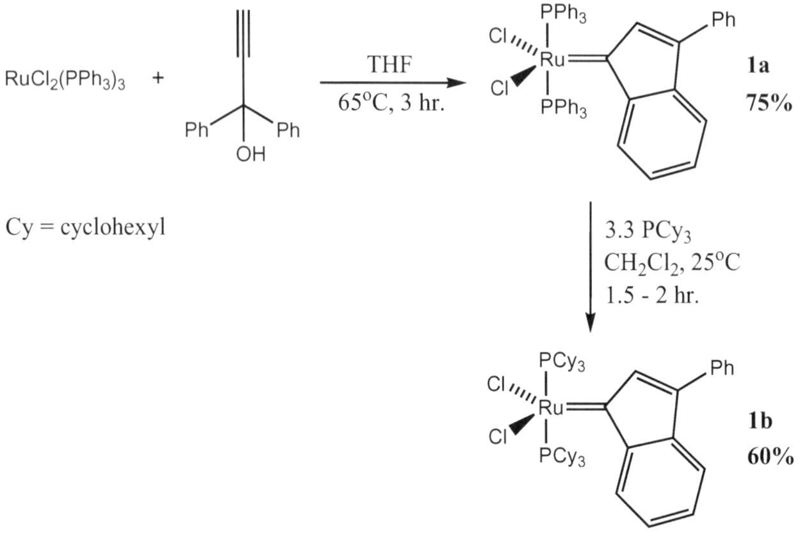

SCHEME 6.26 Synthesis of two ruthenium catalysts for ring-closing metathesis.

found favor in a teaching environment. To address this and expose undergraduates to epoxide rearrangements, Christensen et al. have developed a discovery-oriented experiment using Bi(III)-based catalysts.[83] Despite being a heavy metal, most bismuth compounds are surprisingly nontoxic and noncorrosive. The procedure involves the rapid reaction of *trans*-stilbene oxide with a variety of metal triflate catalysts, in a range of solvents. Student results indicate that bismuth(III) triflate is an effective catalyst in this reaction, and selectively yields the aldehyde product, as shown in Scheme 6.27.

6.2.7 Biocatalysis

In addition to the biocatalyzed reductions discussed in Section 6.2.2.1, there are several other biocatalytic reactions modified for student experimental purposes. As well as the examples presented here, many biocatalytic transformations have been developed for industrial and synthetic applications.[84–86] These examples could be used as instructive case studies in a lecture course, or potentially converted into novel laboratory experiments.

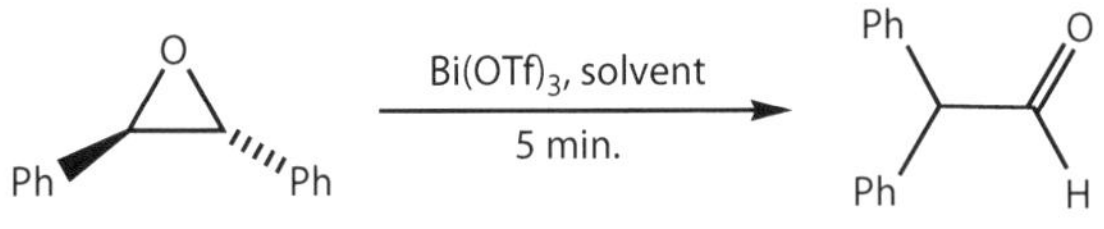

SCHEME 6.27 Ring opening of epoxides via metal triflate catalysis.

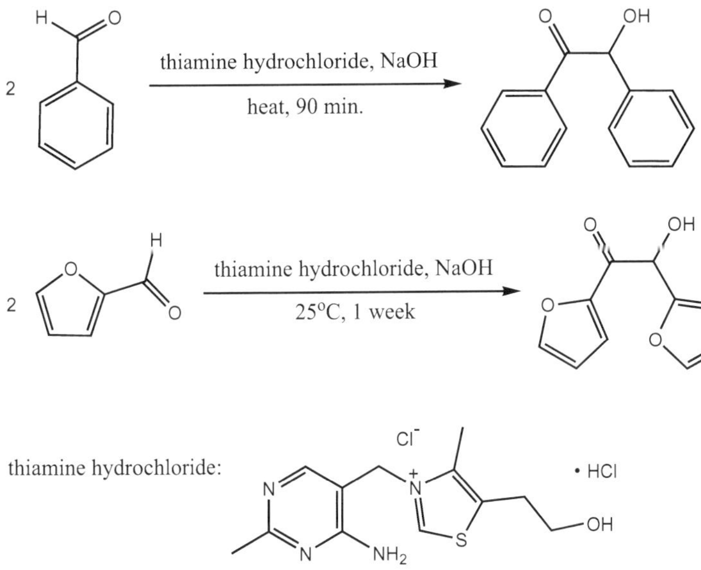

2a

2b

SCHEME 6.28 Thiamine hydrochoride-catalyzed benzoin condensations.

6.2.7.1 Thiamine Hydrochloride-Catalyzed Benzoin Condensation

The benzoin condensation was serendipitously discovered by Wöhler and von Liebig in the nineteenth century. The first example involved condensation of two benzaldehyde molecules in the presence of cyanide to form benzoin (Scheme 6.28, structure 2a). The benzoin condensation represents an efficient method to form carbon-carbon bonds, and has been used to synthesize a wide variety of compounds since its discovery. Unfortunately, use of cyanide as a catalyst presents a serious problem if an aqueous acidic reaction workup is employed. In 1958, Breslow reported the use of thiamine hydrochloride (vitamin B1 hydrochloride) as a catalyst for the benzoin condensation.[87] Application of thiamine hydrochloride as a catalyst in a synthetic organic reaction represents one of the earliest reports of biocatalysis. Warner introduced the benzoin condensation to the undergraduate curriculum, using benzaldehyde as the substrate and thiamine hydrochloride as the catalyst (Scheme 6.28).[88] Following this, Doxsee and Hutchison adapted the benzoin condensation of furfural to suit the undergraduate laboratory environment (Scheme 6.28).[89]

6.2.7.2 Oxidative Coupling

While enzyme-catalyzed carbonyl reductions have become quite popular as teaching tools, there are relatively few examples of other enzyme-catalyzed reactions in the pedagogical literature. Very recently, Nishimura et al. incorporated the horseradish peroxidase-catalyzed dimerization of vanillin and apocynin into the undergraduate practical curriculum.[90] This oxidative dimerization is very rapid under ambient conditions, only requires 3% hydrogen peroxide, and uses water as the solvent

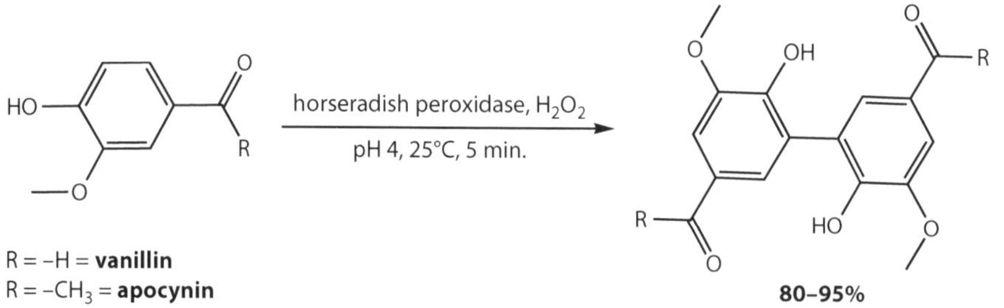

SCHEME 6.29 Oxidative coupling of vanillin and apocynin.

(Scheme 6.29). Traditional preparations of divanillin and diapocynin require stoichiometric inorganic oxidants.[91,92] Therefore, this reaction is an excellent example of an environmentally benign transformation.

6.2.7.3 Enzymatic Resolution

The need for enantiomerically pure chiral compounds in the pharmaceutical industry has driven the development of a variety of methods to obtain single enantiomeric products. Generation of pure enantiomers can be accomplished through asymmetric synthesis, or via enantiomeric resolution. The latter technique has been accomplished through several different strategies: (1) crystallization,[93–94] (2) chiral chromatography,[95] (3) kinetic resolution,[96–98] (4) extraction and membrane-based processes,[99] (5) sublimation,[100] and (6) distillation.[101] Enzymatic resolution has proven to be a particularly robust process toward enantiomerically pure molecules. Monteiro et al. have developed an undergraduate enzymatic resolution using a lipase enzyme.[102] The experiment resolves a racemic mixture of 1-phenylethanol using *Candida antarctica* lipase B (CAL B) and ethyl myristate at 40°C under vacuum for one day (Scheme 6.30). Once the unreacted alcohol is removed by distillation, the transesterification reaction is reversed by adding excess ethanol to the CAL B/esterified alcohol mixture. The reaction is allowed to proceed for one day, and the alcohol is again extracted by distillation. Several green chemistry principles are illustrated in this reaction. The use of an enzyme as the resolving agent showcases a renewable catalyst. Employing a fatty acid ester as the alkylating agent eliminates the need for organic solvents, and provides an additional sustainability advantage. Unfortunately, the reaction must be performed under vacuum and with mild heat. However, these minor issues are outweighed by the nature of the resolving agent, and the lack of organic solvents used in the resolution.

6.2.8 Solid-Supported Reactions

6.2.8.1 Oxidations

Several examples of a greener oxidation of 9-fluorenol using polymer-supported CrO_3 (Scheme 6.31) have appeared in the primary literature and textbooks during the past thirty years.[103–106] These procedures use an Amberlyst® resin to support the CrO_3 oxidizing agent. The reactions routinely involve refluxing the alcohol and

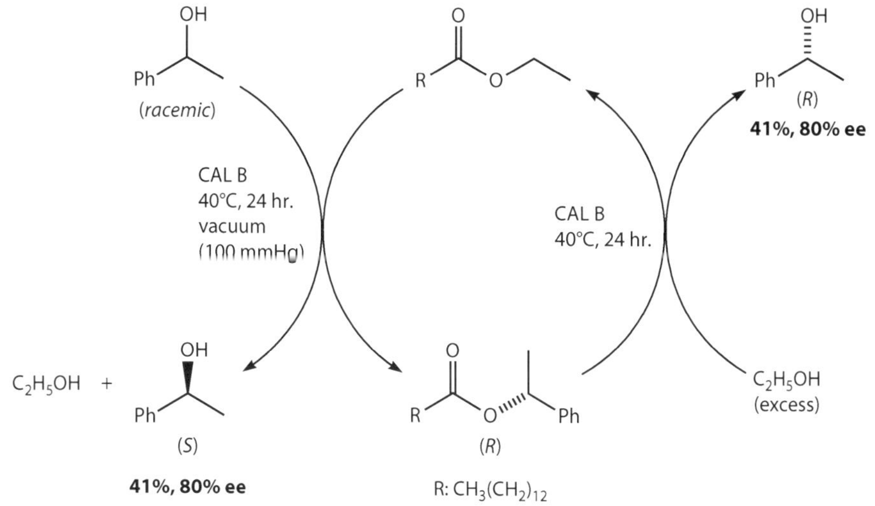

SCHEME 6.30 Enzymatic resolution of racemic 1-phenylethanol.

polymer-supported CrO_3 in toluene for one hour. While Cainelli et al. used this oxidizing agent for a variety of primary and secondary alcohols, the required reaction times for some of the substrates limit their usefulness in a three-hour laboratory period.[103] However, several of the alcohols studied required only one hour of reaction time. Buglass and Waterhouse reported the successful oxidation of 9-hydroxyfluorene, diphenylmethanol, and 1-phenylethanol using a similar polymer (Amberlyst A-26)[107] (Scheme 6.31).

Crumbie has published the use of Magtrieve™ to oxidize benzoin to benzil[108] (Scheme 6.32). Magtrieve is a commercially available, heterogeneous form of chromium dioxide (rather than CrO_3) whose surface is reduced during the alcohol oxidation reaction. Since only the surface of the material is reduced, the oxidant can easily be removed from the reaction mixture by magnetic separation and regenerated by simple heating. These modifications illustrate several green chemistry principles and provide concrete advantages over other oxidation methods. In both the polymer-supported CrO_3 and Magtrieve oxidations, the solid support can be reused and regenerated, therefore demonstrating the recycling of materials. Since the reduced form of CrO_2 remains on the surface in both methods, there is

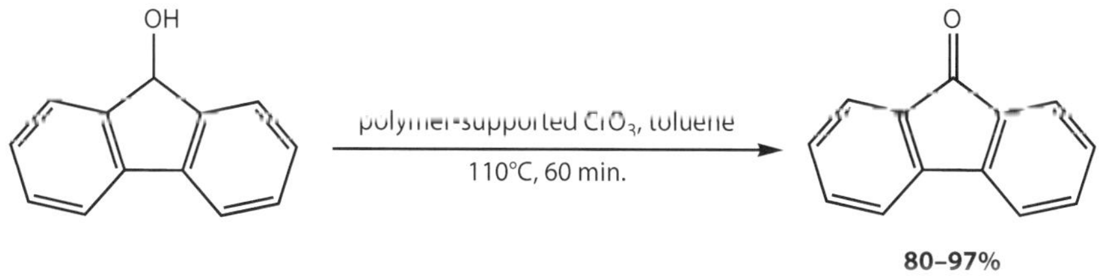

SCHEME 6.31 Polymer-supported oxidation of 9-fluorenol.

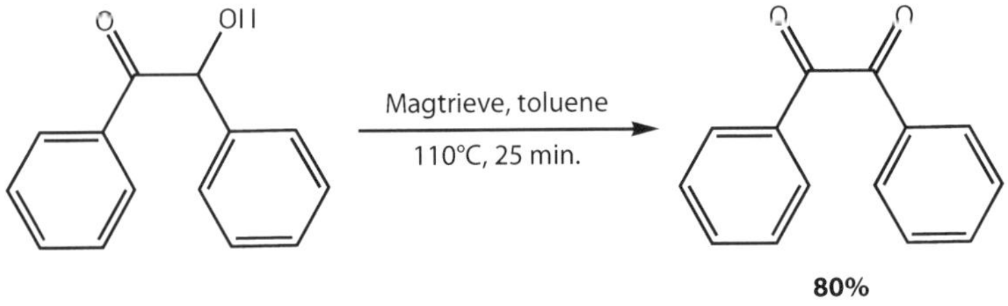

SCHEME 6.32 Magtrieve oxidation of benzoin.

no chromium waste or aqueous waste (from the workup) generated. The use of solid supports allows for simple and easy purification that does not require the use of additional solvents. However, the reaction itself is typically performed in chlorinated solvents or toluene, neither of which is an ideal solvent from a green chemistry perspective.

Crouch et al. have also outlined a benzoin oxidation procedure utilizing alumina-supported manganese dioxide,[109] which can be conveniently extended to oxidize allylic, propargylic, and benzylic alcohols.[110,111] The reaction is performed at ambient temperature for two hours, so the energy requirements of this oxidation are reduced compared to the previously mentioned reactions (Scheme 6.33). Mn(IV) salts are additionally less toxic and less hazardous than chromium trioxide and chromate/dichromate salts. Therefore, this oxidation demonstrates the use of less hazardous reagents. In contrast, dichloromethane is used as a solvent rather than toluene. These two reactions (using either Magtrieve or MnO_2 as the benzoin oxidant) could serve as an interesting class discussion in terms of comparing and contrasting the relative "greenness" of each method.

The preparation of a polymer-supported hypervalent iodine reagent (2-iodoxybenzoic acid (IBX)) in an undergraduate laboratory has recently been described,[112] which is based on a research literature procedure.[113] The synthesis requires five steps with four intermediate compounds generated, and is designed to take place over a number of weeks. The reactions needed to synthesize polymer-bound IBX are diazotization, aromatic iodination, Fischer esterification, Williamson ether synthesis, basic ester hydrolysis, and oxidation, which are all familiar to undergraduates. Upon isolating the hypervalent iodine reagent (bound to Merrifield's resin), students select

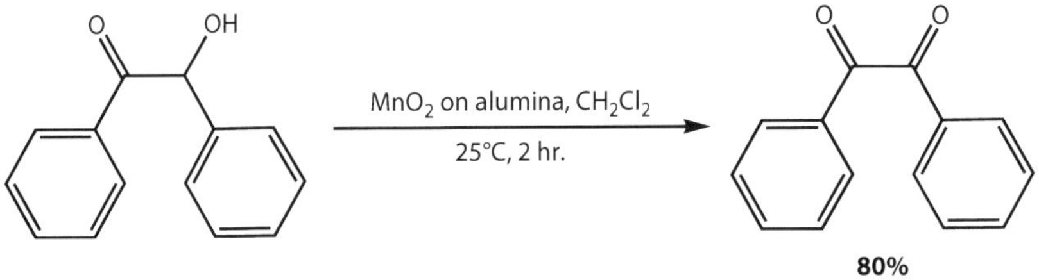

SCHEME 6.33 Manganese dioxide-immobilized oxidation of benzoin.

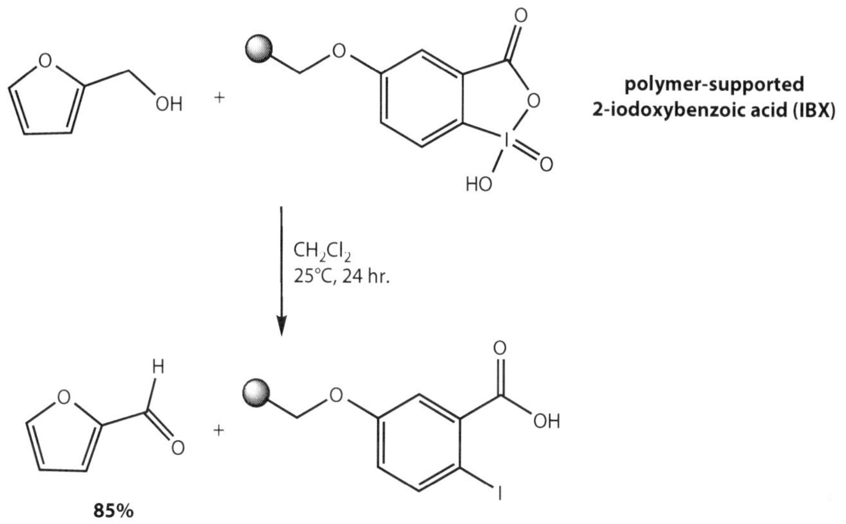

SCHEME 6.34 Polymer-supported IBX oxidation of furfuryl alcohol.

a primary, secondary, allylic, benzylic, or propargylic alcohol to oxidize. A typical reaction is shown in Scheme 6.34, where furfuryl alcohol is oxidized to fufural. The reduced polymer-supported product can be collected, dried, and reoxidized, thus highlighting reagent recycling to students. The oxidations performed also illustrate that hypervalent iodine reagents generally have lower toxicity and operate under much milder conditions than traditional heavy metal oxidizing agents.

6.2.8.2 Alkene Epoxidation

Given the synthetic utility of epoxides and their common use in industrial chemistry, experiments that introduce students to epoxide formation are desirable. The strong oxidizing behavior and explosive nature of organic peracids typically mean that epoxide synthesis is relatively uncommon in the introductory organic laboratory. Reagents such as *m*-chloroperoxybenzoic acid (MCPBA) do permit epoxide preparation from alkenes under mild conditions.[114] Unfortunately, such transformations are less than ideal from a green perspective, as they suffer from low atom economy and generally require chlorinated solvents. However, the use of a polymer-supported peroxide greatly reduces the hazards associated with organic peroxide solutions, as long as the peroxide resin remains wet and swollen. Buglass and Waterhouse reported the successful use of a variety of polymers to support peroxide reagents.[107] The polymer-supported peroxides were prepared by oxidizing dried beads of ion exchange resin in the presence of *p*-toluenesulfonic acid and hydrogen peroxide.[115] The resin is then added to tetrahydrofuran and mixed with cyclohexene (Scheme 6.35). The resin is easily removed by filtration and can be regenerated after the epoxidation has taken place.

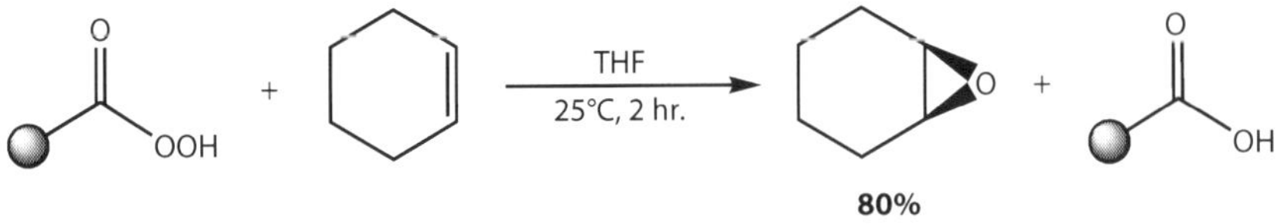

SCHEME 6.35 Polymer-supported epoxidation of cyclohexene.

6.2.8.3 Fischer Esterification

Cioffi has designed an experiment that uses activated carbon as a solid support for *p*-toluenesulfonic acid.[116] Students utilize the H^+-immobilized activated carbon to catalyze the esterification of benzoic acid with 1-hexanol (Scheme 6.36). Using a solid-supported acid catalyst reduces the hazards of using corrosive solutions of mineral acids or solid *p*-toluenesulfonic acid. The reaction also features very rapid microwave heating in place of thermal heating, thus greatly reducing the energy cost of the experiment. Finally, the reaction uses excess alcohol as the solvent. However, an important point to note is that the solid-supported catalyst is not isolated and regenerated after the reaction, thus reducing some of the advantages of solid-support reagents.

6.2.8.4 Alkene Isomerization

Another common reaction that is catalyzed by a strong acid is the isomerization of an alkene. One catalyst system used for this transformation is a homogeneous mixture of $Ni[P(OEt)_3]_4$ and sulfuric acid.[117–119] Seen has developed a catalyst system that replaces sulfuric acid with Nafion-H^+.[120] This greener approach is used to catalyze the isomerization of 1-octene to a mixture of *cis*-2-octene and *trans*-2-octene (Scheme 6.37). One should note this experiment is designed for an upper-level

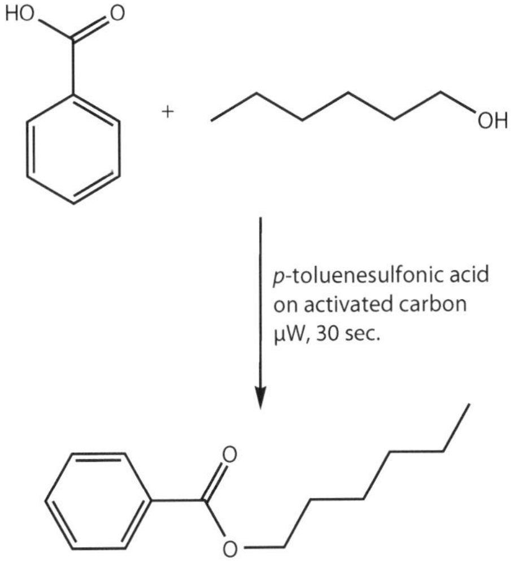

SCHEME 6.36 Esterification using an activated carbon-supported acid catalyst.

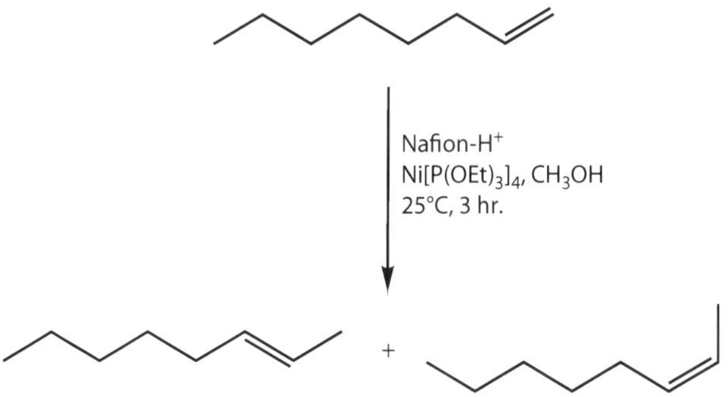

SCHEME 6.37 Alkene isomerization using a solid-supported acid catalyst.

laboratory course, as the $Ni[P(OEt)_3]_4$ is an air-sensitive transition metal complex. Vacuum distillation apparatus and inert atmospheric conditions are essential.

6.2.8.5 Ester and Phosphoester Hydrolysis

Enzymes are finding increased use in the undergraduate synthetic laboratory, as evidenced by Section 6.2.7. Unfortunately, enzymes are often expensive to purchase, and can have limited stability. One way to address the latter challenge is to immobilize the enzyme on an insoluble polymeric support. The immobilization makes the enzyme more stable to deactivation by heat, pH, organic solvents, and oxygen. As with other polymer-supported reagents, the enzyme is easily recovered and reused, thus reducing the cost somewhat. Conlon and Walt have documented the use of immobilized hydrolase enzymes to transform an ester into a carboxylic acid and a phosphoester into an alcohol[121] (Scheme 6.38). While this experiment is a biological assay that is primarily suited to a biochemistry laboratory, the reactions exploited clearly illustrate the synthetic use of immobilized enzymes, and could be readily introduced to an organic curriculum.

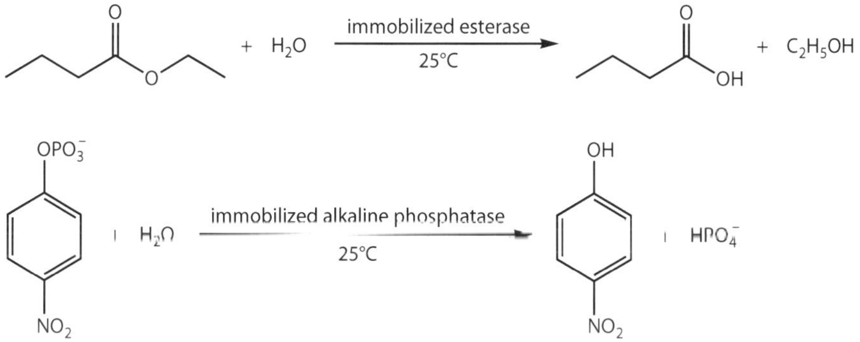

SCHEME 6.38 Enzyme-immobilized ester and phosphoester hydrolysis.

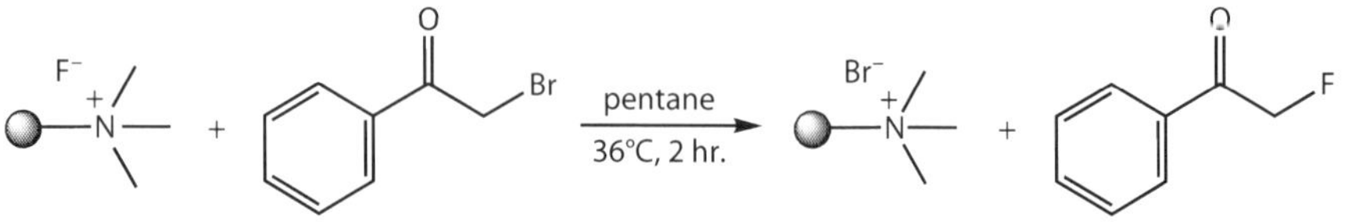

SCHEME 6.39 Solid-supported fluoride α-substitution.

6.2.8.6 Fluorination

Reactions that incorporate fluorine atoms into molecules are challenging to run in the undergraduate laboratory, despite their industrial importance. Pohl and Schwarz have designed an experiment that utilizes polymer-supported fluoride as a reagent for nucleophilic substitutions[122] (Scheme 6.39). The reported reaction converts 2-bromoacetophenone into 2-fluoroacetophenone with the fluoride ion form of Amberlyst A-25, and requires refluxing in pentane for two hours. The resin is easily removed by filtration after reaction is complete, and the pure product is obtained by simple solvent evaporation. Since the reaction proceeds via an S_N2 pathway, primary alkyl halides provide a single product, while secondary alkyl halides typically form a mixture of substitution and elimination products. The possible regeneration or reuse of the Amberlyst resin is not discussed in this article.

6.2.8.7 Amide Formation

Polymer-supported reagents can also be used to generate polymer-supported products. These compounds can be subsequently analyzed or removed from the polymer support, depending on the desired outcome. This method has found particular value in the pharmaceutical industry, where large libraries of related compounds can be quickly assembled using combinatorial methods.[123–125] Hailstone et al. have formulated an experiment that exposes students to the generation of polymer-supported products (Scheme 6.40). Here, a polystyrene-supported amine is reacted with an aromatic carboxylic acid using diisopropylcarbodiimide as the coupling agent, in the presence of 4-(dimethylamino)pyridine (DMAP) and 1-hydroxybenzotriazole.[126] While the reagents used in the experiment are highly flammable or toxic, the reactions are performed on a small-scale, thus reducing potential exposure.

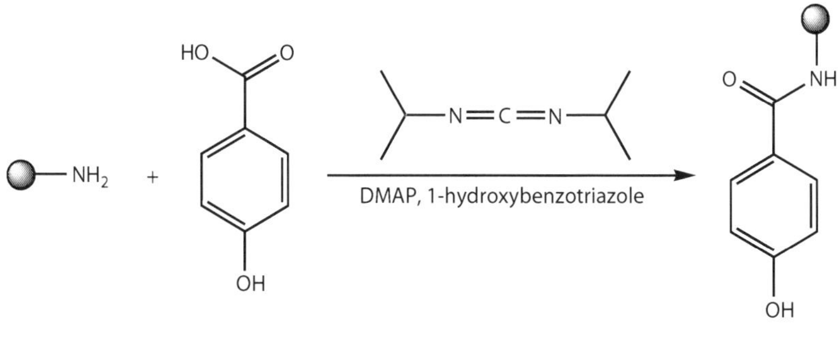

SCHEME 6.40 Solid-supported amide formation.

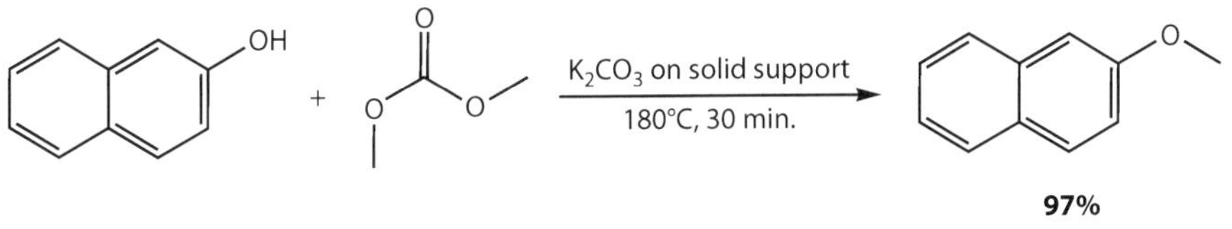

SCHEME 6.41 Methylation of 2-naphthol with dimethyl carbonate.

A related experiment has recently been published that showcases the important aspects of using polymeric supports to work with nanoparticles.[127] These are as follows: attachment of a linker unit to the support, linker modification, binding of nanoparticles to the linker, and cleavage of nanoparticles into solution. Gold nanoparticles are used in this regard, which are conveniently observed due to their purple color in solution.

6.2.8.8 Aromatic Alcohol Alkylation

Tundo et al. recently reported the use of a polyethylene glycol–K_2CO_3-supported catalyst for the alkylation of 2-naphthol using dimethyl carbonate (Scheme 6.41).[128] The experiment setup involves several components that may not typically be found in undergraduate organic laboratories, but is manageable for students if time permits and the equipment is readily available. The procedure involves pumping a mixture of 2-naphthol and dimethyl carbonate through a heating column containing the solid-supported catalyst, so that the reaction takes place in the gas phase. Once the gaseous products are eluted through the column, they are condensed and collected. Dimethyl carbonate is significantly less toxic than traditional methylating agents (methyl halides and dimethylsulfate), readily produced from renewable sources, and biodegradable.[129] Section 5.8 briefly discusses dimethyl carbonate as a potential nontoxic and environmentally friendly reaction solvent.

6.2.8.9 Porphyrin Synthesis and Metallation

Silica gel has been used as a solid support for the solvent-free formation of a tetraphenylporphyrin from benzaldehyde and pyrrole[130–132] (Scheme 6.42). Here, the liquid reactants become adsorbed on the gel surface, which may behave as a catalyst. As described by Doxsee and Hutchison, column chromatography can be employed to purify the porphyrin product, using hexanes/ethyl acetate as the elution solvent. This represents a greener choice than the usual chlorinated hydrocarbons such as dichloromethane. The reaction requires microwave irradiation for ten minutes, and thus exemplifies energy efficiency. The authors have extended this experiment and developed a greener method for porphyrin metallation.[133,134] This approach uses zinc acetate dissolved in dimethylsulfoxide (DMSO) and *N*-methylpyrrolidinone (NMP) to metallate the tetraphenylporphrin formed previously (Scheme 6.42). Halogenated solvents or *N,N*-dimethylformamide have been historically used in the metallation process. In this greener method, these hazardous solvents are replaced with more benign alternatives. DMSO has relatively low toxicity, although care must be taken to avoid skin contact. This version of the reaction also occurs at room temperature, unlike traditional porphyrin metallations, which often require heating.

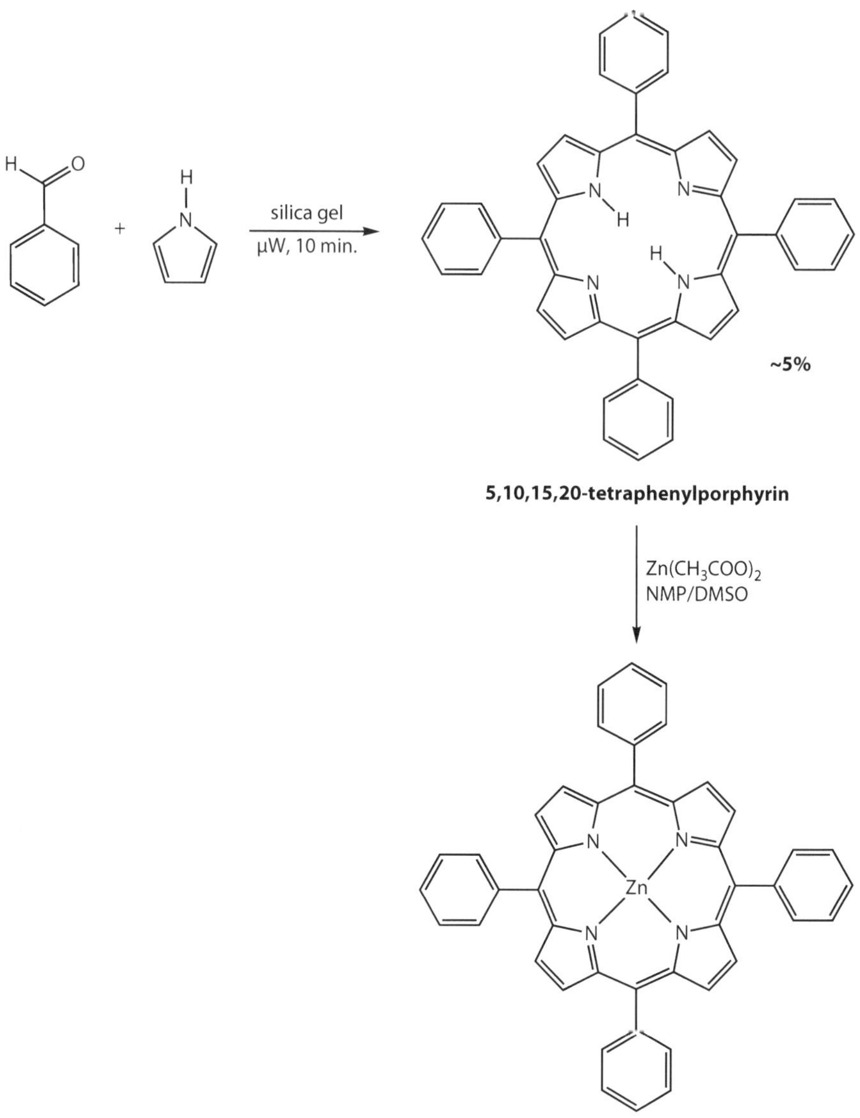

SCHEME 6.42 Formation and subsequent metallation of 5,10,15,20-tetraphenylporphyrin.

6.3 CONCLUSION

This chapter demonstrates the wide range of alternative reagents that can be used to illustrate greener organic reactivity approaches in a laboratory environment. It is clear from the broad scope of "greened" reactions that many are applicable to the introductory organic curriculum. Fundamental oxidation and reduction reactions, which are so pivotal to many functional group transformations, are areas that have received much attention. Laboratory instructors are highly encouraged to critically look at the reagents (and solvents) that their students currently handle. Serious consideration should be given to whether they are completely appropriate within the context of promoting sustainability. It may be the case that greener reagents are already being used, but that this fact is not being emphasized enough to students. A comparison between more classical and contemporary synthetic approaches is often instructive in this regard. A central aspect of green chemistry is informed decision making, and if students can learn to do this in terms of reagent selection, significant progress will have been made in their chemical education.

REFERENCES

1. Stevens, R. V., Chapman, K. T., Weller, H. N. *J. Org. Chem.* 1980, 45, 2030–2032.
2. Zuczek, N. M., Furth, P. S. *J. Chem. Educ.* 1981, 58, 824.
3. Kauffman, J. M., McKee, J. R. *J. Chem. Educ.* 1982, 59, 862.
4. Perkins, R. A., Chau, F. *J. Chem. Educ.* 1982, 59, 981.
5. Mohrig, J. R., Nienhuis, D. M., Linck, C. F., Van Zoeren, C., Fox, B. G., Mahaffy, P. G. *J. Chem. Educ.* 1985, 62, 519–521.
6. dos Santos, A. P. B., Gonçalves, I. R. C., Pais, K. C., Martinez, S. T., Lachter, E. R., Pinto, A. C. *Quim. Nova* 2009, 32, 1667–1669.
7. Straub, T. S. *J. Chem. Educ.* 1991, 68, 1048–1049.
8. Blunt, S. B., Hoffman, V. F. *Chem. Educator* 2004, 9, 370–373.
9. Ballard, C. E. *J. Chem. Educ.* 2010, 87, 190–193.
10. Lehman, J. W. Minilab 32: Air Oxidation of Fluorene to 9-Fluorenone. In *Operational Organic Chemistry*. Prentice-Hall, Upper Saddle River, NJ, 2009, 518–519.
11. Stocksdale, M. G., Fletcher, S. E. S., Henry, I., Ogren, P. J., Berg, M. A. G., Pointer, R. D., Benson, B. W. *J. Chem. Educ.* 2004, 81, 388–390.
12. Gandhari, R., Maddukuri, P. P., Vinod, T. K. *J. Chem. Educ.* 2007, 84, 852–854.
13. North, M. *J. Chem. Educ.* 1998, 75, 630–631.
14. Patterson, J., Sigurdsson, S. T. *J. Chem. Educ.* 2005, 82, 1049–1050.
15. Pohl, N., Clague, A., Schwarz, K. *J. Chem. Educ.* 2002, 79, 727–728.
16. Jayasinghe, L. Y., Kodituwakku, D., Smallridge, A. J., Trewhella, M. A. *Bull. Chem. Soc. Jpn.* 1994, 67, 2528–2531.
17. Rotthaus, O., Krüger, D., Demuth, M., Schaffner, K. *Tetrahedron* 1997, 53, 935–938.
18. North, M. *Tetrahedron Lett.* 1996, 37, 1699–1702.
19. Ravía, S., Gamenara, D., Schapiro, V., Bellomo, A., Adum, J., Seoane, G., Gonzalez, D. *J. Chem. Educ.* 2006, 83, 1049–1051.
20. Baldassarre, F., Bertoni, G., Chiappe, C., Marioni, F. *J. Mol. Catal. B Enzym.* 2000, 11, 55–58.
21. Chadha, A., Manohar, M., Soundararajan, T., Lokeswarl, T. S. *Tetrahedron Asymmetry* 1996, 7, 1571–1572.
22. Comasseto, J. V., Omori, Á. T., Porto, A. L. M., Andrade, L. H. *Tetrahedron Lett.* 2004, 45, 473–476.
23. Yadav, J. S., Nanda, S., Thirupathi Reddy, P., Bhaskar Rao, A. *J. Org. Chem.* 2002, 67, 3900–3903.
24. Yadav, J. S., Thirupathi Reddy, P., Nanda, S., Bhaskar Rao, A. *Tetrahedron Asymmetry* 2001, 12, 3381–3385.
25. Bruni, R., Fantin, G., Medici, A., Pedrini, P., Sacchetti, G. *Tetrahedron Lett.* 2002, 43, 3377–3379.
26. Giri, A., Dhingra, V., Giri, C. C., Singh, A., Ward, O. P., Narasu, M. L. *Biotechnol. Adv.* 2001, 19, 175–199.
27. Mączka, W. K., Mironowicz, A. *Tetrahedron Asymmetry* 2002, 13, 2299–2302.

28. Mączka, W. K., Mironowicz, A. *Tetrahedron Asymmetry* 2004, 15, 1965–1967.
29. Natarajan, K. R. *J. Chem. Educ.* 1991, 68, 13–16.
30. Koga, N., Oliveira, A. H. A., Sakamoto, K. *Chem. Educator* 2008, 13, 344–347.
31. Boykin, D. W. *J. Chem. Educ.* 1998, 75, 769.
32. O'Brien, K. E., Wicht, D. K. *Green Chem. Lett. Rev.* 2008, 1, 149–154.
33. Lawrence, N. J., Drew, M. D., Bushell, S. M. *J. Chem. Soc. Perkin Trans.* 1 1999, 3381–3391.
34. Lehman, J. W. Experiment 29: Borohydride Reduction of Vanillin to Vanillyl Alcohol. In *Operational Organic Chemistry*. Prentice-Hall, Upper Saddle River, NJ, 2009, 246–254.
35. Mohrig, J. R., Hammond, C. N., Schatz, P. F., Morrill, T. C. Experiment 24.1: Reduction of 3-Nitroacetophenone using Sodium Borohydride. In *Modern Projects and Experiments in Organic Chemistry: Miniscale and Standard Taper Microscale*, 2nd ed. W.H. Freeman, New York, 2003, 193–195.
36. Mayo, D. W., Pike, R. M., Trumper, P. K. Experiment 5: Reduction of Ketones Using a Metal Hydride Reagent: Cyclohexanol and cis- and trans-4-tert-Butylcyclohexanol. In *Microscale Organic Laboratory: With Multistep and Multiscale Syntheses*, 4th ed. Wiley, New York, 2000, 133–144.
37. Pavia, D. L., Lampman, G. M., Kriz, G. S., Engel, R. G. Experiment 28: An Oxidation-Reduction Scheme: Borneol, Camphor, Isoborneol. In *Introduction to Organic Laboratory Techniques: A Microscale Approach*, 3rd ed. Brooks/Cole, Pacific Grove, CA, 1999, 266–278.
38. Baru, A. R., Mohan, R. S. *J. Chem. Educ.* 2005, 82, 1674–1675.
39. Lecher, C. S. Sodium Borohydride Reduction of Vanillin: A Low Solvent Synthesis of Vanillyl Alcohol. http://greenchem.uoregon.edu/PDFs/GEMsID90.pdf (accessed December 23, 2010).
40. Gilbertson, R., Parent, K., McKenzie, L., Hutchison, J. Electrophilic Aromatic Iodination of 4′-Hydroxyacetophenone. In *Greener Approaches to Undergraduate Chemistry Experiments*, Kirchhoff, M., Ryan, M. A., Eds. American Chemical Society, Washington, DC, 2002, 1–3.
41. Doxsee, K. M., Hutchison, J. E. Experiment 12: Electrophilic Aromatic Iodination. In *Green Organic Chemistry—Strategies, Tools, and Laboratory Experiments*. Brooks/Cole, Pacific Grove, CA, 2004, 182–188.
42. Doxsee, K. M., Hutchison, J. E. Experiment 13: Palladium-Catalyzed Alkyne Coupling/Intramolecular Alkyne Addition: Natural Product Synthesis. In *Green Organic Chemistry—Strategies, Tools, and Laboratory Experiments*. Brooks/Cole, Pacific Grove, CA, 2004, 189–196.
43. Eby, E., Deal, S. T. *J. Chem. Educ.* 2008, 85, 1426–1428.
44. Moroz, J. S., Pellino, J. L., Field, K. W. *J. Chem. Educ.* 2003, 80, 1319–1321.
45. Monk, K. A., Mohan, R. S. *J. Chem. Educ.* 1999, 76, 1717.
46. Reeves, W. P., King II, R. M., Jonas, L. L., Hatlevik, O., Lu, C. V., Schulmeier, B. *Chem. Educator* 1998, 3, 1–6.
47. Djerassi, C., Scholz, C. R. *J. Am. Chem. Soc.* 1948, 70, 417–418.
48. McKenzie, L. C., Huffman, L. M., Hutchison, J. E. *J. Chem. Educ.* 2005, 82, 306–310.
49. Chandrasekhar, C., Dragojlovic, V. *Green Chem. Lett. Rev.* 2010, 3, 39–47.
50. Pavia, D. L., Lampman, G. M., Kriz, G. S., Engel, R. G. Experiment 22A: Dehydration of 1-Butanol and 2-Butanol. In *Introduction to Organic Laboratory Techniques: A Microscale Approach*, 3rd ed. Brooks/Cole, Pacific Grove, CA, 1999, 219.
51. Mayo, D. W., Pike, R. M., Trumper, P. K. Experiment 9: The E1 Elimination Reaction: Dehydration of 2-Butanol to Yield 1-Butene, trans-2-Butene, cis-2-Butene. In *Microscale Organic Laboratory: With Multistep and Multiscale Syntheses*, 4th ed. Wiley, New York, 2000, 184–192.

52. Lehman, J. W. Experiment 21: Dehydration of Methylcyclohexanol and the Evelyn Effect. In *Operational Organic Chemistry*. Prentice-Hall, Upper Saddle River, NJ, 2009, 181–190.
53. Mohrig, J. R., Hammond, C. N., Schatz, P. F., Morrill, T. C. Experiment 11: Dehydration of Alcohols. In *Modern Projects and Experiments in Organic Chemistry: Miniscale and Standard Taper Microscale*, 2nd ed. W.H. Freeman, New York, 2003, 83–91.
54. Doxsee, K. M., Hutchison, J. E. Experiment 4: Preparation and Distillation of Cyclohexene. In *Green Organic Chemistry—Strategies, Tools, and Laboratory Experiments*. Brooks/Cole, Pacific Grove, CA, 2004, 129–134.
55. Doyle, M. P., Plummer, B. F. *J. Chem. Educ.* 1993, 70, 493–495.
56. Sereda, G. A. *Tetrahedron Lett.* 2004, 45, 7265–7267.
57. Sereda, G. A., Rajpara, V. B. *J. Chem. Educ.* 2007, 84, 692–693.
58. Bhat, R. P., Raje, V. P., Alexander, V. M., Patil, S. B., Samant, S. D. *Tetrahedron Lett.* 2005, 46, 4801–4803.
59. Wetter, E., Levy, I. J., Kay, R. D. Zeolite-Catalyzed Multi-Component Reaction: Preparation of a β-Acetamido Ketone. http://greenchem.uoregon.edu/PDFs/GEMsID92.pdf (accessed December 23, 2010).
60. Bertelsen, S., Jørgensen, K. A. *Chem. Soc. Rev.* 2009, 38, 2178–2189.
61. For examples of aldol condensations designed for the undergraduate laboratory, see: (a) Pavia, D. L., Lampman, G. M., Kriz, G. S., Engel, R. G. Experiment 35: The Aldol Condensation Reaction: Preparation of Benzalacetophenones (Chalcones). In *Introduction to Organic Laboratory Techniques: A Microscale Approach*, 3rd ed. Brooks/Cole, Pacific Grove, CA, 1999, pp. 316–319. (b) Mayo, D. W., Pike, R. M., Trumper, P. K. Experiment 20: Aldol Reaction: Dibenzalacetone. In *Microscale Organic Laboratory: With Multistep and Multiscale Syntheses*, 4th ed. Wiley, New York, 2000, 279–286. (c) Mohrig, J. R., Hammond, C. N., Schatz, P. F., Morrill, T. C. Project 11: Aldol-Dehydration Chemistry Using Unknown Aldehydes and Ketones. In *Modern Projects and Experiments in Organic Chemistry: Miniscale and Standard Taper Microscale*, 2nd ed. W.H. Freeman, New York, 2003, 353–361.
62. Bennett, G. D. *J. Chem. Educ.* 2006, 83, 1871–1872.
63. Wong, T. C., Sultana, C. M., Vosburg, D. A. *J. Chem. Educ.* 2010, 87, 194–195.
64. Kim, H., Yen, C., Preston, P., Chin, J. *Org. Lett.* 2006, 8, 5239–5242.
65. Doxsee, K. M., Hutchison, J. E. Experiment 6: Oxidative Coupling of Alkynes: The Glaser-Eglinton-Hay Coupling. In *Green Organic Chemistry—Strategies, Tools, and Laboratory Experiments*. Brooks/Cole, Pacific Grove, CA, 2004, 142–151.
66. Hay, A. S. *J. Org. Chem.* 1962, 27, 3320–3321.
67. Gilbertson, R., Doxsee, K., Succaw, G., Huffman, L., Hutchison, J. Palladium-Catalyzed Alkyne Coupling/Intramolecular Alkyne Addition: Synthesis of a Benzofuran Product. In *Greener Approaches to Undergraduate Chemistry Experiments*, Kirchhoff, M., Ryan, M. A., Eds. American Chemical Society, Washington, DC, 2002, 4–7.
68. France, M. B., Uffelman, E. S. *J. Chem. Educ.* 1999, 76, 661–665.
69. Casey, C. P. *J. Chem. Educ.* 2006, 83, 192–195.
70. Taber, D. F., Frankowski, K. J. *J. Chem. Educ.* 2006, 83, 283–284.
71. Pappenfus, T. M., Hermanson, D. L., Ekerholm, D. P., Lilliquist, S. L., Mekoli, M. L. *J. Chem. Educ.* 2007, 84, 1998–2000.
72. Greco, G. E. *J. Chem. Educ.* 2007, 84, 1995–1997.
73. Schepmann, H. G., Mynderse, M. *J. Chem. Educ.* 2010, 87, 721–723.
74. Masuda, T., Jitoe, A. *Phytochemistry* 1995, 39, 459–461.
75. Parker, R. E., Isaacs, N. S. *Chem. Rev.* 1959, 59, 737–799.
76. Buchanan, J. G., Sable, H. Z. Stereoselective Epoxide Cleavages. In *Selective Organic Transformations*, Thyagarajan, B. S., Ed., Vol. II. Wiley, New York, 1972, 1–95.
77. Ranu, B. C., Jana, U. *J. Org. Chem.* 1998, 63, 8212–8216.

78. Kulasegaram, S., Kulawiec, R. J. *J. Org. Chem.* 1997, 62, 6547–6561.
79. Rickborn, B., Gerkin, R. M. *J. Am. Chem. Soc.* 1971, 93, 1693–1700.
80. Settine, R. L., Parks, G. L., Hunter, G. L. K. *J. Org. Chem.* 1964, 29, 616–618.
81. House, H. O. *J. Am. Chem. Soc.* 1955, 77, 5083–5089.
82. House, H. O. *J. Am. Chem. Soc.* 1955, 77, 3070–3075.
83. Christensen, J. E., Huddle, M. G., Rogers, J. L., Yung, H., Mohan, R. S. *J. Chem. Educ.* 2008, 85, 1274–1275.
84. Martinez, C. A., Hu, S., Dumond, Y., Tao, J., Kelleher, P., Tully, L. *Org. Process Res. Dev.* 2008, 12, 392–398.
85. Sime, J. T. *J. Chem. Educ.* 1999, 76, 1658–1661.
86. Straathof, A. J. J., Panke, S., Schmid, A. *Curr. Opin. Biotechnol.* 2002, 13, 548–556.
87. Breslow, R. *J. Am. Chem. Soc.* 1958, 80, 3719–3726.
88. Warner, J. Benzoin Condensation Using Thiamine as a Catalyst Instead of Cyanide. In *Greener Approaches to Undergraduate Chemistry Experiments*, Kirchhoff, M., Ryan, M. A., Eds. American Chemical Society, Washington, DC, 2002, 14–17.
89. Doxsee, K. M., Hutchison, J. E. Experiment 15: Carbonyl Chemistry: Thiamine-Mediated Benzoin Condensation of Furfural. In *Green Organic Chemistry—Strategies, Tools, and Laboratory Experiments*. Brooks/Cole, Pacific Grove, CA, 2004, 201–205.
90. Nishimura, R. T., Giammanco, C. H., Vosburg, D. A. *J. Chem. Educ.* 2010, 87, 526–527.
91. Elbs, K., Lerch, H. *J. Prakt. Chem.* 1916, 93, 1–9.
92. Dasari, M. S., Richards, K. M., Alt, M. L., Crawford, C. F. P., Schleiden, A., Ingram, J., Hamidou, A. A. A., Williams, A., Chernovitz, P. A., Luo, R., Sun, G. Y., Luchtefeld, R., Smith, R. E. *J. Chem. Educ.* 2008, 85, 411–412.
93. Collet, A. *Angew. Chem. Int. Ed.* 1998, 37, 3239–3241.
94. Baar, M. R., Cerrone-Szakal, A. L. *J. Chem. Educ.* 2005, 82, 1040–1042.
95. Ward, T. J. *Anal. Chem.* 2002, 74, 2863–2872.
96. Huerta, F. F., Minidis, A. B. E., Bäckvall, J. E. *Chem. Soc. Rev.* 2001, 30, 321–331.
97. Keith, J. M., Larrow, J. F., Jacobsen, E. N. *Adv. Synth. Catal.* 2001, 343, 5–26.
98. Caddick, S., Jenkins, K. *Chem. Soc. Rev.* 1996, 25, 447–456.
99. Afonso, C. A. M., Crespo, J. G. *Angew. Chem. Int. Ed.* 2004, 43, 5293–5295.
100. Acs, M., von dem Bussche, C., Seebach, D. *Chimia* 1990, 44, 90–92.
101. Markovits, I., Egri, G., Fogassy, E. *Chirality* 2002, 14, 674–676.
102. Monteiro, C. M., Afonso, C. A. M., Lourenço, N. M. T. *J. Chem. Educ.* 2010, 87, 423–425.
103. Cainelli, G., Cardillo, G., Orena, M., Sandri, S. *J. Am. Chem. Soc.* 1976, 98, 6737–6738.
104. Wade, Jr., L. G., Stell, L. M. *J. Chem. Educ.* 1980, 57, 438.
105. Doxsee, K. M., Hutchison, J. E. Experiment 14: Resin-Based Oxidation Chemistry. In *Green Organic Chemistry—Strategies, Tools, and Laboratory Experiments*. Brooks/Cole, Pacific Grove, CA, 2004, 197–200.
106. Mayo, D. W., Pike, R. M., Trumper, P. K. Experiment 33A: 9-Fluorenone: CrO_3 Oxidation of 9-Fluorenol. In *Microscale Organic Laboratory: With Multistep and Multiscale Syntheses*, 4th ed. Wiley, New York, 2000, 357–359.
107. Buglass, A. J., Waterhouse, J. S. *J. Chem. Educ.* 1987, 64, 371–372.
108. Crumbie, R. L. *J. Chem. Educ.* 2006, 83, 268–269.
109. Crouch, R. D., Holden, M. S., Burger, J. S. *J. Chem. Educ.* 2001, 78, 951–952.
110. Fatiadi, A. J. *Synthesis* 1976, 65–104.
111. Fatiadi, A. J. *Synthesis* 1976, 133–167.
112. Zhang, J., Phillips, J. A. *J. Chem. Educ.* 2010, 87, 981–984.
113. Sorg, G., Mengel, A., Jung, G., Rademann, J. *Angew. Chem. Int. Ed.* 2001, 40, 4395–4397.

114. Pageau, G. J., Mabaera, R., Kosuda, K. M., Sebelius, T. A., Ghaffari, A. H., Kearns, K. A., McIntyre, J. P., Beachy, T. M., Thamattoor, D. M. *J. Chem. Educ.* 2002, 79, 96–97.
115. Takagi, T. *J. Appl. Polym. Sci.* 1975, 19, 1649–1662.
116. Cioffi, E. Esterification by Microwave Irradiation Using Activated Carbon. In *Greener Approaches to Undergraduate Chemistry Experiments*, Kirchhoff, M., Ryan, M. A., Eds. American Chemical Society, Washington, DC, 2002, 21–22.
117. Cramer, R., Lindsey, Jr., R. V. *J. Am. Chem. Soc.* 1966, 88, 3534–3544.
118. Tolman, C. A. *J. Am. Chem. Soc.* 1970, 92, 4217–4222.
119. Tolman, C. A. *J. Am. Chem. Soc.* 1972, 94, 2994–2999.
120. Seen, A. J. *J. Chem. Educ.* 2004, 81, 383–384.
121. Conlon, H. D., Walt, D. R. *J. Chem. Educ.* 1986, 63, 368–370.
122. Pohl, N., Schwarz, K. *J. Chem. Educ.* 2008, 85, 834–835.
123. Truran, G. A., Aiken, K. S., Fleming, T. R., Webb, P. J., Markgraf, J. H. *J. Chem. Educ.* 2002, 79, 85–86.
124. Birney, D. M., Starnes, S. D. *J. Chem. Educ.* 1999, 76, 1560–1561.
125. Miles, W. H., Gelato, K. A., Pompizzi, K. M., Scarbinsky, A. M., Albrecht, B. K., Reynolds, E. R. *J. Chem. Educ.* 2001, 78, 540–542.
126. Hailstone, E., Huther, N., Parsons, A. F. *J. Chem. Educ.* 2003, 80, 1444–1445.
127. Sereda, G., Rajpara, V. *J. Chem. Educ.* 2010, 87, 978–980.
128. Tundo, P., Rosamilia, A. E., Aricò, F. *J. Chem. Educ.* 2010, 87, 1233–1235.
129. Tundo, P., Selva, M. *Acc. Chem. Res.* 2002, 35, 706–716.
130. Warner, M. G., Succaw, G. L., Hutchison, J. E. *Green Chem.* 2001, 3, 267–270.
131. Warner, M., Succaw, G., Doxsee, K., Hutchison, J. Microwave Synthesis of Tetraphenylporphyrin. In *Greener Approaches to Undergraduate Chemistry Experiments*, Kirchhoff, M., Ryan, M. A., Eds. American Chemical Society, Washington, DC, 2002, 27–31.
132. Doxsee, K. M., Hutchison, J. E. Experiment 8: Microwave Synthesis of 5,10,15,20-Tetraphenylporphyrin. In *Green Organic Chemistry—Strategies, Tools, and Laboratory Experiments*. Brooks/Cole, Pacific Grove, CA, 2004, pp. 159–162.
133. Doxsee, K. M., Hutchison, J. E. Experiment 9: Metallation of 5,10,15,20-Tetraphenylporphyrin. In *Green Organic Chemistry—Strategies, Tools, and Laboratory Experiments*. Brooks/Cole, Pacific Grove, CA, 2004, pp. 163–166.
134. Warner, M., Hutchison, J. Metallation of Tetraphenylporphyrin. In *Greener Approaches to Undergraduate Chemistry Experiments*, Kirchhoff, M., Ryan, M. A., Eds. American Chemical Society, Washington, DC, 2002, 32–34.

7 Organic Waste Management and Recycling

Ms. Amanda R. Edward

CONTENTS

7.1 INTRODUCTION

A primary goal of green chemistry is the adjustment of reaction conditions such that waste generation is minimized *without* compromising product yields.[1] Green modifications of diverse transformations, including the Diels-Alder reaction,[2,3] aldol condensation,[2,4] Wittig reaction,[5] Passerini reaction,[6] and Suzuki reaction,[7] have been reported for the organic teaching laboratory. Such experiments and others address organic waste management by use of an environmentally benign solvent such as water (or sometimes no solvent at all), and are discussed in Chapters 3 and 4. However, waste production is realistically an inevitable part of the laboratory component of undergraduate organic chemistry courses. Williamson has described a waste as a substance simply "declared as a waste by a chemist."[8] In certain instances a waste product might be recycled, or converted into another usable form, thus rendering it a "nonwaste." As such, instructors can introduce strategies to reduce the amount and type of waste produced in undergraduate laboratories, without attempting to achieve total elimination.

Chemical resource management in the teaching laboratory was thoroughly described in 1977 by Neckers et al.[9] This was primarily from a cost perspective, owing to the 1973 Organization of Arab Petroleum Exporting Countries (OAPEC) oil embargo to the United States and various other nations. The yearlong embargo greatly increased the price of oil and petroleum by-products, including organic chemicals. As a result, operational costs of organic laboratories doubled or tripled during a short timeframe, and led to a scaling down of reactions at many institutions. Interestingly, a number of ideas from this paper can be adopted and applied to showcase principles of green chemistry. Three students conducted an analysis of the wastefulness of undergraduate laboratories, and subsequently devised economical modifications. These included solvent recovery and purification from a classical caffeine extraction procedure. Sequential reactions (where the product of one transformation is the starting material for a further reaction) form a considerable component of the article. Use of sunlight as the driver for photochemical reactions is also discussed. All of these methods are in keeping with the modern-day green chemistry movement. Recent pedagogical approaches have been described that foster a sense of environmental awareness among chemistry students.[10–12] These also serve as excellent starting points for resource management in undergraduate laboratories.

One of the twelve principles of green chemistry states it is "better to prevent than to treat or clean up created waste."[13] Established literature methods to reduce organic waste generation form the core of this chapter, along with strategies seeking to render waste less hazardous, or to convert it into useful substances. The recycling component of this chapter initially focuses on reclamation of organic reagents and solvents. Experiments illustrating how consumer products such as newspapers[14–16] and soft drink bottles[17,18] can be recycled are also highlighted. Although not traditionally defined as green reactions, these experiments utilize everyday materials that would ordinarily go to waste, and high-profile substances such as biofuels can be generated. The recycling of some natural products is additionally considered. In undertaking practical work in this field, students can learn and appreciate the value and complexity of industrial recycling operations in the context of real-world applications. As process "greenness" is primarily the responsibility of research chemists in (for example) pharmaceutical venues,[1] creating awareness of chemical waste management and recycling at the undergraduate level *is* demonstrably of utmost importance.

7.2 THREE INDUSTRIAL CASE STUDIES

Principles of waste management and recycling are clearly of enormous significance to the chemical industry, and many corresponding examples highlight progress made. Three such case studies are summarized in this section. First, a greener synthesis of the anti-impotence drug Viagra® is briefly outlined. Following this, preparation of the chemical building block propylene oxide by an award-winning new technology is profiled. Last, the transformation of commercial wastewater containing aromatic sulfonic acids into an important antituberculosis medication is discussed.

7.2.1 Synthesis of Viagra

Pfizer, the pharmaceutical company known worldwide for marketing Viagra (sildenafil citrate; Figure 7.1), has focused heavily on synthesis optimization in terms of resource management.[19] In 1990, the initial medicinal chemistry route to Viagra required 1,300 L of solvent per kg of product, of which nearly 75% was dichloromethane. The 1994 optimized medicinal chemistry route lowered solvent usage to 100 L/kg, with the commercial route (1997) operating at 22 L/kg. By this time, all environmentally unfriendly halogenated solvents had been completely eliminated, along with the highly volatile diethyl ether, methanol, and acetone. In 2004, the four solvents employed were 2-butenone, ethyl acetate, toluene, and *t*-butanol. Recycling of these led to only 7 L of solvent being used per kilogram of synthetic Viagra produced. Further refinements plan to replace *t*-butanol, which is completely water soluble and challenging to recover for reuse. Pfizer has now reached an E-factor of 6 for Viagra synthesis (6 kg total waste per kg of final product, as discussed in Section 1.4). This places the current process in line with bulk chemical syntheses, which typically exhibit E-factors in the range of <1 to 5.[19] As a comparison, the majority of pharmaceutical syntheses have E-factors above 25, with some greater than 100.

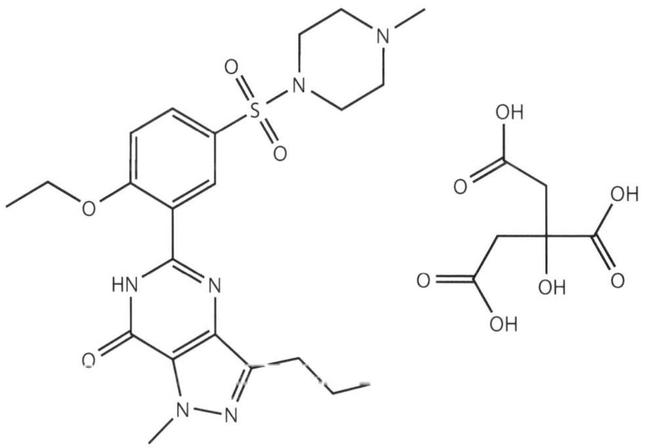

FIGURE 7.1 Structure of sildenafil citrate (Viagra).

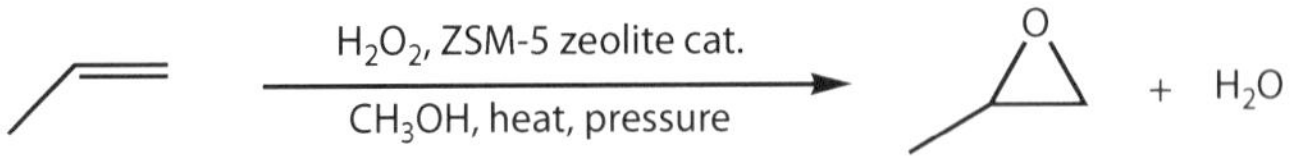

SCHEME 7.1 Propylene oxide synthesis via propene epoxidation.

7.2.2 Propylene Oxide via Propene Oxidation

The Dow Chemical Company and BASF were jointly honored with a Presidential Green Chemistry Challenge Award in 2010 for their "innovative, environmentally benign production of propylene oxide via hydrogen peroxide."[20] Propylene oxide is a major industrial chemical worldwide (more than 14 billion pounds are used annually), and a precursor for many household products, including personal care items, detergents, and furniture. Historically, propylene oxide synthesis (utilizing organic peroxides or chlorohydrin) led to significant coproduct and waste formation. Dow and BASF have developed a high-yielding catalytic reaction between propene and hydrogen peroxide that only generates water as a coproduct (Scheme 7.1). Under the reaction conditions, propene is epoxidized with H_2O_2 in a methanol solvent, using a zeolite catalyst. Significantly, hydrogen peroxide is completely consumed and less is required compared to organic peroxides. Forming water as the sole coproduct leads to cheaper and more efficient manufacturing processes, and wastewater production is lowered by up to 80%.

7.2.3 Wastewater Treatment: Preparation of Paramycin

Aromatic sulfonic acids (ASAs) are produced on a multi-kiloton scale every year and used in a variety of industrial processes. For example, the stilbene derivative 4,4′-dinitrostilbene-2,2′-disulfonic acid (DNS) is required for synthesis of dyes and fluorescent brightening agents.[21] Unfortunately, these compounds are highly water soluble and environmentally persistent, so their presence in wastewater derived from synthesis of DNS is of significant concern to chemists (Figure 7.2). Incineration methods have been employed to degrade such wastewater, but these form acidic gases that require further manipulation before atmospheric release. Biological approaches are ineffective at degradation, and chemical transformations have thus far proved unable to recycle any substances of worth.

Very recent research has demonstrated that ASA wastewater can be comprehensively recycled to generate paramycin (4-amino-2-hydroxybenzoic acid), an antituberculosis drug.[21] Scheme 7.2 illustrates how this can be achieved for *p*-nitrotoluene-*o*-sulfonic acid (NTS), one of the important components of DNS wastewater (Figure 7.2). NTS is initially oxidized with sodium hypochlorite to 4-nitro-2-sulfobenzoic acid (NSBA). Section 6.2.1.1 profiles undergraduate oxidations using NaOCl as a greener oxidizing agent. Following this, reduction of NSBA to 4-amino-2-sulfobenzoic acid (ASBA) is effected with iron metal under weakly acidic conditions. Paramycin is then generated by a high-temperature alkali fusion reaction. Importantly, it should be noted that these reactions can be applied to the wastewater itself, without the need to extract any ASA compounds. The sequence of oxidation-reduction-alkali fusion works on all the significant wastewater ASAs, so

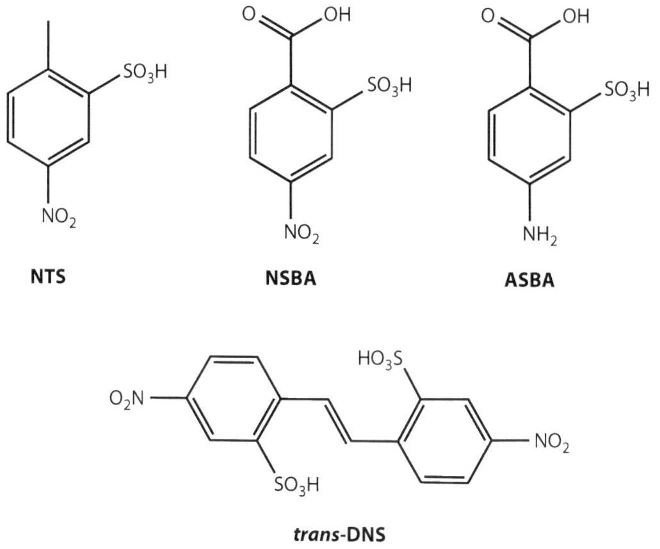

FIGURE 7.2 Representative aromatic sulfonic acids found in industrial wastewater.

that paramycin can be formed in yields of over 85%. This case study is highly appropriate for inclusion in an introductory organic course. The reactions highlighted (benzylic oxidation, aromatic nitro group reduction, and phenol formation) are all ones typically discussed during lectures. Here, they are placed into a real-world context in the setting of both waste management and recycling.

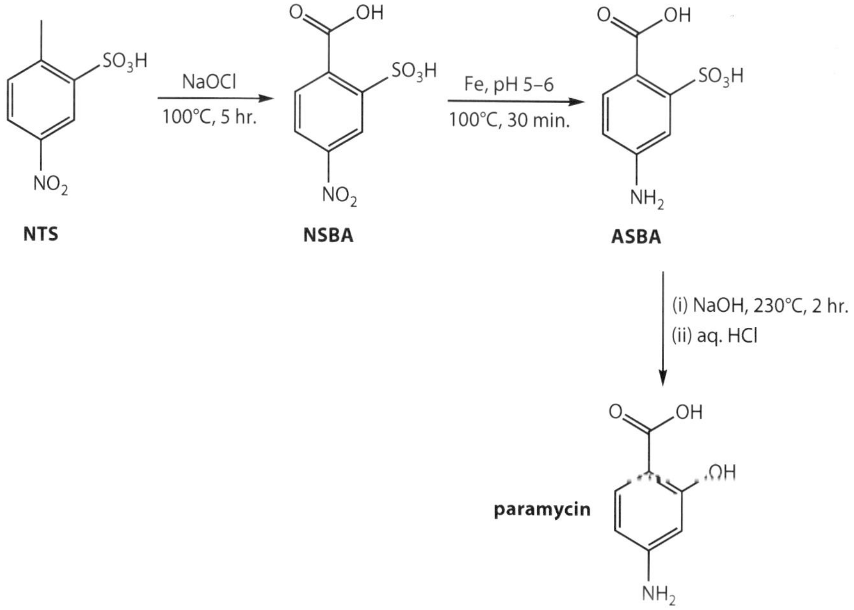

SCHEME 7.2 Synthesis of paramycin from *p*-nitrotoluene-*o*-sulfonic acid (NTS).

7.3 REDUCTION OF WASTE GENERATION

7.3.1 Introduction and Student-Centered Strategies

Organic waste management was described from an environmental perspective by Martin and Waldman in their 1994 landmark article.[22] Their "three R's" of chemical resource management serve as excellent starting points for monitoring waste production. This involves *reducing* the scale of reactions, *recycling* products and reagents, and *rendering* waste products safe for both human handling and the environment. Conducting reactions on a milliscale (0.5–5.0 g of reactant) is touted as a good compromise between the wasteful macroscale and less industrially significant microscale. Shelden has described the widespread availability of semimicroscale glassware and equipment, enabling organic reactions to be easily scaled down.[23] Sequential reactions, solvent and catalyst recycling, and saving organic products for future use have additionally been discussed.[22] Waste treatment has also been introduced in chemistry courses since the early 1990s,[24] and thorough operating procedures according to specific waste properties have been outlined (Section 7.4).

7.3.1.1 Constructing Ecological Diagrams

An ecological diagram has been reported that requires undergraduates to list, identify, and quantify all waste from every reaction step, including purification and product analyses.[10] Once identified, students then label and outline appropriate handling methods for the waste. An extensive database of appropriate treatments for various waste residues is compiled into their laboratory manuals. In doing so, reactions can be modified so waste production is minimized. An ecological diagram outlining preparation of anthraquinone from anthracene using chromium trioxide is provided, and the waste is identified, including unreacted starting material. Upon appreciation of the toxicity and corrosive nature of various reaction components, students devise less hazardous and less wasteful procedures. These include anthraquinone synthesis from *o*-benzoylbenzoic acid via an acidic dehydration, and separately employing microwave irradiation.[25]

7.3.1.2 Reaction Component Quantification

Van Arnum and Savers has outlined a procedure where students quantify all components of an organic reaction (solvent-free synthesis of 3-acetyl-5-methylisoxazole from 2,5-hexanedione).[11,26] This includes starting materials, products, by-products, and all solvents utilized. The molecular weight of each substance and its mass, volume, and density (where applicable) are tabulated, along with calculated mole quantities. This analysis enables reaction optimization to eliminate unwanted by-products and waste, and to increase the desired product yield. Comparison of these data tables proves an efficient method for making reaction modifications. Both the ecological diagram and reaction quantification approaches to minimize waste generation are student directed. The shift of responsibility from the instructor to the undergraduate encourages and develops a sense of environmental responsibility, which is a key attribute for all practicing chemists.

7.3.2 Microscale and Reduced-Scale Reactivity

Microscale reactivity represents a movement to reduce waste output and exposure to hazardous chemicals, by using smaller quantities of all reaction components. The final product of a microscale reaction should amount to no more than approximately 100 mg. Many organic reactions were historically conducted under relatively harsh conditions, using large amounts of reagents and solvents. An impressive range of experiments have subsequently been scaled down to be performed at the undergraduate level. A result of this shift is the conversion of entire laboratory curricula to accommodate microscale reactivity. Microscale chemistry kits have been designed and patented in response to this reduction in scale.[27] The components of initial kits (e.g., pipettes, syringes) were mostly made of plastic. This equipment was highly versatile and less prone to breakage than regular glassware. While useful for analysis and milder reaction conditions, however, problems could occur with high temperatures or corrosive solvents. As such, glassware and other equipment suitable for carrying out microscale reactions were required.[28] As an example, a Soxhlet-like microscale solvent extractor uses less than 4 mL of solvent for extraction of 100–500 mg samples.[29] This continuous extractor is more efficient than a traditional Soxhlet owing to its smaller size. The precise boiling point of a solvent can be reached, 99% less solvent is used, and better product recovery is observed. This extractor is ideal for microscale reactions and reduces solvent waste.

Although the advent of microscale chemistry initially arose due to financial difficulties (Section 7.1), modern microscale reactivity reflects green and environmentally friendly principles. Significantly, it has been noted that microscale chemistry and green chemistry are pedagogically complementary.[30] Use of a less hazardous solvent such as water, or employing no solvent, often accompanies the scaling down of a reaction, with a concomitant reduction in generated waste. It was previously thought that conducting experiments on a microscale was only suitable for the skilled and trained chemist, requiring finesse and extensive practical knowledge. Pike et al. outlined a microscale Wittig reaction for undergraduate students as long ago as 1986,[31] and Williamson published a laboratory manual describing experiments from both microscale and macroscale perspectives in 1989.[32]

An example of how a teaching tool has been "greened" in terms of waste management is shown by the classical Blue Bottle demonstration (Scheme 7.3). This popular visual is widely used at secondary and postsecondary levels to introduce reduction-oxidation and kinetic principles. It involves oxidation of methylene blue by oxygen

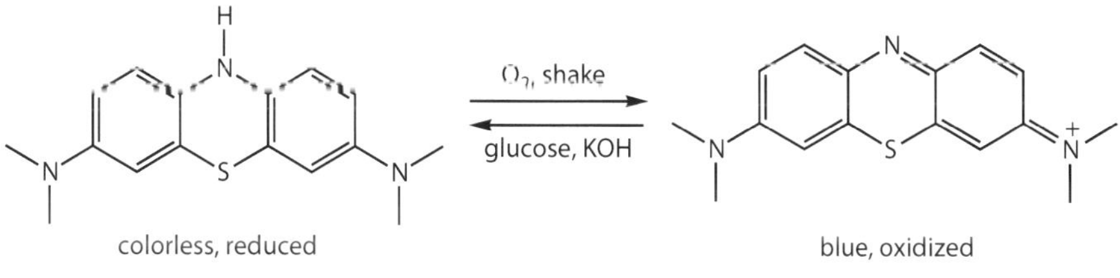

SCHEME 7.3 Blue Bottle demonstration using methylene blue.

in a basic environment, to turn the original colorless solution light or bright blue. Glucose then reduces the oxidized methylene blue to its original state, observed by a slow fade from blue to colorless. An early procedure described by Campbell in 1963 requires copious amounts of corrosive base, which must be disposed of after a lecture demonstration.[33] Several modifications have scaled down the demonstration and rendered it more environmentally friendly. Use of other dyes and sugars for a more efficient and visually appealing experiment has been reported.[34,35]

Extensive modifications of the Blue Bottle demonstration from a green perspective have been discussed, owing in large part to the amount of corrosive hydroxide base required in the original procedure.[36] Ascorbic acid is now used in lieu of glucose, and corrosive potassium hydroxide replaced with less harmful sodium bicarbonate. Copper(II) sulfate is added to catalyze oxidation of methylene blue by oxygen, although other Cu^{2+} compounds, such as copper(II) chloride, are equally effective. Sodium chloride catalyzes the reduction process in these cases. Use of household products for the demonstration has also been outlined.[36] Commercial vitamin C powder is now the source of ascorbic acid, and an additive to home aquaria, (MethyBlu), serves as a source of 5% methylene blue. Another aquarium additive (Had-A-Snail) provides Cu^{2+} ions from a dilute solution (1.61%) of copper(II) sulfate pentahydrate. Sodium chloride is obtained from table salt, and tap water used instead of distilled water. All modern reports on this demonstration are inherently less wasteful than the original procedure.

7.3.3 Multistep Syntheses

Multistep organic syntheses use the product(s) from one reaction step as the starting material(s) for another. In simple terms, these syntheses adhere to the recycling products and reagents proposal by Martin and Waldman.[22] Student products are not disposed of, but are "recycled" as starting materials for another step or reaction. A representative example is the three-step preparation of dulcin, an artificial sweetener, from acetaminophen (Tylenol®) via the intermediate compound phenacetin[37] (Scheme 7.4). This is very well suited to a second-year undergraduate organic curriculum, with each reaction step taking place in good yield. Both acetaminophen and phenacetin exhibit analgesic properties and afford opportunities to discuss structure-activity relationships with links to biochemistry and the pharmaceutical industry. From other green perspectives, each reaction is performed in the absence of toxic solvents under mild reaction conditions, without the need for column chromatography or even recrystallization as purification techniques. The synthesis can be undertaken starting with a single Tylenol tablet, which significantly reduces reagent use and waste formation.

Oxidation of vanillyl alcohol to vanillin using 2,2,6,6-tetramethyl piperidine -1-oxyl (TEMPO) has recently been reported.[38] Although this does not represent part of a linear multistep transformation, it illustrates how a reaction product can be converted back into the important starting material for another process, thus minimizing waste. In this instance, vanillin is a popular substrate on which to perform reduction with sodium borohydride[39] *(*Scheme 7.5).

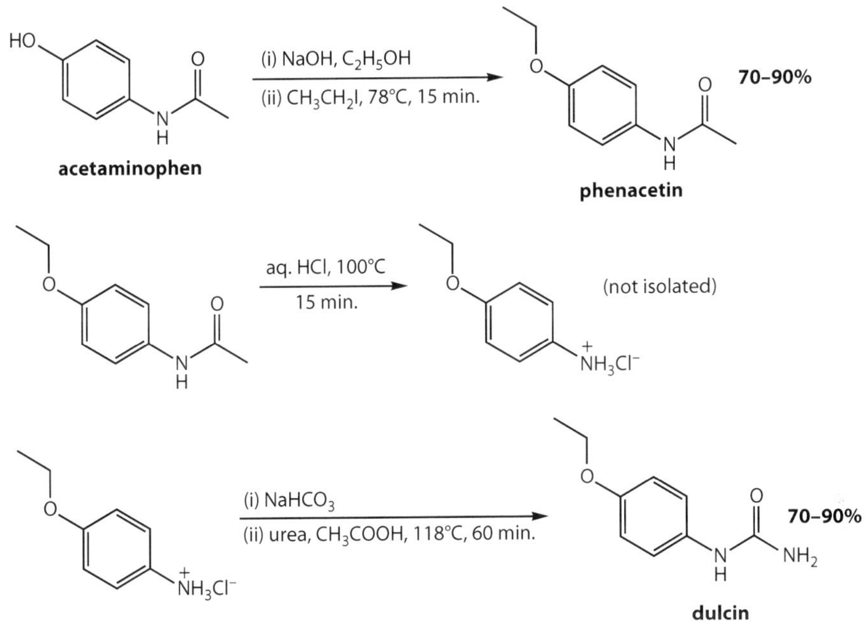

SCHEME 7.4 Synthesis of dulcin from acetaminophen.

7.3.4 Single-Step, High Atom Economy Reactions

High atom economy reactions incorporate most (if not all) starting material atoms into the final desired product(s). A high atom economy reaction adheres to one of the twelve principles of green chemistry, and can simultaneously prevent excess waste generation.[13] Many articles report the high atom economy nature of a reaction in conjunction with another green approach, such as a solventless method, or utilizing alternative greener solvents. This section highlights some selected single-step, high atom economy reactions that are carried out sans solvent or in an environmentally benign solvent.

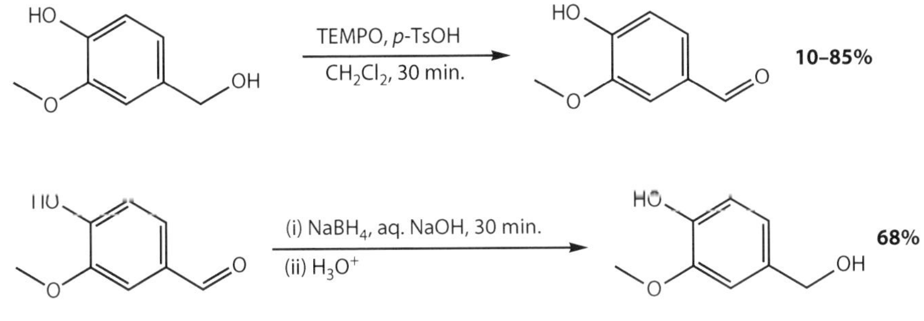

SCHEME 7.5 Oxidation of vanillyl alcohol and reduction of vanillin.

The Diels-Alder reaction is taught in many introductory organic chemistry courses to illustrate key concepts, such as stereoselectivity and enantioselectivity, electron-withdrawing and -donating groups, and highest-occupied molecular orbitals (HOMO) and lowest-unoccupied molecular orbitals (LUMO).[40] It is not always highlighted as a green reaction in the literature, but has an intrinsic atom economy of 100%, and therefore generates very little (if any) waste. Correspondingly, excellent didactic examples of green Diels-Alder reactions have been described under solvent-free conditions, in water and in the presence of polyethylene glycol.[2] These reactions are discussed in more detail in Sections 3.5.2.2, 4.3.2, and 5.6.2, along with microwave Diels-Alder transformations in Section 8.8.5.

The aldol condensation is a carbon-carbon bond-forming reaction taught in undergraduate courses to highlight important mechanistic steps, including enolate ion formation and nucleophilic addition. If an α,β-unsaturated carbonyl product is formed, water is a "waste" product from the dehydration step, so that special disposal systems are not necessary. The base required to generate the nucleophilic enolate ion is neutralized with acid during product workup, giving water and a nontoxic salt. Recent reports have outlined greener aldol condensations[2,4] that proceed without solvent, with noteworthy intrinsic atom economies (>90%). Hutchison et al. (Scheme 7.6) reported a modification on an aldol condensation where the two reactants are ground with a small amount of solid sodium hydroxide.[2,41] This reaction was undertaken by over 250 high school students at the Forty-First International Chemistry Olympiad held in Cambridge, United Kingdom, during July 2009.[42] Further solventless aldol condensation reactions are featured in Section 3.5.2.1.

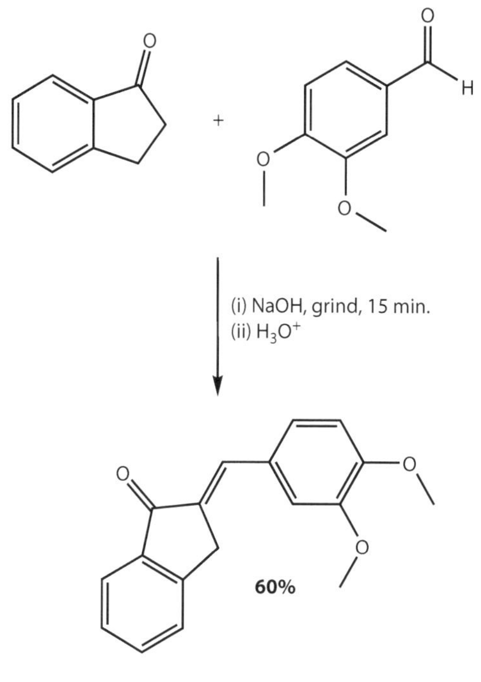

SCHEME 7.6 Solventless aldol condensation of 1-indanone and 3,4-dimethoxybenzaldehyde.

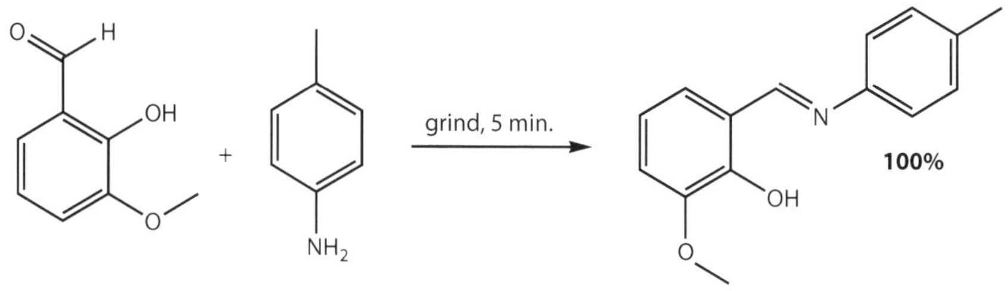

SCHEME 7.7 Solventless imine synthesis.

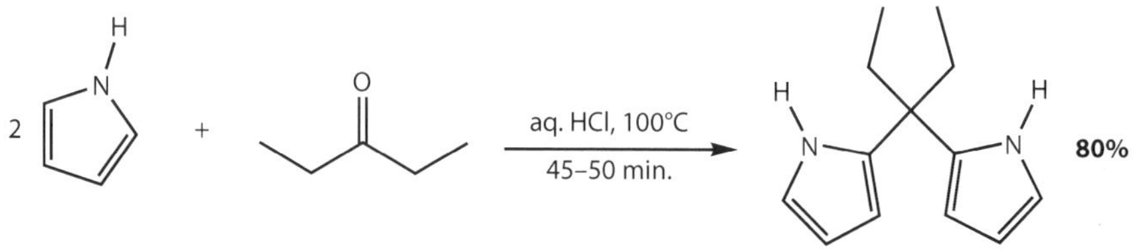

SCHEME 7.8 Aqueous synthesis of *meso*-diethyl-2,2′-dipyrromethane.

Imines have also been synthesized in a similar manner, as discussed by Touchette (Scheme 7.7).

This exothermic reaction is quantitative and requires five minutes of grinding and product recrystallization from ethanol.[43] The intrinsic atom economy in this instance is 93%. Sobral has introduced an aqueous synthesis of *meso*-diethyl-2,2′-dipyrromethane, consistent with green principles outlined previously, and an intrinsic atom economy of 92%. This methodology affords facile product collection, as the dipyrromethane separates as a solid from the aqueous solution and does not require purification by recrystallization[44] (Scheme 7.8).

7.4 MANAGING GENERATED WASTE

As discussed previously, total waste elimination in the organic teaching laboratory is a lofty goal. Methods clearly exist to reduce waste generation, such as performing reactions solvent-free or undertaking chemistry with high atom economy. However, many organic reactions do not lead to complete product formation, meaning that unreacted starting materials are present. Alternatively, unwanted by-product formation may be significant, which needs to be treated as waste. As such, it is imperative that there are systems in place to manage formed waste, regardless of the amount and type produced. Use of ecological diagrams in this regard is outlined in Section 7.3.1.1.[10] In addition, a very thorough account of waste management specific to its identity has been published.[45] Waste treatment and disposal procedures reported by Zimmer reduce harmful environmental effects, or render the waste safer for commercial disposal, and are briefly outlined below.

Collected wastes are first categorized as (1) acidic (inorganic), (2) basic (inorganic), (3) oxidizing material, (4) reducing material, (5) halogenated organic solvent, (6) nonhalogenated organic solvent, (7) organic acid or base, (8) heavy metal solution, or (9) miscellaneous material. These wastes are then disposed of according to their identity, as follows, for selected categories:

1. Acid waste is neutralized using calcium carbonate or sodium bicarbonate blocks to a pH between 6 and 9. This is applicable to most acids except nitric acid, which must first be reduced. This overnight process is done in a fume hood, as CO_2 gas is generated. The amount generated is negligible and not considered to be a contributor to the greenhouse effect. A method to produce sodium bicarbonate blocks has been described.[45] A slurry of sodium bicarbonate is dried overnight in an oven in large crucibles. The blocks are easily removed and can neutralize acidic wastes at a slow enough rate so overflow does not occur. Calcium carbonate blocks can also be produced in this manner. The powder form of either calcium carbonate or sodium bicarbonate can be used for neutralization purposes. The powder is added slowly to the acidic waste until the desired pH is reached.
2. Basic waste is neutralized using hydrochloric acid. This is an exothermic process that must be undertaken slowly and with care. Alternatively, HCl may be placed in a large separatory funnel and slowly dripped into the basic solution with stirring. This is done overnight in a fume hood. Both neutralized solutions can then be disposed down the drain and flushed with large amounts of water.
3. Oxidizing material is first detected using HCl-soaked potassium iodide/starch paper. An oxidizing agent is present if the sample drop turns the indicator purple. These wastes are reduced to the lowest stable state, or to a safer form if toxic heavy metals are present. Heavy metals have distinctive colors that are indicative of their oxidation states. These colors can be used to determine if the wastes have been reduced to a safer form. For example, the less toxic chromium(III) is green, whereas toxic hexavalent chromium(VI) compounds are orange or yellow. If a color test is not possible for a given set of wastes, the KI/starch paper test will suffice. Once reduced, oxidizing material without toxic metals can be flushed down the drain with large amounts of water. If toxic metals are present, the ions are precipitated out according to the solubility table of metal ions. Cr, Ag, and Cu ions are precipitated as either the oxide or hydroxide, and Pb ions as the silicate. The precipitate is collected, dried, and stored for commercial disposal.
4. Disposal of organic solvents is dependent on if they are halogenated or nonhalogenated. Proper collection and storage of these solvents is of utmost importance to prevent a potential hazardous mix-up of waste. To avoid this, a waste container must be labeled with a list of allowed wastes. This careful listing of the types of wastes allowed in a given container will provide a much more efficient collection and disposal system.
5. Experiments that historically used compounds containing toxic heavy metals such as Cd, Cr, Pb, and Hg should be modified so that these are either

completely eliminated or reduced so that the amount of waste produced is negligible. This is an easy step that undergraduate laboratories can undertake to prevent the generation of unnecessary toxic waste.

The Department of Chemistry at the Federal University of Rio de Janeiro, Brazil, has a *compulsory* waste management and treatment course for undergraduate chemists.[12] Students learn and appreciate waste management processes at both academic and industrial levels, broadening their views and obtaining a bigger picture on the course importance. The findings from a survey conducted by the department show a gradual increase in student environmental awareness from 1998 to 2007. The number of students unfamiliar with environmental legislation for waste disposal in Brazil decreased from eighty-five students in 1998 to fifty-eight students in 2007. Students confident in their knowledge of this subject rose from one student in 1998 to fourteen students in 2004. Another interesting trend is observed when students were asked about their expectations from the course. Most students listed a "better chance of employment" in 1998, whereas in 2007, a majority of the students listed "environmental awareness."

7.5 REAGENT RECYCLING

Use of catalysis is an application of one of the twelve principles of green chemistry, where catalytic reagents are noted to be "superior to stoichiometric reagents."[13] Many greener catalytic reactions are detailed in Sections 6.2.2.2 and 6.2.4–6.2.7. The selected examples described here are chosen to profile catalytic, selective, and recyclable properties.

Nafion® NR50 is a perfluorinated solid phase resin produced by the DuPont Chemical Company. This superacid catalyst contains a sulfonic acid group in the reactive chambers. Approximately 50% of the chambers are large enough to catalyze common organic reactions. Nafion NR50 serves as a replacement for concentrated mineral acids, and can be used in lieu of boiling chips. Used Nafion NR50 can be reactivated by stirring in 25% nitric acid for four hours, followed by washing thoroughly with water. Doyle and Plummer have described a series of reactions conducted using Nafion NR50, including dehydration of cyclohexanol (Scheme 7.9) and esterification of isoamyl alcohol (Scheme 7.10).[46] Other reactions reported to be successfully catalyzed using Nafion NR50 are the pinacol-to-pinacolone rearrangement and preparation/hydrolysis of dimethyl acetals.

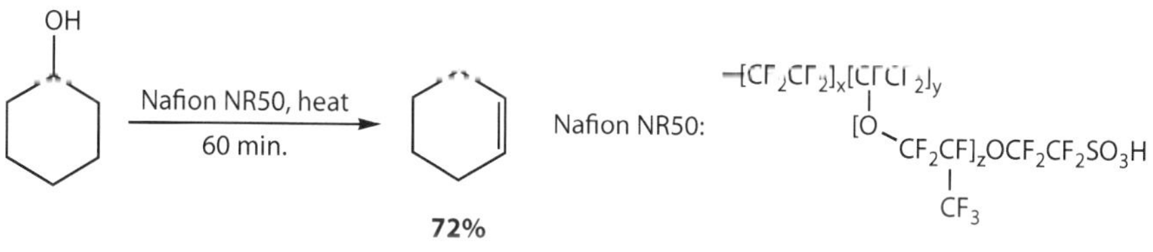

SCHEME 7.9 Dehydration of cyclohexanol to cyclohexene.

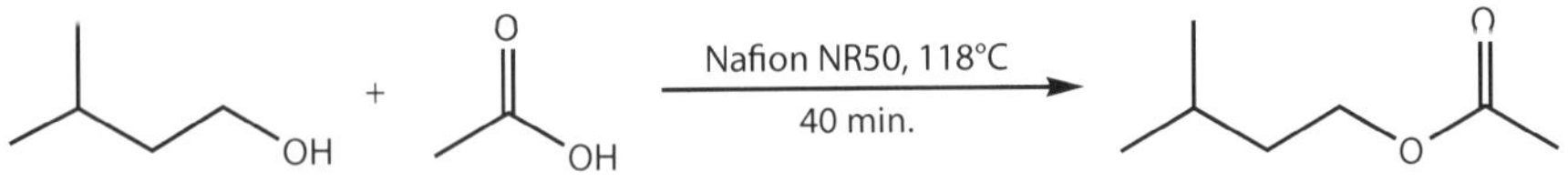

SCHEME 7.10 Esterification of isoamyl alcohol.

Another resin capable of catalyzing dehydration of cyclohexanol is Amberlyst® 15, consisting of cross-linked polystyrene chains with sulfonic acid active sites.[47] Miles and Connell have reported use of this resin to synthesize Methyl Diantilis, a fragrance with an odor reminiscent of vanillin and carnation flowers[48] (Scheme 7.11). The chemoselective etherification step uses Amberlyst 15 in place of $NaHSO_4$.[49] A comparison of this resin vs. concentrated acids indicates the resin to be more effective in producing higher yields of the desired product. Other fragrant derivatives of Methyl Diantilis can be synthesized using Amberlyst 15, and the industrial importance of greener syntheses of commercial fragrances discussed with students.

Azo dyes are derivatives of the diimide bond with substituted aryl rings or other extensive π-delocalized cyclic structures.[50] These dyes are synthesized on an industrial scale due to their importance in the food, cosmetics, and textile industries. As a consequence, a greener method to synthesize these dyes has been developed by Caldarelli et al. using an Amberlyst 26 resin.[51] This method produces less waste owing to use of a reusable resin to couple diazonium salts with phenols or aromatic amines, and avoids workup and purification steps (Scheme 7.12). Yields are good to excellent and purities are high. This report is from an industrial perspective, although applications to the undergraduate lecture and laboratory are potentially possible owing to the curricular relevance, practical simplicity, and green aspects of the procedure.

All solid phase resin conditions described are excellent substitutes to concentrated mineral acids, as they can be reused multiple times, and do not present a threat to the environment. They are able to catalyze a variety of processes efficiently and bypass the required neutralization step for many reactions.

An aqueous oxidation of cyclohexene to produce adipic acid, using a recyclable sodium tungstate catalyst and the phase transfer catalyst Aliquat® 336, has been

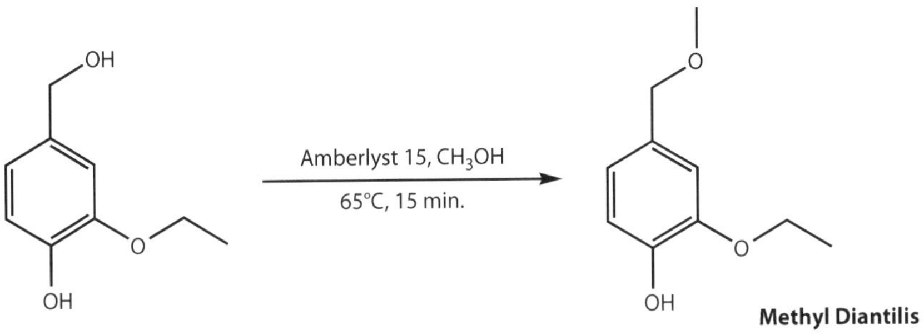

SCHEME 7.11 Synthesis of Methyl Diantilis using Amberlyst 15 resin.

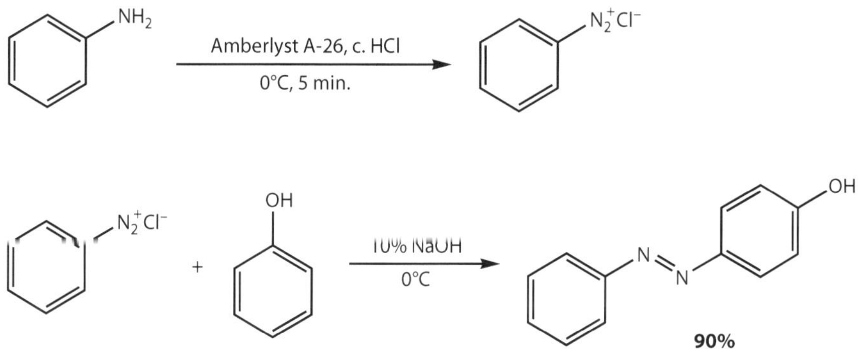

SCHEME 7.12 Polymer-supported azo dye synthesis.

devised[52] (Scheme 7.13). This represents the earliest example of a green, metal-catalyzed reaction for the undergraduate laboratory. Adipic acid is usually synthesized by cyclohexene oxidation with a hot, basic potassium permanganate solution. Potassium permanganate is a harsh and hazardous oxidizer, and the reaction produces a large quantity of manganese dioxide waste. This greener alternative uses hydrogen peroxide as the ultimate oxidant. The sodium tungstate catalyst is easily isolated from the filtrate for further use, and adipic acid is produced in good yields.

Use of another recyclable transition metal catalyst under aqueous conditions is apparent in a green modification of the Suzuki reaction. In a discovery-oriented experiment, Novak et al. reacted phenylboronic acid with a range of halogenated phenols under conventional and microwave heating conditions.[7] A typical example is shown in Scheme 7.14. The palladium catalyst can be recovered by filtration and reused. Other pedagogical Suzuki reactions have been designed with water as the sole solvent and are described in Section 4.3.1.

The Diels-Alder reaction of furan and maleic anhydride is used as a classical introduction to [4 + 2] cycloadditions, and occurs readily due to the highly reactive diene and electron-withdrawing ester moieties of the dienophile. The adduct from this reaction can be ring opened and polymerized under aqueous conditions using a recyclable ruthenium catalyst.[53] This reaction is also discussed in more detail in Section 4.3.1.

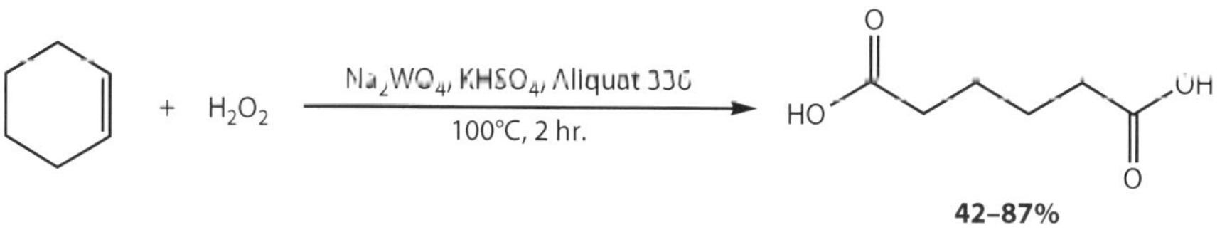

SCHEME 7.13 Adipic acid synthesis using Aliquat 336 phase transfer catalyst.

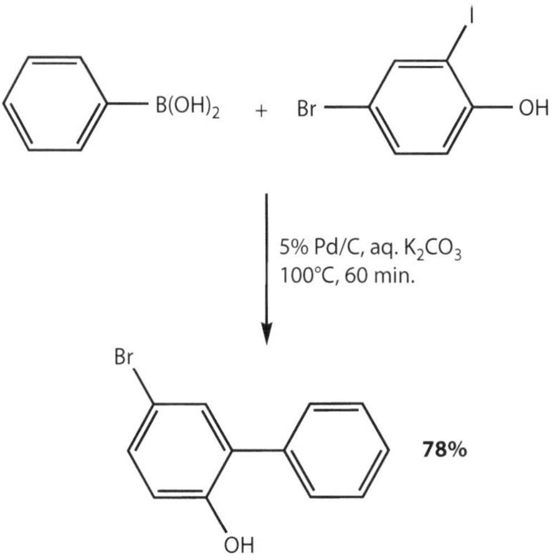

SCHEME 7.14 Aqueous Suzuki biaryl synthesis.

7.6 RECYCLING SOLVENTS

Systems and procedures to recycle organic solvents have been established in an attempt to reduce unnecessary waste output. Acetone is a solvent routinely used to rinse remaining chemical residues from glassware prior to regular washing, in both undergraduate and research laboratories. This "wet" acetone (approximately 20% water content) is commonly stored for disposal along with other waste organic solvents. Dussault and Woller outlined a mechanism to recover acetone with insignificant water content (2%–3%), using a simple distillation apparatus.[54] The distilled acetone can subsequently be used again to rinse glassware. Using this procedure, recovered acetone is reported to have reduced waste production in student laboratories by 50%.

Ethyl acetate and hexanes can additionally be recycled by distillation methods. Large quantities of these solvents are commonly mixed, and used to elute reaction products via column chromatography. Smaller amounts are often used for thin-layer chromatography (TLC) analysis of reaction progress. Mixtures of ethyl acetate and hexanes will undergo simple distillation without solvent separation, due to boiling point similarities (hexanes = 69°C, ethyl acetate = 77°C). The distillate will be a mixture of solvents, and its composition can be determined via TLC. The R_f value of a known, pure solid using the distillate as an eluent can be compared with the R_f values of the same solid using various known ethyl acetate:hexanes mixtures. For example, the R_f value of a known solid using a 50:50 mixture of ethyl acetate:hexanes is easily determined experimentally. If a distillate sample is calculated to produce a similar R_f value, then it can be concluded that it is, or is very near to, a 50:50 composition of both solvents. Wilkinson described a related procedure after separation of acetylferrocene and 1,1′-diacetylferrocene with ethyl acetate:hexanes via column chromatography, following a ferrocene acetylation reaction.[55] The R_f values of these two products are plotted against mixtures of ethyl acetate:hexanes of known composition. The R_f values of the distillate sample are compared with the values in this graph, and the composition of ethyl acetate and hexanes determined.

A similar procedure for distilling ethyl acetate:hexanes mixtures after predrying with $MgSO_4$ has also been discussed.[54] A simple distillation apparatus is used, and the percentage of ethyl acetate in the distillate calculated with a formula using the density values of both solvents. All recycling methods are reported to effectively reduce organic solvent waste output by half. The purity of the solvents in question is not of utmost importance owing to their roles in the laboratory. Recycling systems for these solvents therefore reduce unnecessary waste output. A recovery process for waste sodium metal used for drying organic solvents is also possible.[56] Slow melting of used sodium chunks in mineral oil affords recycled sodium of higher purity below the oil and lower purity at the bottom of the reaction vessel. The recycled sodium can be removed as a block and stored in petroleum oil for reuse.

An environmentally friendly extraction of lycopene from tomato sauce has recently been published.[57] A copolymer surfactant is synthesized and used as a recyclable emulsion to perform extractions in the absence of organic solvents. This approach has been extended to the technique of foam fractionation, which operates under milder conditions.[58]

7.7 RECYCLING CONSUMER AND NATURAL PRODUCTS

Public environmental awareness has increased dramatically in the past twenty years, with recycling facilities and systems to collect materials for this purpose expanding worldwide. Recycling receptacles have been created to collect specific waste and facilitate this process. The most common recyclable household items are paper, glass, and metal products. Recycling plastic products is of considerable interest, as about 80% of all plastic waste ends up in landfills.[59] Reactions that mimic recycling processes are often not specifically mentioned in the chemical education literature as green reactions. However, such experiments often demonstrate key concepts and fundamental reaction mechanisms taught in undergraduate organic courses. The real-world application of these processes piques student interest in both chemistry and environmental responsibility.

7.7.1 Polylactic Acid

Polylactic acid (PLA; also referred to as polylactide) is a condensation polymer formed from lactic acid monomers. It is well known as a preferred material for commercial packaging (such as cold beverage cups and salad trays) owing to its synthesis from a renewable resource, and just as importantly because it is completely biodegradable. Because lactic acid exists in two enantiomeric forms, several types of polylactic acid exist that have different mechanical properties. Some exhibit thermal stability and can be used to make microwavable trays and items of clothing. However, poly-L-lactic acid (derived from naturally occurring L-lactic acid) has a relatively low glass transition temperature, and is unsuitable to hold hot drinks, for example.

A procedure developed for the undergraduate laboratory based on the PLA degradation process has been devised.[60] PLA can be hydrolyzed into lactic acid monomers

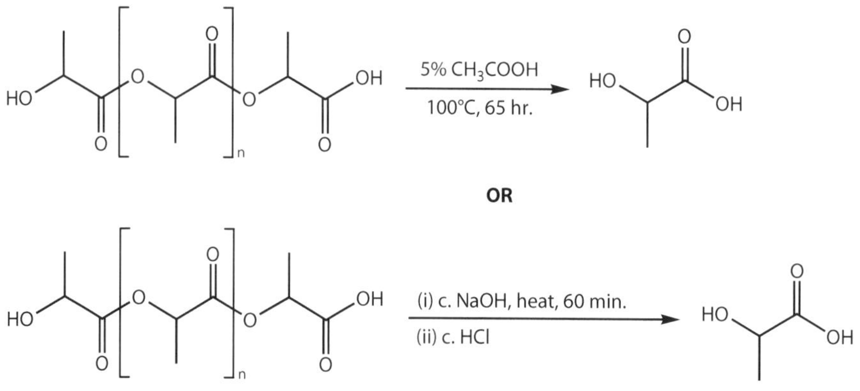

SCHEME 7.15 Hydrolysis of polylactic acid.

under acidic or basic conditions (Scheme 7.15). It is possible to effect the basic hydrolysis by heating in a microwave oven. The more time-consuming acidic hydrolysis employs use of a common household ingredient (vinegar), and takes about one week of heating under reflux to proceed to completion. The lactic acid product is a soap that students can test on soiled bathroom tiles.

Robert and Aubrecht have designed an experiment involving the ring-opening polymerization of lactide to synthesize polylactic acid[61] (Scheme 7.16). The reaction uses tin(II) bis(2-ethylhexanoate) to catalyze the ring opening of lactide and initiate polymerization to PLA. This procedure minimizes the use of hazardous reagents and starting materials, reduces waste, showcases a reaction with high atom economy, and emphasizes production of a biodegradable polymer. This synthesis might be combined with the previously described degradation experiment[60] in a multiweek project. This would illustrate how an important polymeric material can be both generated and decomposed in the laboratory.

7.7.2 Polyethylene Terephthalate

Polyethylene terephthalate (PET) is a plastic polymer used to manufacture 2 L soft drink bottles. Industrial methanolysis of PET used by DuPont Chemicals occurs readily under conditions of high temperature and pressure, which are not amenable to

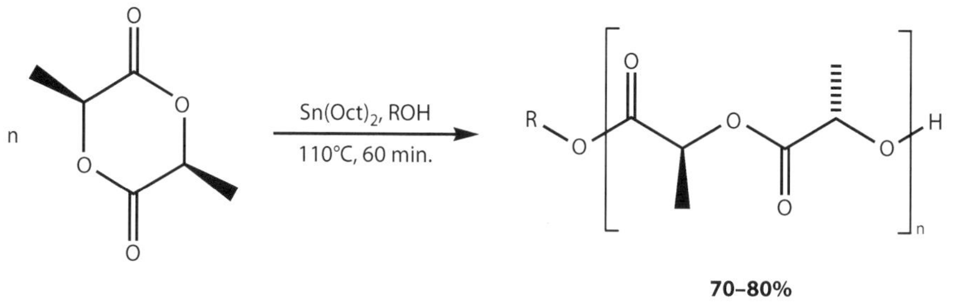

SCHEME 7.16 Ring-opening polymerization synthesis of polylactic acid.

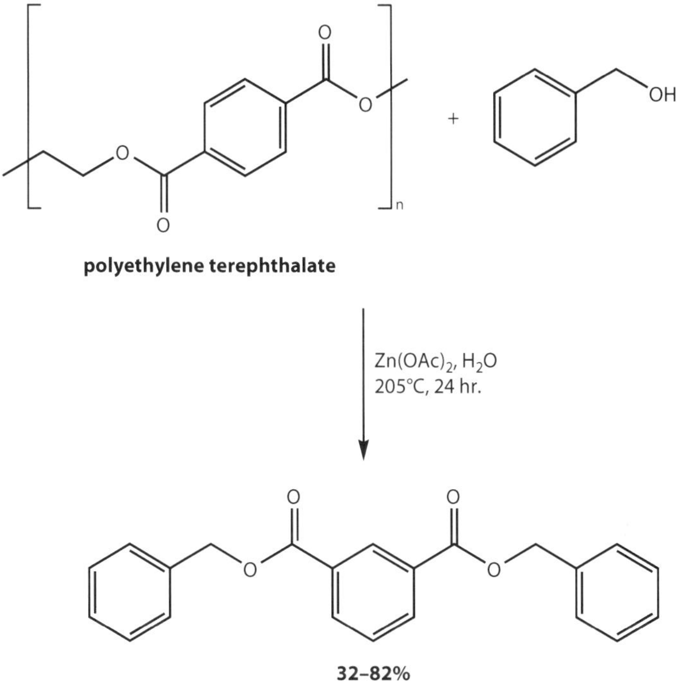

SCHEME 7.17 Synthesis of dibenzyl terephthalate from PET plastics.

an undergraduate laboratory environment. A modified microscale reaction reported by Donahue et al. employs benzyl alcohol in lieu of methanol, with zinc acetate as catalyst[17] (Scheme 7.17). The overall transesterification readily occurs by nucleophilic attack on the electrophilic ester carbonyl carbon in PET. The reaction product (dibenzyl terephthalate) can be used to synthesize a new batch of PET.

An alternative approach to polymer degradation involves refluxing PET samples in potassium hydroxide or potassium *t*-butoxide, using pentanol as solvent.[18] A recent report uses *N*-heterocyclic carbenes to catalyze PET degradation[62] (Scheme 7.18). These carbenes are derived from ionic liquids, and the degradation procedure can be achieved in a three-hour laboratory period.

7.7.3 Recycling Coffee Grounds

Used and discarded coffee grounds provide an environmental sample from which to extract oil (about 10% of the anhydrous grounds) and degreased residue.[63] The extracted oil can be analyzed for free fatty acid content, saponification number, peroxide value, iodine value, insaponifiable residue, and fatty acid composition. Following this, the oil is hydrolyzed under basic conditions to form soap, and degreased residue is evaluated for organic, nitrogen, phosphorus, and heavy metal content. Coffee oil displays a very high saponification number, and produces soap with good detergent and foaming characteristics. The degreased residue exhibits significant organic and

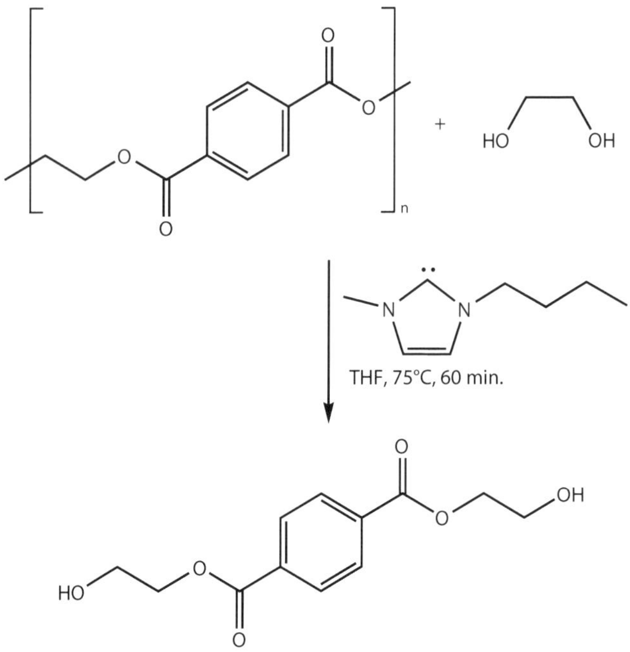

SCHEME 7.18 PET degradation using an *N*-heterocyclic carbene.

nitrogen content, with negligible heavy metals present, making it a viable compost in areas of poor soil fertility.

7.7.4 Biofuel Synthesis

The major component of newspaper and other paper products is cellulose, a polymer of glucose monomers. Acid-catalyzed hydrolysis of cellulose yields both glucose and cellobiose. Glucose produced in this manner can be further degraded by yeast action to synthesize carbon dioxide and ethanol (also known as bioethanol in this context). Ethanol production has garnered much attention recently due to its use as a cleaner, alternative biofuel. It is widely used in the United States and particularly in Brazil as an additive to gasoline (up to 15%), as it increases the fuel octane rating and improves vehicle emissions. Ethanol can also be used as a vehicular fuel in its pure state. Paper products can be extensively recycled and the industrial process mimicked at a microscale level with good results. Approaches have been described from both environmental and biochemical perspectives.[14,15] Mascal and Scown outlined degradation of a small newspaper sample (4 g) using 75% w/w sulfuric acid in water[14] (Scheme 7.19). Hot water is added to create a suspension and heated to near boiling for ninety minutes. The solution is cooled and filtered through a layer of Celite, then neutralized with $Ca(OH)_2$. Calcium sulfate is removed by filtration, and the filtrate adjusted to a neutral pH. Excess water is evaporated until 25 mL of brown syrup is obtained. The resulting mixture of glucose and cellobiose can be analyzed by TLC, and glucose converted to ethanol by fermentation. In a similar manner, the enzyme

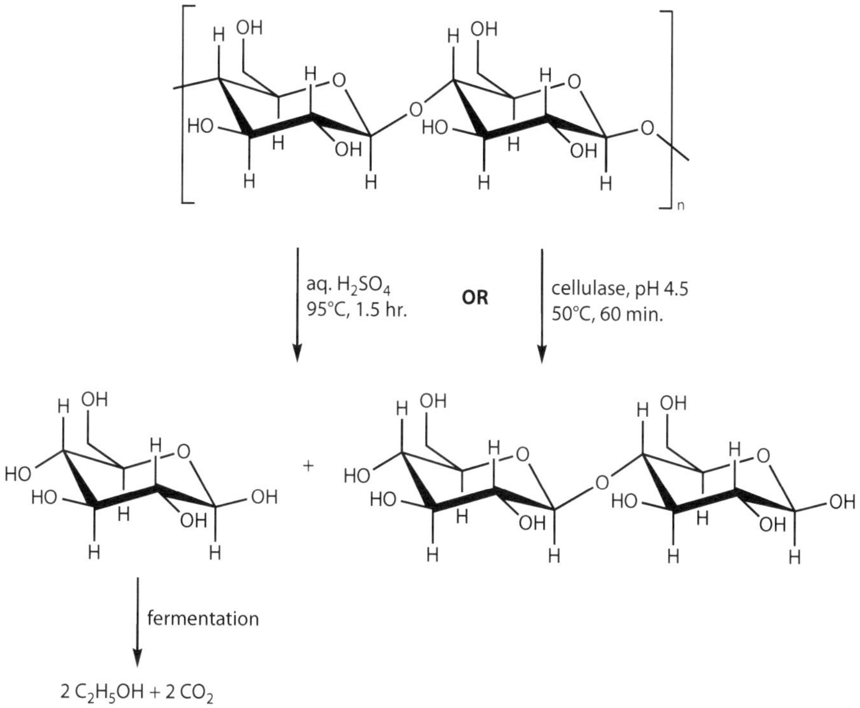

SCHEME 7.19 Degradation of cellulose to glucose and cellobiose by different approaches.

cellulase has been used to degrade the β-glycosidic linkages that connect glucose monomers together[15] (Scheme 7.19). This enzymatic approach is laudable, as cellulase has negligible toxicity levels and exhibits specific activity, and the hydrolysis reaction does not produce harmful waste products. Cellulase is typically extracted from microorganisms (e.g., *Penicillium funiculosum* and *Trichoderma viride*) and can be purchased commercially. Cellulose has been saccharified from various paper sources, including filter paper, foolscap, newspaper, and office papers.

Recently, biofuel production has also been described by fermentation of grass, fruit juice, and grains.[64] In all of these cases, one week is required for alcohol synthesis, and the ethanol is collected by fractional distillation. Students discover that although the percentage of ethanol is highest for fruit (grape) juice, more ethanol per gram of biomass comes from the cellulose and starch sources. Thompson has developed an experiment that produces ethanol from the yeast fermentation of molasses.[65] Molasses contains two enzymes (invertase and zymase) that hydrolyze disaccharides and monosaccharides, respectively (Scheme 7.20). The reaction takes about one week for 70 mL of molasses to yield about 10 mL of ethanol of 90%–95% purity, after simple and fractional distillations.

SCHEME 7.20 Biosynthesis of ethanol from molasses.

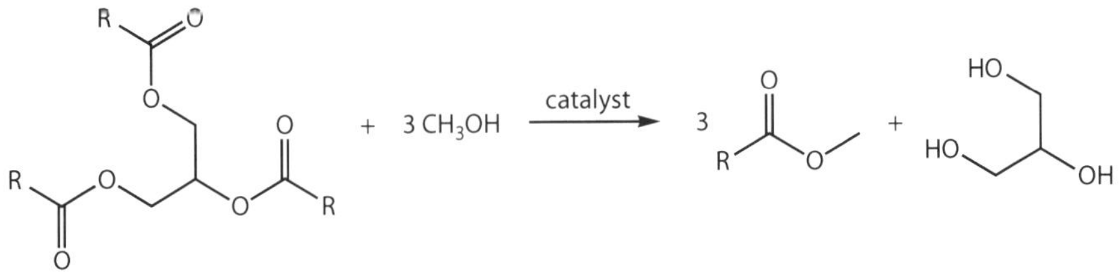

SCHEME 7.21 General biodiesel synthetic scheme.

There has been a strong emphasis on the production of renewable fuels during the past decade. One of the fuels receiving a lot of attention is biodiesel, which can be used directly in a diesel engine. Biodiesels are alkyl ester fuels derived from animal fats or vegetable oils. They are generally synthesized by reacting triglycerides obtained from these sources with an alcohol (Scheme 7.21). This transesterification process forms a methyl ester (if methanol is the reacting alcohol) and glycerol. The preparation and study of biodiesel has been researched extensively, and well-defined methods of production and analysis have been developed. Unsurprisingly, therefore, the formation of biodiesel has become one of the most common green experiments in both high school and college chemistry curricula.

Several pedagogical approaches have been adopted to synthesize and analyze biodiesel, which employ the fundamental reaction shown in Scheme 7.21.[66–71] Meyer and Morgenstern describe an experiment for the general chemistry laboratory where biodiesel is formed on a small scale from vegetable oil.[66] Other experiments differ in the manner by which the biodiesel is analyzed. First, Bucholtz requires students to analyze the glycerol content of their biodiesel by iodometric titration, following synthesis from simulated waste vegetable oil.[67] Alternatively, viscosity measurements are used to compare transesterification of new and old vegetable oil in an inquiry-based experiment.[68] In a third strategy, students synthesize biodiesel under either microwave or conventional heating. This experiment maintains a discovery-based modus operandi where different catalytic conditions are considered, and the biodiesel product is analyzed by proton nuclear magnetic resonance (NMR) spectroscopy.[69] It is possible for students to use infrared (IR) spectroscopy to note the production of glycerol during biodiesel formation.[70] Lastly, the heat of combustion of biodiesel synthesized from peanut cooking oil can be measured using bomb calorimetry.[71]

As a final point regarding biodiesel, Stout has made an important argument about its synthesis from "fresh" oils compared to that from "fryer" oils.[72] Oils that have been used several times can prove problematic for biodiesel production on a laboratory scale. This is due to the formation of emulsions, which are thought to be composed of soap. It is urged that this fact be discussed with students, with the inclusion of washes with vinegar, if necessary, to destabilize such emulsions.

7.7.5 Miscellaneous Examples

Parajó et al. have designed an experiment that nicely illustrates the processing of a natural product to form a polymeric glue.[73] Students experience the extraction,

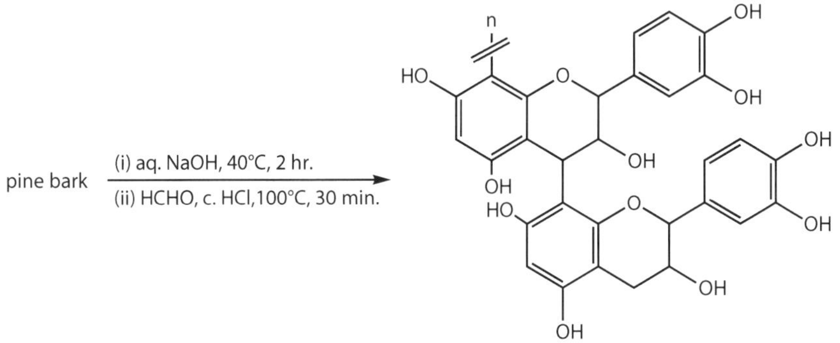

SCHEME 7.22 Liquid extraction of pine bark and subsequent polymeric adhesive formation.

concentration, and polymerization processes required to convert pine bark into a phenolic adhesive (Scheme 7.22). Dry bark samples are ground and homogenized, then extracted with alkali and reacted with aqueous acidic formaldehyde to form a solid polymer. This substance is analyzed for its properties as an adhesive (including strength, color, viscosity, and water repellency). The experiment can be used to introduce students to sustainable development, in addition to solid-liquid extraction techniques and spectrophotometric/gravimetric analysis. The process could additionally be expanded to involve lignin oxidation and cellulose hydrolysis, in order to incorporate additional green chemistry principles, organic reactions, and techniques.

A recent article outlines industrial recycling processes for old and used batteries.[74] A thorough analysis of the composition of batteries that determine their recycling method is included. Their effects on the environment are discussed, and provide an insight into how improper battery disposal is inappropriate and dangerous. Finally, an expansion on newspaper recycling has been described by Venditti, where removal of ink from newspapers and other paper products in an undergraduate organic laboratory is achieved.[75]

7.8 CONCLUSION

This chapter serves as an overview of methods that seek to reduce or manage waste production in an organic teaching laboratory environment. In addition, some processes that highlight recycling of reagents and solvents are discussed. Reactions illustrating industrial consumer product recycling are often not described as green in the pedagogical literature, but foster sustainability awareness among students by connecting environmental, green, and organic chemistry principles. Waste management and recycling issues strike at the heart of industrial technology, and an understanding of them is critical to the education of chemists in general.

REFERENCES

1. Tucker, J. L. *Org. Process Res. Dev.* 2006, 10, 315–319.
2. Hutchison, J. E., Huffman, L. M., McKenzie, L. C., Goodwin, T. E., Rogers, C. E., Spessard, G. O. *J. Chem. Educ.* 2009, 86, 488–493.

3. Delaude, L., Sauvage, X. *J. Chem. Educ.* 2008, 85, 1538–1540.
4. Doxsee, K. M., Hutchison, J. E. Experiment 1: Solventless Reactions: The Aldol Reaction. In *Green Organic Chemistry—Strategies, Tools, and Laboratory Experiments*. Brooks/Cole, Pacific Grove, CA, 2004, 115–119.
5. Angel, S. A., Leung, S. H. *J. Chem. Educ.* 2004, 81, 1492–1493.
6. DeBoef, B., Hooper, M. M. *J. Chem. Educ.* 2009, 86, 1077–1079.
7. Novak, M., Wang, Y.-T., Ambrogio, M. W., Chan, C. A., Davis, H. E., Goodwin, K. S., Hadley, M. A., Hall, C. M., Herrick, A. M., Ivanov, A. S., Mueller, C. M., Oh, J. J., Soukup, R. J., Sullivan, T. J., Todd, A. M. *Chem. Educator* 2007, 12, 414–418.
8. Williamson, K. L. *Can. Chem. News* 1991, 43, 14–15.
9. Neckers, D. C., Duncan, M. B., Gainor, J., Grasse, P. B. *J. Chem. Educ.* 1977, 54, 690–692.
10. Gomez, E. F. L., Garcia, I. C. G., Santos, E. S. *J. Chem. Educ.* 2004, 81, 232–238.
11. Van Arnum, S. D. *J. Chem. Educ.* 2005, 82, 1689–1692.
12. Alfonso, S. A. J. C., Leite, Z. T. C. *Quim. Nova* 2008, 31, 1892–1897.
13. Anastas, P. T., Warner, J. C. *Green Chemistry: Theory and Practice*. Oxford University Press, New York, 1998, 29–56.
14. Mascal, M., Scown, R. *J. Chem. Educ.* 2008, 85, 546–548.
15. van Wyk, J. P. H. *Chem. Educator* 2000, 5, 315–316.
16. Dawson-Andoh, B. E., Filson, P. B., Schwegler-Berry, D. *Green Chem.* 2009, 11, 1808–1814.
17. Donahue, C. J., Exline, J. A., Warner, C. *J. Chem. Educ.* 2003, 80, 79–82.
18. Engel, J., Kaufman, D., Kroemer, R., Wright, G. *J. Chem. Educ.* 1999, 76, 1525–1526.
19. Dunn, P. G., Galvin, S., Hettenbach, K. *Green Chem.* 2004, 6, 43–48.
20. Presidential Green Chemistry Challenge, 2010 Greener Synthetic Pathways Award. www.epa.gov/gcc/pubs/pgcc/winners/gspa10.html (accessed December 23, 2010).
21. Peng, W., Chen, Y. Fan, S., Zhang, F., Zhang, G., Fan, X. *Environ. Sci. Technol.* 2010, 44, 9157–9162.
22. Martin, N. H., Waldman, F. S. *J. Chem. Educ.* 1994, 71, 970–971.
23. Shelden, H. R. *J. Chem. Educ.* 1989, 66, 74.
24. Dhawale, S. W. *J. Chem. Educ.* 1993, 70, 395–397.
25. Bram, G., Loupy, A. *Chem. Ind.* 1991, 396–397.
26. Sauers, R. R., Van Arnum, S. D. *J. Heterocycl. Chem.* 2003, 40, 665–668.
27. Bradley, J. D. *Pure Appl. Chem.* 1999, 71, 817–823.
28. (a) Crouch, R. D., Nelson, T. D., Kinter, C. M. *J. Chem. Educ.* 1993, 70, A203–A204. (b) Silberman, R. G. *J. Chem. Educ.* 1994, 71, A140–A141.
29. Wesolowski, S. S., Mulcahy, T., Zafoni, C. M., Wesolowski, W. E. *J. Chem. Educ.* 1999, 76, 1116–1117.
30. Singh, M. M., Szafran, Z., Pike, R. M. *J. Chem. Educ.* 1999, 76, 1684–1686.
31. Pike, R.M., Mayo, D. W., Butcher, S. S., Butcher, D. J., Hinkle, R. J. *J. Chem. Educ.* 1986, 63, 917–918.
32. Williamson, K. L. *Macroscale and Microscale Organic Experiments*. Houghton Mifflin, Boston, MA, 1989.
33. Campbell, J. A. *J. Chem. Educ.* 1963, 40, 578–583.
34. Cook, A. G., Tolliver, R. M., Williams, J. E. *J. Chem. Educ.* 1994, 71, 160–161.
35. Mosher, M., Vandaveer, W. R. *J. Chem. Educ.* 1997, 74, 402.
36. Healy, T., Noble, M. E., Wellman, W. E. *J. Chem. Educ.* 2003, 80, 537–549.
37. Williams, B. D., Williams, B., Rodino, L. *J. Chem. Educ.* 2000, 77, 357–358.
38. Zablowsky, E., Gordon, P., Jarowek-Lopes, C. H. *Chem. Educator* 2010, 15, 115–116.
39. Fowler, R. G. *J. Chem. Educ.* 1992, 69, A43–A46.
40. Diels, O., Alder, K. *Liebigs Ann. Chem.* 1928, 460, 98–112.

41. Rothenberg, G. D., Downie, A. P., Raston, C. L., Scott, J. L. *J. Am. Chem. Soc.* 2001, 123, 8701–8708.
42. 41st International Chemistry Olympiad, Cambridge, United Kingdom, July 18–27, 2009. www.icho2009.co.uk (accessed December 23, 2010).
43. Touchette, K. M. *J. Chem. Educ.* 2006, 83, 929–930.
44. Sobral, A. J. F. N. *J. Chem. Educ.* 2006, 83, 1665–1666.
45. Zimmer, S. W. *J. Chem. Educ.* 1999, 76, 808–811.
46. Doyle, M. P., Plummer, B. F. *J. Chem. Educ.* 1993, 70, 493–495.
47. Moeur, H. P., Swatik, S. A., Pinnell, R. P. *J. Chem. Educ.* 1997, 74, 833.
48. Miles, W. H., Connell, K. B. *J. Chem. Educ.* 2006, 83, 285–286.
49. Miles, W. H. *J. Chem. Educ.* 2008, 85, 917.
50. Moss, G. P., Smith, P. A. S., Tavernier, M. *Pure Appl. Chem.* 1995, 67, 1307–1375.
51. Caldarelli, M., Baxendale, I. R., Ley, S. V. *Green Chem.* 2000, 2, 43–46.
52. Reed, S. M., Hutchison, J. E. *J. Chem. Educ.* 2000, 77, 1627–1629.
53. Viswanathan, T., Jethmalani, J. *J. Chem. Educ.* 1993, 70, 165–167.
54. Dussault, P. H., Woller, K. R. *Chem. Educator* 1996, 1, 1–6.
55. Wilkinson, T. J. *J. Chem. Educ.* 1998, 75, 1640.
56. Hubler-Blank, B., Witt, M., Roesky, H. W. *J. Chem. Educ.* 1993, 70, 408–409.
57. Zhu, J., Zhang, M., Liu, Q. *J. Chem. Educ.* 2008, 85, 256–257.
58. Wang, Y., Zhang, M., Hu, Y. *J. Chem. Educ.* 2010, 87, 510–511.
59. Snyder, C. H. In *The Extraordinary Chemistry of Ordinary Things*, 4th ed. Wiley, Hoboken, NJ, 2003, p. 548.
60. Boice, J. N., King, C. M., Higginbotham, C., Gurney, R. W. *J. Mat. Educ.* 2008, 30, 257–280.
61. Robert, J. L., Aubrecht, K. B. *J. Chem. Educ.* 2008, 85, 258–260.
62. Kamber, N. E., Tsuji, Y., Keets, K., Waymouth, R. M., Pratt, R. C., Nyce, G. W., Hedrick, J. L. *J. Chem. Educ.* 2010, 87, 519–521.
63. Orecchio, S. *J. Chem. Educ.* 2001, 78, 1669–1671.
64. Epstein, J. L., Vieira, M., Aryal, B., Vera, N., Solis, M. *J. Chem. Educ.* 2010, 87, 708–710.
65. Thompson, J. E. Biosynthesis of Ethanol from Molasses. http://greenchem.uoregon.edu/PDFs/GEMsID86.pdf (accessed December 23, 2010).
66. Meyer, S. A., Morgenstern, M. A. *Chem. Educator* 2005, 10, 130–132.
67. Bucholtz, E. C. *J. Chem. Educ.* 2007, 84, 296–298.
68. Clarke, N. R., Casey, J. P., Brown, E. D., Oneyma, E., Donaghy, K. J. *J. Chem. Educ.* 2006, 83, 257–259.
69. Miller, T. A., Leadbeater, N. E. *Chem. Educator* 2009, 14, 98–104.
70. Thompson, J. E. Biodiesel Synthesis. http://greenchem.uoregon.edu/PDFs/GEMsID87.pdf (accessed December 23, 2010).
71. Akers, S. M., Conkle, J. L., Thomas, S. N., Rider, K. B. *J. Chem. Educ.* 2006, 83, 260–262.
72. Stout, R. *J. Chem. Educ.* 2007, 84, 1765.
73. Parajó, J. C., Domínguez, H., Santos, V., Alonso, J. L., Garrote, G. *J. Chem. Educ.* 2008, 85, 972–975.
74. Smith, M. J., Gray, F. M. *J. Chem. Educ.* 2010, 87, 162–167.
75. Venditti, R. A. *J. Chem. Educ.* 2004, 81, 693–694.

8 Greener Organic Reactions under Microwave Heating

Dr. Marsha R. Baar

CONTENTS

8.1 INTRODUCTION

This chapter discusses the benefits of microwave heating over traditional thermal heating, both to the environment and to pedagogy in the organic chemistry laboratory. It is written as a practical guide for educators who may wish to transition to microwave technology, from the perspective of an organic chemistry instructor of undergraduates. The specifics of microwave ovens suitable for the undergraduate laboratory are described, without making a detailed comparison of commercially available ovens and accessories. A foundation is provided in microwave techniques, and the physics of microwave-matter interaction that generates reaction rate enhancements. Also included is a brief history outlining the development of microwave technology, and what qualifies it as a "greener" heating strategy. It is not intended as a comprehensive review of all microwave-accelerated reactions, but transformations that have proven useful for the organic laboratory are highlighted. The chapter offers answers to the following five important questions: (1) What chemistry will most likely undergo microwave rate acceleration? (2) Which type of microwave oven should one buy? (3) How much will an oven cost? (4) Which solvents should be tried? (5) What "cook" temperatures and heating periods are appropriate?

8.2 MICROWAVE HEATING AS A GREENER TECHNOLOGY

As presented in Section 1.3, the twelve principles of green chemistry focus on a number of varied sustainability concepts that can be highlighted in the introductory organic chemistry laboratory. A principle that has not been discussed in detail in the preceding chapters is the one titled "design for energy efficiency." Although some reactions will proceed readily under ambient conditions, many require a heat source in order to be complete within the timeframe of an undergraduate laboratory period. In this regard, microwave technology affords a significant advantage over more conventional heating methods, and can improve the environmental impact of experimental work. Microwave heating accelerates reaction rates and eliminates long refluxes, which require water-cooled condensers and greater energy consumption. Higher yields and fewer side products usually accompany microwave-enhanced reactions. This allows for the scale-down of reagents/solvents and easier purifications, which contribute to waste reduction and lower costs. As microwaves are absorbed by solids as well as liquids, reactions can be run without solvent, with the reagents ground together, or preabsorbed onto clays. Water is a good microwave absorber, so it can serve as an environmentally safe and renewable solvent. Microwave heating can eliminate the need for a catalyst. These features qualify microwave heating as an established greener technology.

The major advantage of microwave reactions for a laboratory instructor is the replacement of lengthy reflux periods with "cook times" that take minutes. This creates opportunities for students to undertake additional chemistry or to perform greater in-depth analyses of existing work. The rate enhancement also enables access to chemistry whose long reflux times or need for toxic/expensive catalysts has prevented incorporation into the laboratory curriculum. Even reactions involving metals can be performed in a microwave oven, as long as they remain in solution. Students additionally gain hands-on experience of a technique routinely used in industry to cleanly generate a myriad of organic substances.

8.3 HISTORICAL BACKGROUND TO MICROWAVE CHEMISTRY

Using microwaves to heat and accelerate reactions owes its discovery to the military's need for a fixed-frequency generator of microwaves required for RADAR. In the 1940s, the magnetron was developed as such a power source. Legend has it that while working on RADAR equipment, Percy Spencer at Raytheon Corporation observed that a candy bar in his pocket melted.[1] This serendipitous discovery eventually led to the manufacture of domestic microwave ovens for cooking food. In the 1950s, industrial applications of microwave heating included digestion of biological and geological samples prior to analysis, enhancement of solvent extractions, drying of products, vulcanization of rubber, and desulfurization of coal, to name a few.

It was not until the mid-1980s that the first extensions of microwave enhancement to organic reactions were reported in the research literature.[2,3] Using a 720 W domestic oven, Gedye et al. investigated amide hydrolysis reactions, esterifications, aromatic side chain oxidations, and S_N2 etherifications under microwave heating.[2] The authors found that in all cases, microwaves accelerated the reaction rates from hours to minutes, and yields were comparable to those observed under conventional reflux. The reactions were performed in sealed Teflon® vessels to exceed the boiling points of the solvents and to isolate ignitable vapors. However, these higher temperatures led to increased pressures, which on occasion led to vessel rupture. In comparison, Giguere et al. monitoerer Diels-Alder cycloadditions and Claisen rearrangements.[3] Sealed Pyrex® tubes or screw-cap pressure tubes surrounded by a vermiculite bed within a Nalgene® desiccator were used to absorb the reactions in the event of an explosion. As a 600 W domestic oven was employed, temperatures were approximated by placing sealed capillaries containing compounds of known melting point either within or affixed to the reaction container.

Encouraged by these early successes, other researchers attempted to adapt reactions to the safety limitations of domestic microwave ovens. Transformations were performed without solvent, using pulverized solid reagents or compounds preabsorbed on solid supports. High-boiling solvents such as diglyme (bp 162°C) and *o*-dichlorobenzene (bp 180°C) for nonpolar reagents, and *N,N*-dimethylformamide (bp 153°C) for polar reagents were used in open-vessel refluxes. Unfortunately, ignition of vapors by electronics remained a real possibility. Several experiments performed in domestic microwave ovens that are suitable for the organic chemistry laboratory have since been published,[4–13] where samples are usually submitted to the oven one at a time. The explosion hazard from flammable solvents and the lack of the ability to both monitor temperature and obtain uniform heating within dry media remain problematic with domestic ovens.

Industry has responded to these safety and measurement concerns with laboratory grade microwave ovens. The first industrial units had their design based on domestic multimode appliances. Ovens are now built with temperature, pressure monitoring, and feedback controls that ensure automatic shutoff if values exceed set limits. They are constructed with reinforced stainless steel, corrosion resistant walls, and a special door-locking mechanism in the event of vessel rupture. Units use the same magnetron power source, and have slotted rotating carousels that run dozens of samples simultaneously. In these research grade ovens, conditions can be easily monitored for reproducibility, safety, and translation to other ovens. The multimode ovens are appropriate as heat sources for the undergraduate organic laboratory, as they can run 14–24 reaction vessels simultaneously. Safety concerns are minimized, and the need to modify standard experiments is reduced. Volatile and flammable materials can be used without fear of explosion or need to change purification methods. These laboratory grade microwave ovens are more expensive than their domestic counterparts (~$20,000 for a base model with academic discount). However, the benefits outweigh the initial instrumental cost, particularly if expensive reflux glassware, heating mantles, and transformers are replaced. There will also be concomitant reduced chemical and waste disposal costs.

With concerns regarding safety and reaction conditions ameliorated by laboratory grade ovens, the application of microwave heating to synthetic organic chemistry has escalated. The number of research articles describing reactions accelerated by microwave heating has grown exponentially in the past twenty years, which has been primarily driven by the pharmaceutical industry. Pharmaceutical companies require a large number of new compounds to be produced in a short timeframe for testing as potential drugs. Microwave heating means that more reactions can be performed in less time, which greatly increases productivity and leads to significant cost savings. Microwave technology has been coupled with high-throughput systems, automation, and computer-aided control, which all improve efficiency, but it is only microwave heating that speeds up the chemistry itself.

8.4 MICROWAVE VS. CONVENTIONAL THERMAL HEATING

In order for a reaction to occur, molecules must collide with the correct geometric orientation and with sufficient energy. The relationship between temperature, activation energy, and rate is described by the Arrhenius expression ($k = Ae^{-Ea/RT}$). Heat provided to a reaction increases molecular kinetic energy, be it from a heating mantle, oil bath, or microwaves. This raises the temperature and the number of collisions. An increase in temperature results in a faster reaction rate, as more molecules possess sufficient energy to overcome the activation energy barrier. Conventional heating methods depend on heat transfer from an external power source, such as a hot plate or heating mantle through a reaction vessel (e.g., a glass beaker or round-bottom flask). The efficiency of this transfer is dependent on the thermal conductivity of the vessel material. Within the reaction vessel, the solution closer to the wall is hotter, so stirring is employed to disperse the heat throughout the reaction mixture. The reaction vessel is hotter than the solution, and a temperature gradient can develop within the sample, with hotter regions potentially leading to product or reagent degradation. This thermal conductivity is a slow and inefficient method for transferring energy.

Microwaves are a form of low-energy electromagnetic radiation possessing long wavelengths (1 m–1 mm) and low frequencies (300–300,000 MHz). All forms of electromagnetic radiation possess oscillating electric and magnetic fields that propagate

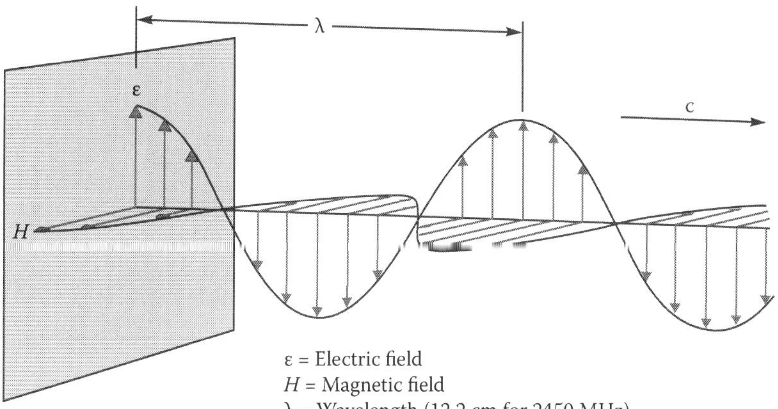

FIGURE 8.1 A microwave. (From Hayes, B. Microwave Synthesis—Chemistry at the Speed of Light. CEM Publishing, Matthews, NC, 2002. Reprinted with permission from CEM Publishing.)

in perpendicular planes (Figure 8.1). It is a molecular interaction with the electric field component of the microwave that transfers energy and generates heat. Microwave energy is not powerful enough to cleave covalent bonds; rather, its absorption by organic molecules only leads to rotational changes, nor are higher rotation levels responsible for the rapid rise in temperature observed on microwave heating.

There are two methods by which microwave energy is transferred to molecules: dipole rotation and ionic conduction. Dipole rotation is a result of polar molecules trying to align their dipole with the oscillating electric field of the microwave. The rotation of the molecule cannot quite keep up with the rapidly changing electric field, which generates a molecular friction as the faster spinning molecules collide. Ionic conduction occurs when there are free ions or ionic species being heated. Again, as the ionic species oscillate back and forth with the changing electric field, they collide and generate heat. Of the two, ionic conduction creates a greater rise in temperature.

Microwave heating, therefore, involves the conversion of electrical energy into kinetic (thermal) energy. As such, only solids and liquids experience this phenomenon. Gases do not generate a microwave heating effect, as the molecules are too far apart to interact. Microwaves are transparent to glass and Teflon, so they pass through reaction vessels and are absorbed directly by polar reagents and solvents, which leads to instantaneous localized heating and a rapid rise in temperature. Since the energy is directly delivered to the reaction mixture where it is converted to heat, there is no reliance on the thermal conductivity of a glass vessel. Consequently, using microwaves is a more efficient heating method than conventional refluxing.

There is much discussion within the scientific community as to whether there is a "microwave effect" in addition to thermal heating. It has been suggested that non-thermal effects may result from a microwave electric field, causing an orientation of a molecular dipole, which may alter the entropy of the system. For reactions that progress through more polar transition states, there may be lowering of the activation energy due to stabilization arising from electrostatic interactions of polar species with the electric field. In addition, it has been shown that a simultaneous cooling of reaction mixtures while being irradiated with microwave energy leads to rate enhancements. Whatever the reason, there is scientific evidence to give credence to an additional microwave effect. Reaction selectivity differences (chemo-, regio-, and stereoselectivity) have been observed when comparing microwave and thermal heating, but the debate rages on. Several books and articles that discuss the physics of microwave heating have been written.[14–19] They are also excellent resources for a discussion of the different types of microwave ovens and conditions, and provide many examples of accelerated reactions.

8.5 SOLVENTS FOR MICROWAVE HEATING

Most organic reactions are performed in solvents whose polarity and inertness play critical roles. Their boiling points are also taken into consideration when reflux is required to provide thermal energy to accelerate reaction rates. With microwave heating, the ability of the solvent to couple to the electrical energy is the most important criterion. This is because effective coupling leads to a rapid temperature rise, which usually corresponds to a faster reaction rate. As microwave reaction vessels are designed to withstand high pressures and temperatures, low boiling solvents that are traditionally not used for reflux purposes can be employed, regaining the benefit of their easier removal during purification. For example, methanol (bp 65°C) and ethanol (bp 76°C) can be heated to twice their boiling points in pressurized reaction vessels within seconds, even at the lowest power setting (400 W). These relatively volatile alcohols are removed rapidly during rotary evaporation, suction filtration, or oven drying. One would have to use 1-pentanol (bp 137°C) and 1-hexanol (bp 157°C) to achieve similar temperatures under reflux.

The suitability of a solvent for microwave heating is dependent upon its dielectric properties, particularly its dielectric constant and dielectric loss. The dielectric constant measures how well a solvent can store electric charges, and is a reflection of its dipole moment. Dielectric loss is the amount of microwave energy that is lost by being dissipated as heat. The ratio of dielectric loss to dielectric constant, or tangent delta (tangent loss), describes the efficiency of a solvent to absorb microwave energy and convert it into thermal energy. It is the tangent delta that is most helpful in judging the coupling efficiency and ability of a solvent to generate heat. For example, one might conclude that water, which is very polar with a high dielectric constant (80.4), should be among the best microwave solvents. However, its dielectric loss of only 9.9 indicates a relatively poor ability to convert microwave energy into thermal energy. The tangent delta value of water (0.123; Table 8.1) reflects this inefficiency, and ranks this solvent as a moderate absorber.

So which solvents are efficient absorbers of microwave energy? Excellent absorbers are polar solvents, including ethanol, methanol, and dimethylsulfoxide (DMSO). Moderate absorbers are water, *N,N*-dimethylformamide (DMF), acetonitrile, and acetone. Particularly poor absorbers are chloroform, dichloromethane, tetrahydrofuran (THF), toluene, pentane, and hexane.

An important message here is that whatever type of solvent has traditionally been used for a particular transformation (e.g., a polar aprotic solvent to promote an S_N2 or

TABLE 8.1
Dielectric Constant (ε′), Tan δ, and Dielectric Loss (ε″) for Thirty Common Solvents (Measured at Room Temperature and 2,450 MHz)

Solvent (bp °C)	ε′	Solvent	Tan δ	Solvent	ε″
Water (100)	80.4	Ethylene glycol	1.350	Ethylene glycol	49.950
Formic acid (100)	58.5	Ethanol	0.941	Formic acid	42.237
DMSO (189)	45.0	DMSO	0.825	DMSO	37.125
DMF (153)	37.7	2-Propanol	0.799	Ethanol	22.866
Acetonitrile (82)	37.5	1-Propanol	0.757	Methanol	21.483
Ethylene glycol (197)	37.0	Formic acid	0.722	Nitrobenzene	20.497
Nitromethane (101)	36.0	Methanol	0.659	1-Propanol	15.216
Nitrobenzene (202)	34.8	Nitrobenzene	0.589	2-Propanol	14.622
Methanol (65)	32.6	1-Butanol	0.571	Water	9.889
NMP (215)	32.2	Isobutanol	0.522	1-Butanol	9.764
Ethanol (78)	24.3	2-Butanol	0.447	NMP	8.855
Acetone (56)	20.7	2-Methoxyethanol	0.410	Isobutanol	8.248
1-Propanol (97)	20.1	*o*-Dichlorobenzene	0.280	2-Butanol	7.063
MEK (80)	18.5	NMP	0.275	2-Methoxyethanol	6.929
2-Propanol (82)	18.3	Acetic acid	0.174	DMF	6.070
1-Butanol (118)	17.1	DMF	0.161	*o*-Dichlorobenzene	2.772
2-Methoxyethanol (124)	16.9	1,2-Dichloroethane	0.127	Acetonitrile	2.325
2-Butanol (100)	15.8	Water	0.123	Nitromethane	2.304
Isobutanol (108)	15.8	Chlorobenzene	0.101	MEK	1.462
1,2-Dichloroethane (83)	10.4	Chloroform	0.091	1,2-Dichloroethane	1.321
o-Dichlorobenzene (180)	9.9	MEK	0.079	Acetone	1.118
Dichloromethane (40)	9.1	Nitromethane	0.064	Acetic acid	1.079
THF (66)	7.4	Acetonitrile	0.062	Chloroform	0.437
Acetic acid (113)	6.2	Ethyl acetate	0.059	Dichloromethane	0.382
Ethyl acetate (77)	6.0	Acetone	0.054	Ethyl acetate	0.354
Chloroform (61)	4.8	THF	0.047	THF	0.348
Chlorobenzene (132)	2.6	Dichloromethane	0.042	Chlorobenzene	0.263
o-Xylene (144)	2.6	Toluene	0.040	Toluene	0.096
Toluene (111)	2.4	Hexane	0.020	*o*-Xylene	0.047
Hexane (69)	1.9	*o*-Xylene	0.018	Hexane	0.038

Source: From Hayes, B. Microwave Synthesis—Chemistry at the Speed of Light. CEM Publishing, Matthews, NC, 2002. Reprinted by permission from CEM Publishing.

E2 reaction), it can likely remain the same for its microwave counterpart. DMSO, DMF, acetonitrile, and acetone can still be used; there is often no need to rethink solvent choice when adapting a reaction to microwave heating. However, sometimes it may be advantageous to switch to a better absorber to shorten the ramp time (the period of time required to heat to a desired cooking temperature), and the cook time itself.

From an environmental perspective, water is an excellent solvent choice for microwave-accelerated reactions. It is readily available, renewable, nontoxic, inexpensive, and a reasonable microwave absorber. Its polarity does not allow it to solvate hydrophobic organic compounds particularly well. Despite this, Section 4.3 details a number of aqueous reactions that have successfully been undertaken in undergraduate laboratories. Interestingly, at supercritical temperatures (~300°C), the properties of water alter significantly. It experiences a drop in dielectric constant, which enables it to solvate nonpolar materials more effectively. Although supercritical temperatures cannot be achieved in a microwave oven, near supercritical temperatures are possible. This adds an intriguing aspect to aqueous reactivity under microwave conditions.[20]

As discussed in Section 5.5, ionic liquids are a new class of environmentally benign solvents, which couple very efficiently with microwave energy through ionic conduction. They are organic salts comprised of an organic cation and an organic or inorganic anion. Their usefulness comes from their low melting points, which give them a large liquid temperature ranging from –96 to 200°C. They readily dissolve both organic and inorganic compounds, and possess negligible vapor pressure, which minimizes hazards from overpressurization upon heating. They are commercially available (although expensive), which necessitates their recovery and reuse. However, recent articles indicate that the addition of a few drops of an ionic liquid to a microwave reaction mixture can facilitate a rate increase.[21]

Common ionic salts, such as NaCl or Na_2CO_3, can be added to act as heat sinks during microwave reactions. Indeed, using a magnetic stir bar adds energy to the reaction mixture, because the iron core of the bar absorbs energy. The microwave manufacturer Milestone, Inc. sells Weflon®-coated stir bars that are excellent absorbers of microwave energy, as a way to get more heat into a reaction. It is the graphite within Weflon that serves as the heat sink in this instance.

8.6 A COMPARISON OF MULTIMODE AND MONOMODE MICROWAVE OVENS

There are four major manufacturers of microwave ovens: Anton Paar, Biotage, CEM Corporation, and Milestone, Inc.[22–25] CEM and Milestone have both had a long history designing units for the academic laboratory, while Anton Paar has only recently targeted this market. I have personally tested both the multimode MARS® and monomode Discover® units marketed by CEM, as well as Milestone's multimode START system. Milestone does not produce a monomode unit. The MARS standard reaction vessels can withstand the temperature and pressures that are routinely encountered during undergraduate experiments. The Milestone START system basic vessel set is more limited, and requires an upgrade to the more resilient research quality vessels at additional expense. An important factor to take into account when purchasing an oven is the knowledge of the sales personnel. Their ability to discuss the many options and help with the selection of needed features at the most affordable price is critical. The quality of the technical support provided over the telephone, and the availability of repair personnel are other considerations. The following discussion is therefore relevant to the CEM MARS and Discover systems that were purchased. An additional perspective on incorporating microwave heating into undergraduate curricula, and potential ovens to consider, has recently been published.[26]

The magnetrons in all commercial microwave ovens irradiate at 2.45 GHz (12.3 cm), to avoid interference with RADAR, telecommunication, and cellular phone frequencies. They can be operated at different power settings, have built-in magnetic stirring, and feature efficient post-run air cooling. The domestic microwave ovens sold for home cooking are multimode units, and served as the prototype for the first laboratory grade ovens. Multimode units have multiple wavelengths of microwave energy within the cavity of the oven. These microwaves reflect off the oven walls, creating complex wave patterns. Wherever the waves reinforce one another, there is a greater intensity of microwave energy, or hot spots. Conversely, wherever the waves negate one another, energy is minimized, producing cold spots. It is for this reason that multiple-sample holding carousels are rotated within the cavity to even out exposure to varying levels of microwave energy.

Monomode ovens have a small cavity, which holds only one wavelength of microwave energy in which the sample is positioned at the crest. This provides greater field homogeneity and density, with sufficient microwave energy for even small samples to absorb, so reaction scale-down is possible. The higher-power microwave field produces faster reaction rates. Field homogeneity is important to ensure reproducible reaction conditions and consistent yields. The less dense and less homogeneous microwave fields found within multimode units allow for more variation in reaction conditions and yields. However, increasing cook time can compensate for the irregularity in field energy, thus keeping yields comparable between vessels. Multimode ovens are the ones most often employed in the undergraduate organic laboratory, as many samples can be irradiated simultaneously and one can conveniently run experiments on a miniscale or a macroscale.

The CEM MARS unit can be purchased with temperature and pressure monitoring. Temperature is measured by either an infrared sensor located in the floor of the unit that measures the sample vessels as they rotate over it, or a more accurate (and more expensive) fiber optic thermocouple placed in the reference vessel. Pressure is measured by placing a detection device in a special reference cell. Units can be purchased with sensors that measure an unsafe buildup of solvent vapor within the cavity. The reactions vessels are thick walled to withstand high temperatures and pressures. Vessel caps have individual pressure release devices to allow safe venting.

When ordering the MARS oven, one must select the size, type, and number of reaction vessels, as this determines the accompanying rotating carousel. Vessels are available in both glass and Teflon. It should be noted that Teflon vessels stain, and glass ones are easily washable and reusable. There is a carousel that holds fourteen HP 100 mL reaction vessels, and another style that holds sixteen GlassChem 20 mL vessels. The larger vessels have a slightly more complicated locking system. They are loosely capped, slipped into Kevlar® sleeves, then screwed into Teflon modular frames and placed in slots on the carousel. The smaller vessels are easier to use, as they have their own screw tops and the Kevlar sleeves are already in positioned slots on the carousel. Once all the samples are loaded, an additional Teflon circular plate covers and locks them in place.

Student lab enrollment and the scale on which instructors plan to work are critical factors in deciding which reaction vessel and carousel set to purchase. The laboratory sections at Muhlenberg College hold a maximum of sixteen students. Reactions are scaled to between 1.0 and 2.0 g of substrate dissolved in 5.0–7.0 mL of solvent, so the smaller vessels are appropriate and less expensive. When the MARS oven was initially purchased, only the 100 mL glass vessels were available. These larger vessels and their pressure-locking modular frames are somewhat cumbersome, so the smaller 20 mL vessels were acquired due to their increased ease of operation and greater number of vessel slots on the carousel. All the microwave-accelerated reactions that were originally performed in the 100 mL vessels were rerun in the smaller vessels. The reaction conditions and results were transferable by following this power settings guide for percentage of the carousel occupied: 400 W power for up to 25% occupation, 800 W for 25%–50%, and 1,600 W for 50%–100%.

The price of ~$20,000 (with academic discount) for the microwave oven includes the glass reaction vessels, caps, Kevlar sleeves, carousel, stir bars, fiber optic temperature probe, glass thermowell, and wrenches. There are minimum and maximum volumes for each size of reaction vessel, as well as a minimum amount of solvent required within the oven for a reaction. The minimum and maximum solvent volume levels are 5 mL/50 mL and 3 mL/15 mL for the 100 and 20 mL vessels, respectively. The thermowell and fiber optic temperature probe are interchangeable between different sizes of reaction vessels and carousel sets. A reasonable budget for one year is ~$1,000 for replacement of glass reaction vessels ($85), fiber optic probes ($290), and thermowells ($225).

As the monomode Discover unit runs one sample at a time, it may not be as flexible for use in the organic chemistry laboratory, unless enrollments are small or reaction times are short. Scale-up is also limited in monomode units due to the necessarily small cavity size, but 10 and 35 mL glass reaction vessels can hold amounts of solvent typical in the organic chemistry laboratory (0.25 to 5–7 mL in the 10 mL vessel). The 35 mL vessel can hold 15–25 mL, depending on the absorbing ability and boiling point of the solvent. These reaction vessels are much less expensive than those used in the MARS unit. (A box of 100 vessels (10 mL each) costs $94, with 100 caps costing a further $110. A box of twenty-five vessels (35 mL each) costs $170, with twenty-five caps costing a further $85.) The vessels can be reused, although the caps cannot. The importance of filling the reaction vessels appropriately must not be overstated. Sufficient headspace must be left to prevent undue pressure buildup. In addition, if the sample entirely vaporizes, gases do not absorb microwave energy, and the temperature will drop. The oven will interpret this as a need to keep heating in order to restore the set temperature. This will lead to an unsafe rise in pressure, and the run will most likely be aborted.

All Discover microwave ovens simultaneously measure temperature, pressure, and power. The Discover system can be safely heated to higher temperatures and pressures than the MARS. The basic Discover model, the Labmate, costs ~$10,000. It has a pressure-sealing device that the operator places on top of the capped 10 mL reaction vessel once it is loaded into the cavity. The more versatile (and therefore more expensive) SP model (~$15,000) has an automated pressure cover, and can accommodate both 10 and 35 mL vessels. For both Discover models, the pressure-sealing technology allows for safety venting during the reaction and cooldown. The SP model can be expanded to include the Explorer robotic autosampler (additional cost of $15,000). This feature requires computer control, which has the flexibility to subject each reaction vessel to different conditions, including switching vessel size between runs. The robotic system is excellent for method development and optimization of reaction conditions. All Discover models permit retrieval of cooked samples for immediate workup, whereas the MARS system retains all vessels until the entire run is finished. If reaction conditions can be devised so that the entire ramp, cook, and cooldown time is ~2 minutes, then the Explorer may find use in larger laboratory sections. The sequential operation will help to reduce student congestion during workup and analyses. Colleges that have small laboratory enrollments may opt for a nonrobotic monomode unit, as it is less expensive than the MARS unit, provides better reproducibility, and affords earlier access to vessels.

Some final considerations include replacement costs for easily broken and disposable items. If a MARS oven equipped with temperature monitoring is purchased, then the thermowell and fiber optic probe are exposed parts that are regularly handled and sometimes broken. One should budget ~$500/year to replace them. The Discover temperature detector is protected in the base of the cavity, and there are no exposed or handled items, but students can still break the glass reaction vessels! The cost of replacing these is outlined earlier in this section.

8.7 MICROWAVE-ACCELERATED REACTIONS FOR THE UNDERGRADUATE LABORATORY

8.7.1 Searching for Examples

If one does not want to develop a microwave-enhanced reaction himself or herself, the scientific literature is full of examples from which to choose. One approach is to use the "Research Topic" search function in SciFinder® by combining the word *microwave* with a specific reaction name, e.g., "microwave-enhanced Diels-Alder reactions." Abstracts from professional meetings can provide access to more experiments. A few practical manuals devoted to microwave-accelerated reactions have been published to augment the few examples reported in common laboratory texts.[27,28] Certain faculty-posted student experimental handouts are available online, some as part of the Greener Educational Materials for Chemists (GEMs) database.[29] Microwave oven manufacturers also have online libraries that can be freely searched. As outlined in Section 1.5.2, the *Journal of Chemical Education* is an excellent resource where many microwave-accelerated reactions have appeared. These are almost all geared toward the conditions of a typical undergraduate organic chemistry laboratory. A number of representative experiments are discussed in Section 8.8.

8.7.2 Adapting a Reaction for Microwave Acceleration

When I initially considered using microwave heating to accelerate reactions with lengthy reflux times, I had hoped to find experiments from the literature to adopt. Unfortunately, at that time, only a few examples of microwave reactions commonly taught in the organic chemistry laboratory were reported. For those procedures, the inconsistency and lack of clarity concerning reaction conditions were frustrating. Many reactions were performed in domestic ovens, which did not measure

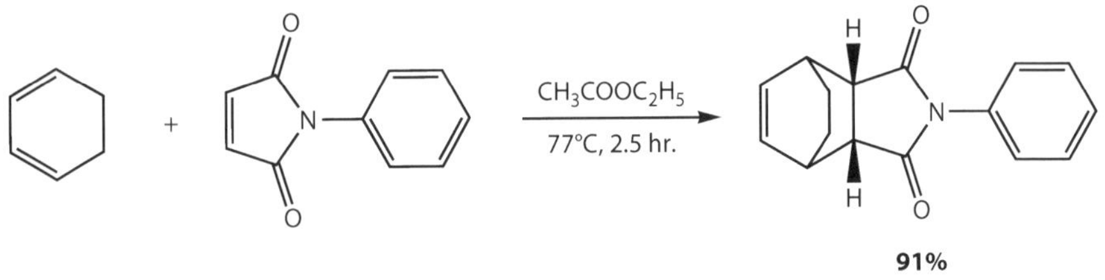

SCHEME 8.1 Diels-Alder cycloaddition between 1,3-cyclohexadiene and *N*-phenylmaleimide under conventional reflux heating.

temperature or pressure. The different oven brands, variations between models, and their power setting information were virtually impossible to interpret.

Therefore, I decided to develop my own rate enhancement methods for a few familiar reactions with long heating times. My approach to microwave acceleration was very simple: *don't change known reaction conditions unless absolutely necessary*. If there was an established reflux procedure for a reaction specifying the solvent choice, concentration of reagents, and purification method, I did not want to have to "reinvent the wheel." Rather, my intent was to simply increase the reaction rate by safely heating at a higher temperature in a closed vessel. The first reaction investigated was a known Diels-Alder cycloaddition performed in ethyl acetate, which is a poor absorber of microwave energy.[30] Under reflux conditions, the cycloaddition between 1,3-cyclohexadiene and *N*-phenylmaleimide takes 2.5 hours to proceed in excellent yield, with a useful visual cue when the reaction is over. The starting dienophile is a yellow solid, and the cycloadduct is a white solid that precipitates out of the reaction mixture, so the transformation can be judged as completed when the yellow color disappears (Scheme 8.1).

The reaction was initially performed in the CEM MARS 5 multimode oven using a 100 mL glass reaction vessel. The reagents and solvent were scaled to the lowest permissible limits, namely, no less than 5 mL of an effective absorbing solvent per reaction vessel. Although ethyl acetate is not a good absorber of microwave energy, choosing the lowest power setting (400 W) ensured that there would not be too much excess microwave energy within the oven cavity, which could do some damage. The experimental considerations were now as follows: (1) what temperature to heat to, (2) the solvent pressure response (would it be within the vessel safety limits?), and (3) the length of the cook time. A pressure and temperature response graph provided by the manufacturer showed that ethyl acetate (bp 77°C) could be heated close to the maximum oven temperature of 200°C, and still be within the glass vessel pressure safety limit of 150 psi (Figure 8.2). However, this graph indicated the response for a pure solvent, not for a reaction mixture.

Reagents can contribute to the temperature and pressure if they also absorb microwave energy. To leave a safety pressure cushion, a temperature of 130°C was selected, because the corresponding pressure (~40 psi) would not force a safety venting of the solvent or the diene (cyclohexadiene, bp 80°C). Different samples were cooked for five minutes, and then ten minutes, therefore doubling the reaction period until the yellow color disappeared. The stirring option was utilized in all runs. Increasing the reagent concentrations relative to the solvent was also investigated, and showed a

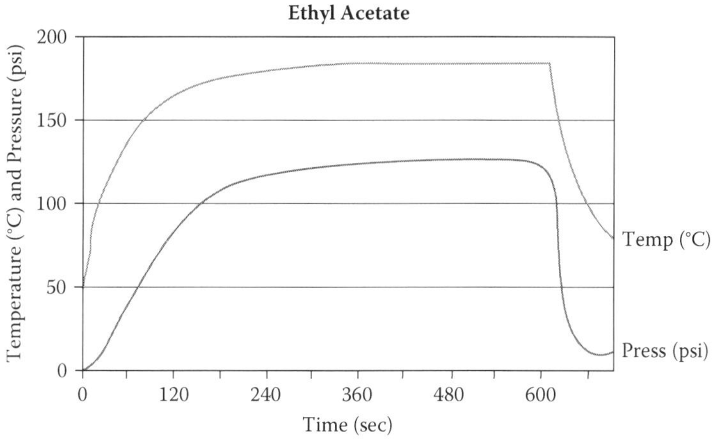

FIGURE 8.2 Temperature and pressure response curve for ethyl acetate. (From Hayes, B. Microwave Synthesis—Chemistry at the Speed of Light. CEM Publishing, Matthews, NC, 2002. Reprinted with permission from CEM Publishing.)

rate increase. The ramp time was recorded along with the percent wattage needed to keep the reaction mixture at 130°C. This information was useful for determining the total time the sample remained in the oven and therefore inaccessible. If the ramp, cook, and cooldown times add up to a lengthy period, then it clearly diminishes the benefit of microwave heating in the organic chemistry laboratory. It also provides an indication of the ability to scale up the reaction. If the power must remain on to keep a few samples at the cook temperature, it does not bode well for running a full carousel, even at a higher power setting.

Once the heating period finished, the sample was allowed to cool within the oven below the boiling point of the solvent, to reduce pressure and prevent ejection of vessel contents upon opening. This usually took several minutes, and often as long as the cook time. The reaction workup consisted of further cooling the vessel in a tap water bath, followed by vacuum filtration and washing with chilled ethyl acetate. The dried product was weighed and its melting point measured to verify high yield and purity. The best conditions were a balance between temperature, cook, and cooldown times. It is desirable to accelerate the reaction at the lowest possible temperature without increasing the cook time, so that the cooldown period is short. For this Diels-Alder reaction, the best result was achieved with a ten-minute cook time at 130°C. The ramp time for a fully loaded carousel (fourteen samples) at full power (1,600 W) was nine minutes, and cooldown took ten minutes, so from start to finish the entire microwave process was twenty-nine minutes.

In an attempt to shorten the overall time, the cycloaddition was repeated in absolute ethanol, which is a better absorber of microwave energy (Figure 8.3). The ramp time shortened to forty seconds, which reduced the overall time by eight minutes (Scheme 8.2). The product precipitated as before, so there was no need to change purification conditions. The boiling point of ethanol (76°C) is comparable to ethyl acetate, so cooldown was not affected. It was interesting to note that although the

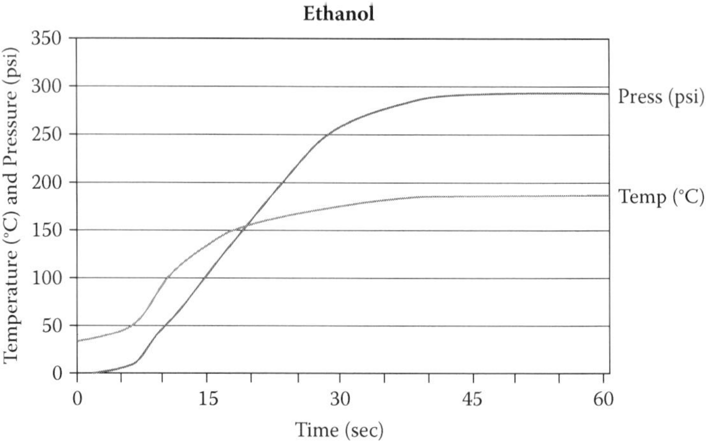

FIGURE 8.3 Temperature and pressure response curve for ethanol. (From Hayes, B. Microwave Synthesis—Chemistry at the Speed of Light. CEM Publishing, Matthews, NC, 2002. Reprinted with permission from CEM Publishing.)

starting dienophile was insoluble in ethanol, this did not seem to affect the efficiency of the reaction. The reagents were measured directly into the reaction vessels, which were then capped and placed on the carousel. As there is no water-cooled condenser or stirring plate required, setup and cleanup involved simply cleaning the reaction vessel, its cap, and stir bar.

Highly encouraged by this first successful microwave-accelerated reaction, an attempt was made to enhance the rate of a sluggish Wittig salt formation known to require seven hours of reflux in xylenes (bp ~144°C). This S_N2 reaction is included in several laboratory textbooks, but with a shorter reflux period of ninety minutes, which only produces 30–40% yields.[31] The choice of xylenes, a high-boiling solvent, speaks to the large energy requirement. The low yield requires a scale-up of reagents, so that a sufficient quantity results for further reactions. The Wittig salt is converted into an ylide, which is then usually reacted with (*E*)-cinnamaldehyde to produce (*E*,*E*)-1,4-diphenyl-1,3-butadiene. The xylenes are among the worst absorbers of microwave energy (Table 8.1), but as their pressure is quite low at 200°C, the reaction was tried at the maximum temperature setting of the microwave oven. There

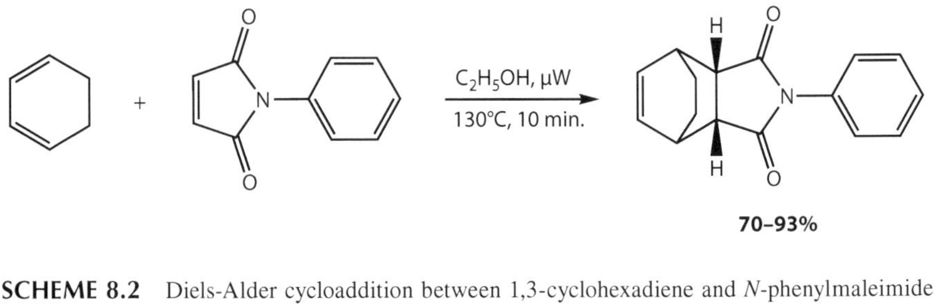

SCHEME 8.2 Diels-Alder cycloaddition between 1,3-cyclohexadiene and *N*-phenylmaleimide under microwave heating.

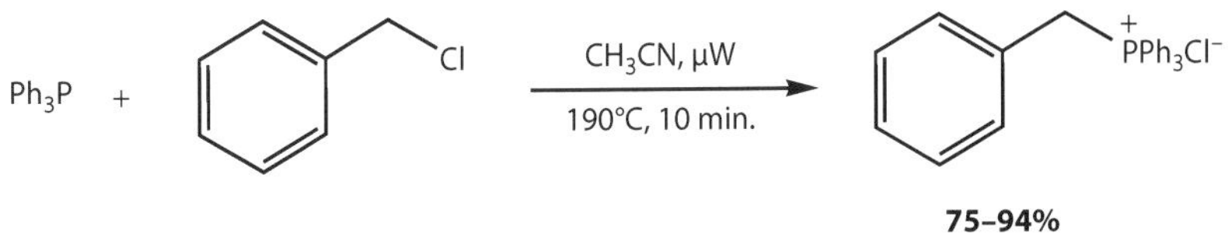

SCHEME 8.3 Wittig salt formation from benzyl chloride and triphenylphosphine.

was quite a lengthy ramp time (fifteen minutes with a full carousel at full power) and cooldown (twelve minutes) involved, but the reaction progressed well, with ~56% product recovery within ten minutes. This was an impressive result, as neither the reagents nor solvents were good microwave absorbers, yet the reaction time decreased from hours to minutes. In an attempt to shorten both ramp and cooldown times, the solvent was switched to acetonitrile, a moderate absorber of microwave energy (Table 8.1). The ramp time decreased to 6.5 minutes and yields improved to over 90% (even with a reduction in cook temperature to 190°C) (Scheme 8.3). As often happens, switching solvents affected the reaction workup. The Wittig salt, which precipitated fully in the nonpolar xylenes, was slightly soluble in acetonitrile. Ligroin was therefore added to the cooled reaction mixture to further increase precipitation of the salt.

What have I learned from developing these two accelerated reactions? First, it is a good idea to initially change nothing except the heat source and temperature, and to run the reaction in the same solvent and at the same reagent concentrations. Second, it is important to check the literature for the pressure response of a solvent at various temperatures. If the solvent pressure response permits, a cook temperature ~50°C above the boiling point of the solvent can be tried as a starting point. If the reflux period is known for the reaction and it is within a few hours, then a microwave heating period of five to fifteen minutes is a good place to begin. For reflux periods that span days, an initial cook time of thirty minutes may be necessary. If that is not sufficient, then the heating interval should be doubled. Once the optimum temperature is determined, attempts can be made to reduce the cook times. If time still remains a factor after this, a better-absorbing solvent can be attempted.

I am currently investigating these Diels-Alder and Wittig reactions and others that have shown microwave rate enhancements in a multimode unit in the monomode Discover oven. As previously mentioned, monomode units possess a microwave field of greater density. This provides for better reproducibility, effective scale-down, and also the possibility of increasing reaction rates even further. Initial results indicate that this is indeed the case. For example, in the MARS unit, the Diels-Alder reaction took ten minutes to cook, but with ramp and cooldown the samples were in the oven for a total of ~21 minutes. In the Discover unit (at one-third of the scale) it also took less than one minute to ramp to 130°C. The reaction was complete upon reaching the set temperature, and cooldown took another minute for a total of two minutes before the sample was available. In this instance, the greater amount of microwave energy provided by the monomode unit shortened the reaction time to one-tenth of the previous time observed with a multimode oven. If this rate enhancement is typical, then the monomode unit may find greater usage in the undergraduate laboratory.

8.8 LITERATURE EXAMPLES OF MICROWAVE-ACCELERATED REACTIONS

The remaining portion of this chapter provides literature examples of microwave-accelerated reactions that an instructor of undergraduates may wish to incorporate into an introductory or advanced organic chemistry laboratory sequence. Some reactions cited come from laboratory textbooks or articles published in the *Journal of Chemical Education* that have been test driven by students. Others are from recent articles performed in laboratory grade ovens with reported conditions, which facilitates their transfer to other ovens. Some experiments that proceed in domestic microwave ovens are included. These are ones where dry media or solvent-free conditions are employed, in order to avoid ignition of flammable solvents or unsafe pressures. Katritzky et al. mention some of these reactions and profile others that are performed by students using industrial monomode ovens.[32] There are many fine examples not covered in this section, and the reader is referred to an article that summarizes some earlier work.[33]

The selections described here span the gamut, including nucleophilic substitutions, heterocyclic ring formations, carbonyl condensations, transition metal-catalyzed reactions, pericyclic reactions, esterifications, oxidations, reductions, and hydrolyses. Reactions that complement microwave heating with another green technique (e.g., catalysis, utilizing water as a solvent, solvent-free reactivity, and using solid supports) have been purposefully chosen. Many more examples of water-based microwave reactions and those performed in dry media or under solvent-free conditions are described in three excellent reviews.[33–35] Along with the chemistry, the examples include cook temperatures, reaction times, and percentage yields. Whenever possible, the type of oven employed and power setting are also listed. Unfortunately, not all of the above information has been routinely reported. Refereed journals are now requiring computer printouts from authors that list temperature, pressure, and power profiles, along with necessary cook times. Some interesting earlier chemistry performed in domestic ovens without rigorous safeguards and temperature measuring devices is additionally mentioned. This is to provide the reader with a range of reactions that are known to undergo microwave rate enhancement, with the hope that they can be reworked in a laboratory grade oven without much modification.

8.8.1 Nucleophilic Substitutions

8.8.1.1 Williamson Ether Syntheses

Many reactions performed in the organic chemistry teaching laboratory utilize an S_N2 step in which oxygen, phosphorus, nitrogen, or carbon nucleophiles are alkylated. These transformations have been historically performed in solvents, particularly polar aprotic ones. A classical oxygen nucleophile is an alkoxide ion employed in Williamson ether syntheses. Many such reactions have exhibited rate accelerations under microwave heating. The two examples presented here produce solid and liquid ether products, respectively. Ethyl 2-naphthyl ether (mp 37°C) can be prepared in a multimode oven[36] (Scheme 8.4), and butyl *p*-tolyl ether synthesized in both multimode and monomode ovens[27] (Scheme 8.5). The latter reaction is performed in

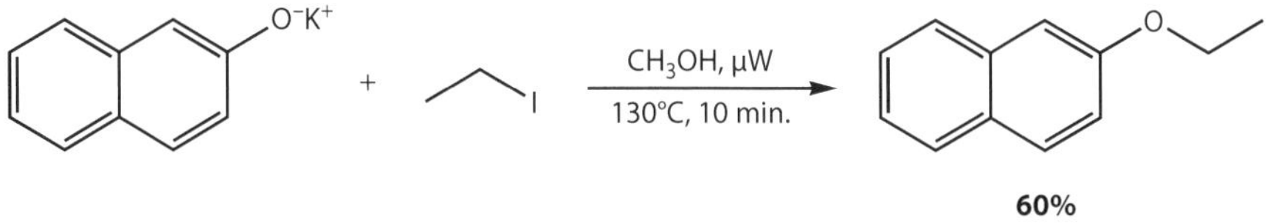

SCHEME 8.4 Williamson synthesis of ethyl 2-naphthyl ether.

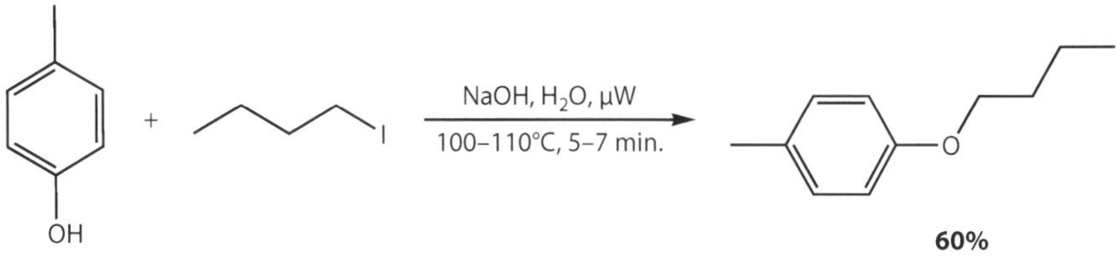

SCHEME 8.5 Williamson synthesis of butyl *p*-tolyl ether.

water, which requires the addition of a phase transfer catalyst (tetra-*n*-butylammonium bromide (TBAB)). In the multimode unit, the cook temperature is 100°C for five minutes, and in the monomode oven it is 110°C for seven minutes. Both methods produce butyl *p*-tolyl ether in yields of around 60%.

8.8.1.2 A Wittig Reaction

The most frequently used phosphorus nucleophiles in the organic teaching laboratory are phosphines, as part of a Wittig reaction. A Wittig salt is formed in excellent yield by reaction between triphenylphosphine and benzyl chloride in acetonitrile, when heated to 200°C for ten minutes in a multimode oven (Section 8.7.2). Its conversion to the ylide using sodium ethoxide and subsequent reaction with (*E*)-cinnamaldehyde is a staple in many laboratory texts. Martin and Kellen-Yuen have employed heating in a domestic microwave oven to generate the ylide from a preformed Wittig salt, in a solvent-free environment. The Wittig salt is simply mixed with potassium carbonate and one of five assigned aromatic aldehydes, then heated for a short time.[37] Observed percentage yields obtained by students mirror results from the research literature (Scheme 8.6).

8.8.1.3 *N*- and C-Alkylations

Ju and Varma have reported the synthesis of tertiary amines via *N*-alkylation of primary and secondary amines with an alkyl halide.[38] Water is the solvent for these

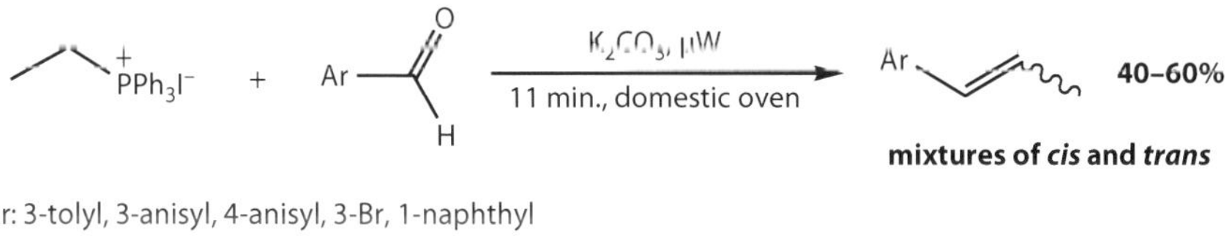

SCHEME 8.6 Generation of an ylide and subsequent Wittig reaction.

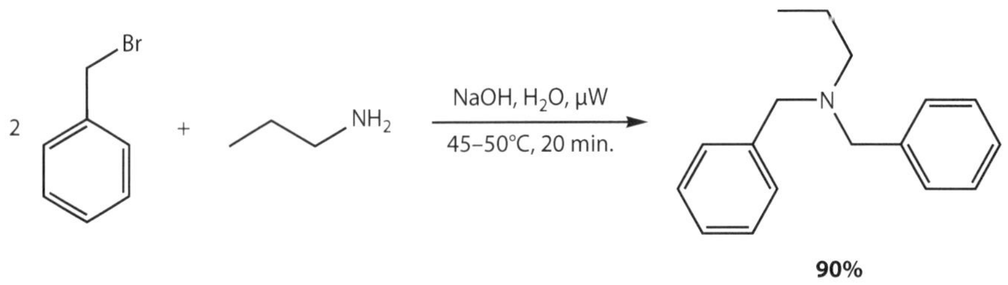

SCHEME 8.7 *N*-Alkylation of primary amines to form tertiary amines.

reactions, which are performed in an open vessel within a multimode oven, at temperatures ranging from 45 to 100°C. The reaction time is twenty minutes, compared to twelve hours of traditional reflux (Scheme 8.7). Impressive yields, coupled with the suppression of secondary amine side products from primary amine reactants, illustrate the benefits of microwave heating.

Alkylation of enolate anions is an important topic covered in introductory organic chemistry courses. Experiments designed at Muhlenberg College allow preparation of diethyl *n*-butylmalonate by alkylation of the anion derived from malonic acid (unpublished results; Scheme 8.8). Currently, the best condition for this reaction in the MARS oven is a solvent-free environment, utilizing a phase transfer catalyst (TBAB) with potassium carbonate as the base.

8.8.1.4 Nucleophilic Aromatic Substitutions

Nucleophilic aromatic substitution reactions are difficult to promote by conventional heating, as they require high temperatures and long reflux times. Additionally, the aryl component must typically contain at least one electron-withdrawing group. McGowan and Leadbeater have described the reaction between 2,4-dinitrobromobenzene

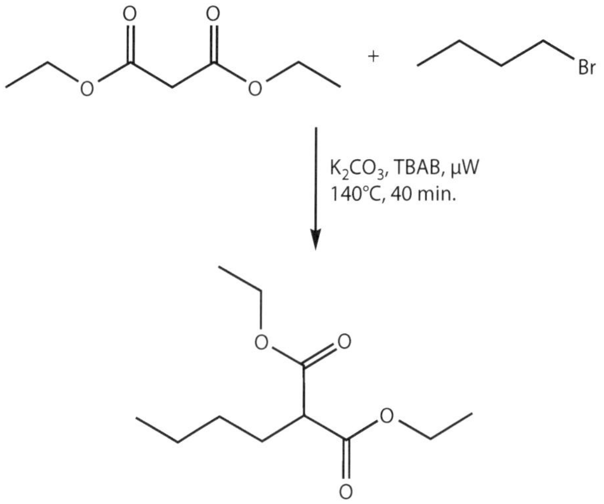

SCHEME 8.8 C-Alkylation of diethylmalonate with *n*-butyl bromide.

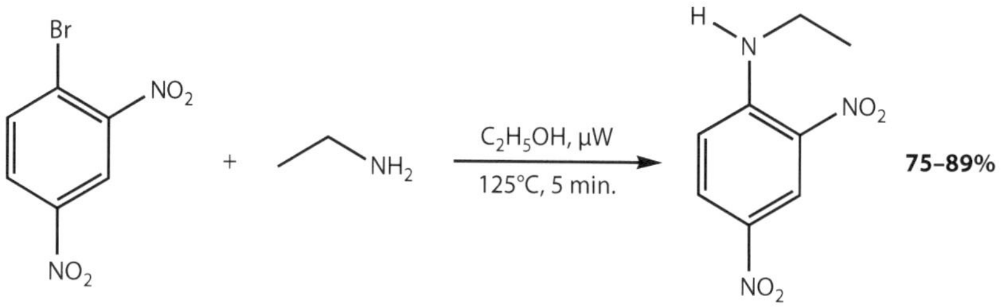

SCHEME 8.9 Nucleophilic aromatic substitution of 2,4-dinitrobromobenzene.

and ethylamine with consistent results in both multimode and monomode ovens[27] (Scheme 8.9).

The water-based reaction outlined by Cherng (which employs chiral amino acids as nucleophiles) produces *N*-aryl α-amino acids in under a minute of microwave heating in a monomode unit, vs. several hours at room temperature[39] (Scheme 8.10). Sodium bicarbonate is required to ensure the amino groups are not protonated, so that they can act as nucleophiles. As a comparison, no products are formed after one minute of oil bath heating at 95°C.

8.8.2 Heterocyclic Syntheses

Barbiturate syntheses are commonplace in most organic chemistry laboratory textbooks.[40] Urea is usually employed as the nitrogen nucleophile, but reactions often require several hours of reflux. Majetich and Hicks have described a successful condensation between urea and diethyl *n*-butylmalonate to produce a barbiturate[41] (Scheme 8.11). This is a noteworthy improvement on the lower yields obtained after conventional heating. This urea condensation and the previously discussed alkylation of diethyl *n*-butylmalonate (Section 8.8.1.3) could easily be combined to form a two-step synthesis of a barbiturate to be accomplished in two laboratory periods.

Hydantoins, which are structurally related to barbiturates, are another class of pharmaceutically important compounds. Coursindel et al. have recently reported an undergraduate solid-supported microwave-assisted synthesis of a library of hydantoins.[42] Commercially available resins preloaded with protected amino acids are

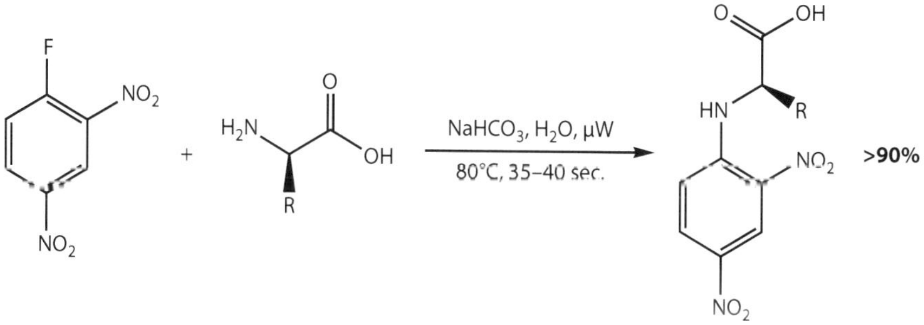

SCHEME 8.10 Water-based nucleophilic aromatic substitution with chiral amino acids.

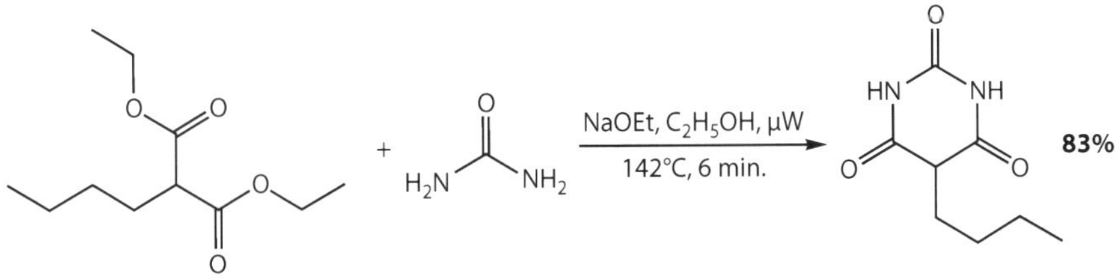

SCHEME 8.11 Reaction of urea and diethyl *n*-butylmalonate to form a barbiturate.

deprotected, and then reacted with phenylisocyanate under microwave heating to promote an additive coupling. The addition of triethylamine and microwave heating subsequently completes the cyclization to form the hydantoin and simultaneously cleave it from the resin (Scheme 8.12).

The microwave-mediated preparation of lophine (2,4,5-triphenylimidazole) has been described in a monomode Milestone START oven.[43] This system permits sixteen students to perform the reaction in under twenty minutes. Benzaldehyde, aqueous ammonia, and benzil are irradiated in glacial acetic acid to form the product (Scheme 8.13). Adaptations have been made to the protocol to facilitate operation in a domestic oven. Musiol et al. have developed a series of heterocyclic reactions for use in the undergraduate laboratory performed either without solvent, where the reactants are mixed together neatly, or first adsorbed onto a carrier surface (e.g., alumina). Irradiation is performed in an open vessel in a domestic oven for no more than three minutes of heating to produce high yields of cyclic anhydrides, hydantoins, imidazoles, and phthalimides, among others.[44]

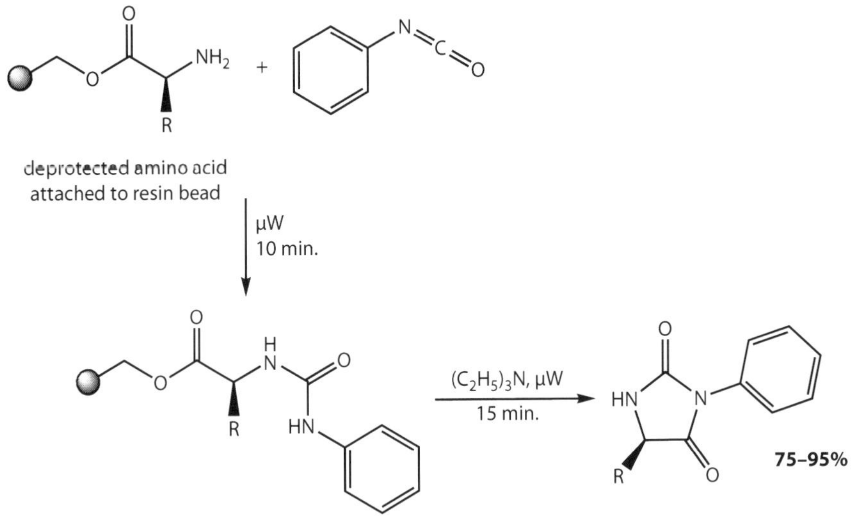

SCHEME 8.12 Solid phase microwave synthesis of hydantoins.

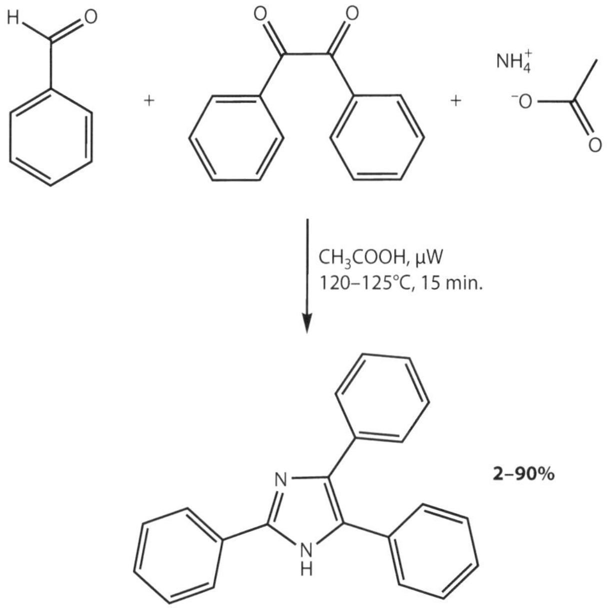

SCHEME 8.13 Synthesis of lophine.

8.8.3 Carbonyl Condensation Reactions

Aldol condensations and Michael additions are other classes of reactions that can be accelerated by microwave heating. A cyclic self-aldol condensation performed in aqueous NaOH[27] and a crossed-aldol reaction promoted in a solvent-free environment[45] are both illustrated in Scheme 8.14. The latter process uses potassium carbonate as a base with the assistance of Aliquat® 336 (Section 3.3.1). This facilitates preparation of jasminaldehyde, an artificial violet scent used in perfumes.

Rao and Jothilingam have performed a series of Michael additions in both multimode and monomode units, using ovens not equipped with temperature monitoring.[46] With microwave irradiation, strong bases often used for enolate generation are replaced with potassium carbonate, with resulting minimization of side reactions (Scheme 8.15).

8.8.4 Transition Metal-Catalyzed Reactions

Reactions that create carbon-carbon bonds are of supreme importance in synthetic planning. In addition to the previously described enolate and alkylation chemistry, palladium-catalyzed cross-couplings, such as Heck and Suzuki reactions, offer another avenue. Palladium-catalyzed vinylic substitutions (Heck reactions) have been performed by Wali et al. in a multimode oven using iodobenzene and 1-decene as substrates, among others[47] (Scheme 8.16). Palladium is supported on alumina for these reactions and can be recycled. Microwave heating proves comparable to the usual lengthy reflux of fourteen hours in terms of reaction yield.

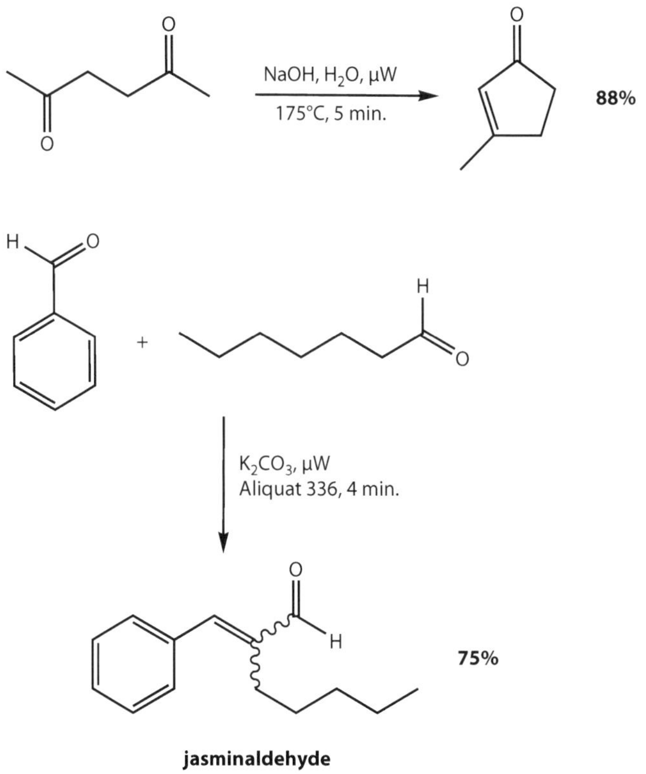

SCHEME 8.14 Aldol condensations under aqueous and solvent-free phase transfer catalytic conditions.

Numerous Suzuki couplings have been reported for the teaching laboratory under aqueous conditions (Sections 4.3.1 and 7.5). McGowan and Leadbeater have described a Suzuki reaction between 4-bromoacetophenone and phenylboronic acid in water that is catalyzed by ligand-free $Pd(OAc)_2$, with the aid of a phase transfer catalyst (TBAB)[27] (Scheme 8.17). Another solvent-free Suzuki coupling using the same palladium catalyst was performed on potassium fluoride/alumina in a monomode oven.[48] In this article the authors also report similar success with Heck and Stille reactions.

8.8.5 Pericyclic Reactions

One of the first microwave-enhanced reactions cited in several laboratory textbooks is the Diels-Alder cycloaddition of anthracene and maleic anhydride, developed by Bari et al.[4] (Scheme 8.18).

The popularity of this reaction is certainly due to its ability to be performed in a domestic microwave oven. It involves the use of diglyme as solvent (bp 162°C) in an open beaker. The two solid reagents are first ground together in a mortar and transferred to a beaker, diglyme is added, and the mixture is irradiated. Open-vessel

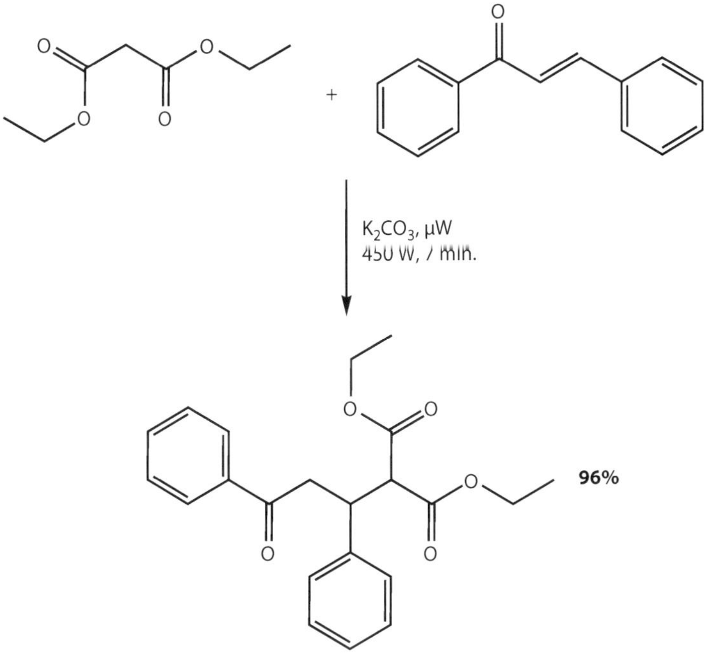

SCHEME 8.15 Michael reaction of chalcone and diethyl malonate.

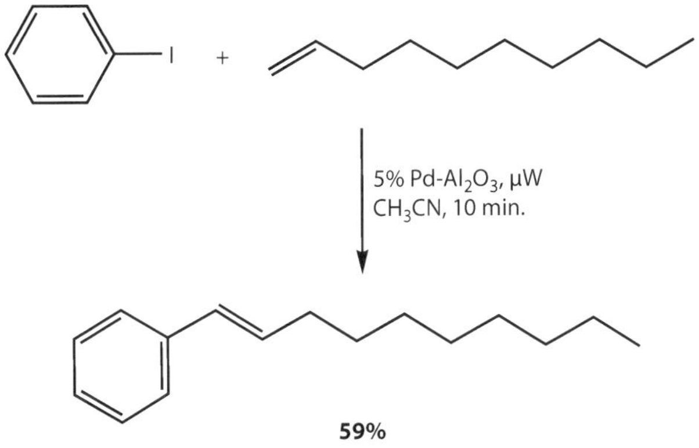

SCHEME 8.16 Heck reaction involving 1-decene and iodobenzene with Pd-Al_2O_3.

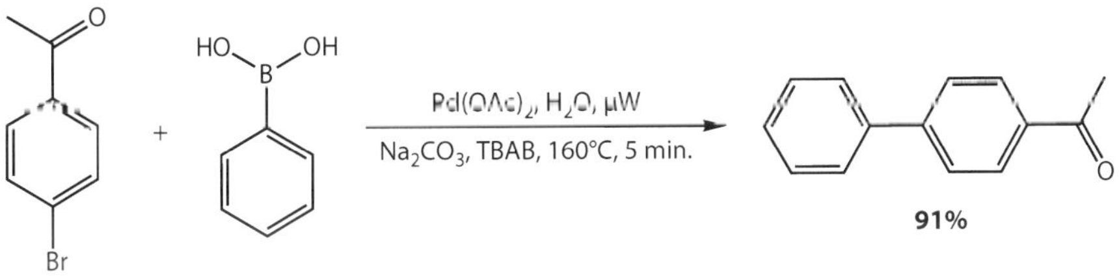

SCHEME 8.17 Suzuki coupling of 4-bromoacetophenone and phenylboronic acid.

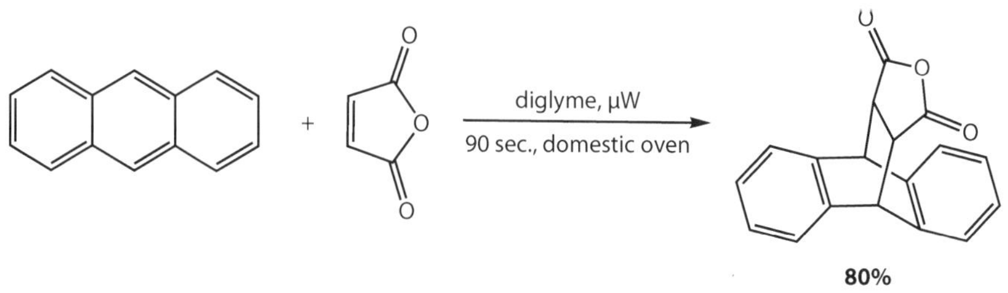

SCHEME 8.18 Diels-Alder cycloaddition of anthracene and maleic anhydride.

reactions are limited to starting materials with high melting points. Greater flexibility is possible in closed vessels employed in laboratory grade ovens.

The Diels-Alder reaction described in Section 8.7.2 between 1,3-cyclohexadiene and *N*-phenylmaleimide is performed in ethanol (Scheme 8.2). Excellent yields are achieved upon heating to 130°C for ten minutes in a multimode oven.[36] The cycloaddition of 2,3-dimethyl-1,3-butadiene and *trans*-dimethyl fumarate to produce the *trans* cycloadduct can be performed in a monomode oven.[27] A nice feature of this reaction is that it is performed under aqueous conditions. Heating in a multimode unit at 140°C for five minutes produces an 89% yield, whereas a larger-scale reaction in the monomode oven requires ten minutes at 140°C, but generates a yield of 96% (Scheme 8.19). Reacting the same diene with a less reactive dienophile (methyl acrylate) requires eighteen to twenty-four hours of heating in toluene or xylene, or five hours if using Lewis acid catalysts. Toluene and xylene are poor absorbers of microwave energy, but the addition of one drop of an ionic liquid means the reaction proceeds within minutes[21] (Scheme 8.20).

An intriguing Diels-Alder reaction described by Dintzner et al. forms the insecticide methylenedioxyprecocene (MDP) on basic montmorillonite K 10 clay in a commercial oven.[49] The electrophilic addition of 3-methyl-2-butenal to sesamol, subsequent dehydration, and intramolecular Diels-Alder reaction rapidly occur (Scheme 8.21). No solvent is required for the reaction itself, and the clay can be recycled. Ethyl acetate, a relatively green solvent required for product isolation, can also be recycled, meaning that several sustainability principles are highlighted by this procedure.

The pericyclic Cope and Claisen rearrangements usually require temperatures in excess of 190°C, and long reflux times to promote the reactions. Majetich and

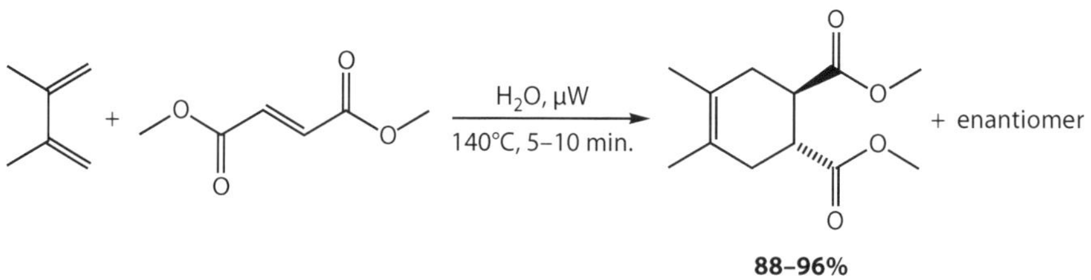

SCHEME 8.19 Diels-Alder cycloaddition of 2,3-dimethyl-1,3-butadiene and *trans*-dimethyl fumarate in water.

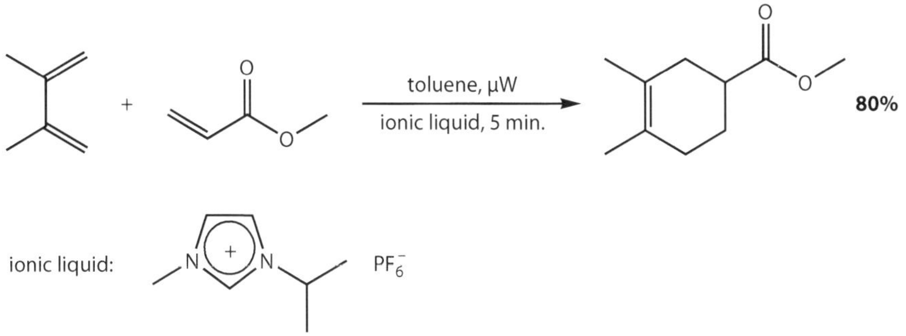

SCHEME 8.20 Diels-Alder cycloaddition of 2,3-dimethyl-1,3-butadiene and methyl acrylate promoted by an ionic liquid.

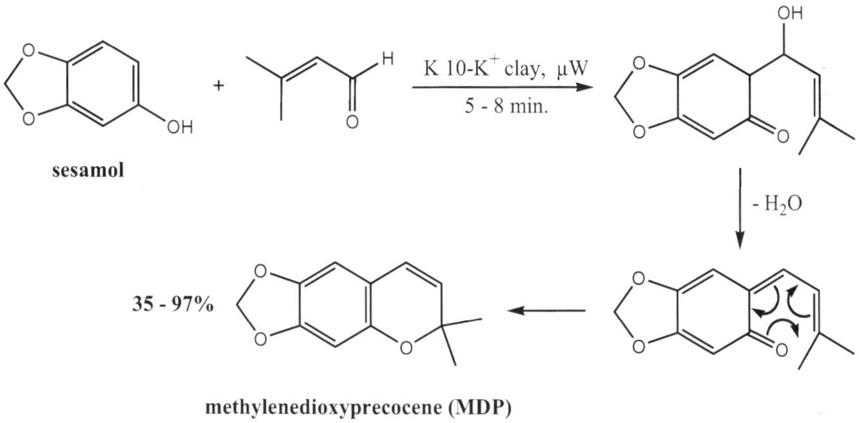

SCHEME 8.21 Intramolecular Diels-Alder synthesis of an insecticide on K 10 clay.

Hicks have studied the Claisen rearrangement of allyl phenyl ether to 2-allylphenol in *N,N*-dimethylformamide at ~190°C.[50] Under microwave conditions the reaction is complete within five hours, as compared to eighty hours at reflux. Raner et al. have undertaken the same reaction in near-critical water in ten minutes with good yields[51] (Scheme 8.22).

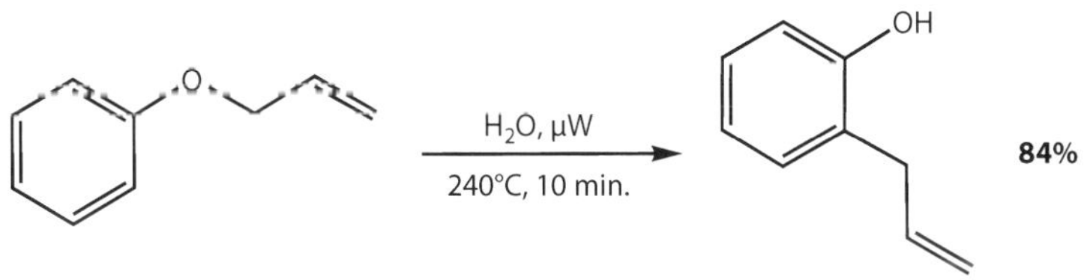

SCHEME 8.22 Claisen rearrangement in near-critical water.

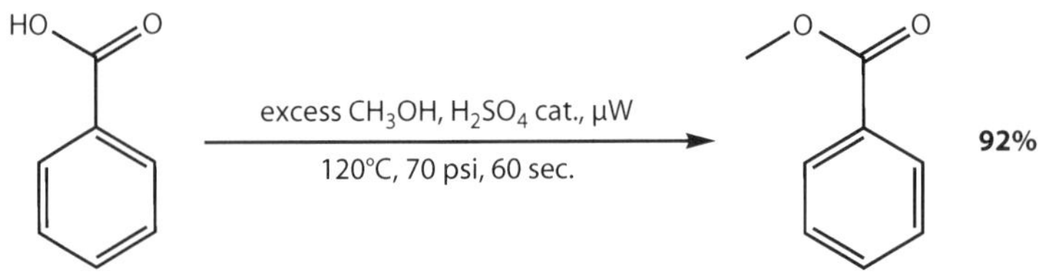

SCHEME 8.23 Esterification of benzoic acid and methanol to produce methyl benzoate.

8.8.6 Esterifications

Microwave-accelerated esterifications abound in the research literature, but there is little apparent consistency in reaction conditions. The ratio of alcohol to carboxylic acid and amount of acid catalyst vary greatly, and often the microwave specifics of the esterification have not been reported. However, esterification reactions work well under most microwave conditions, and they are favorites among students due to the product fragrances and commercial applications. Majetich and Hicks reported preparation of methyl benzoate in a multimode oven using benzoic acid, a large excess of methanol, and a catalytic amount of sulfuric acid[41] (Scheme 8.23). Specific reagent amounts are not reported in this article, but the microwave yield matches that obtained on refluxing for eighty minutes.

Mirafzal and Summer have prepared aspirin using an open beaker as the reaction vessel in a domestic microwave oven[9] (Scheme 8.24). They have also described the synthesis of two other analgesics (phenacetin and acetaminophen) by the corresponding acetylation of amines. Montes et al. have additionally discussed student aspirin synthesis via microwave irradiation under different catalytic conditions.[13] This approach is covered in more detail in Section 3.2.

The preparation of three esters (isoamyl acetate (banana), octyl acetate (orange), and benzyl acetate (peach)) has been designed in both monomode and multimode ovens.[27] In these experiments, silica gel beads are added to the alcohol and acetic acid to absorb the water formed as a coproduct. In the multimode unit, the reactions require holding the temperature at 120°C for five minutes. In the monomode oven, heating at 130°C for the same length of time produces similar results. Student yields of the three esters typically range from 57 to 92%, 70 to 98%, and 80 to 99%, respectively.

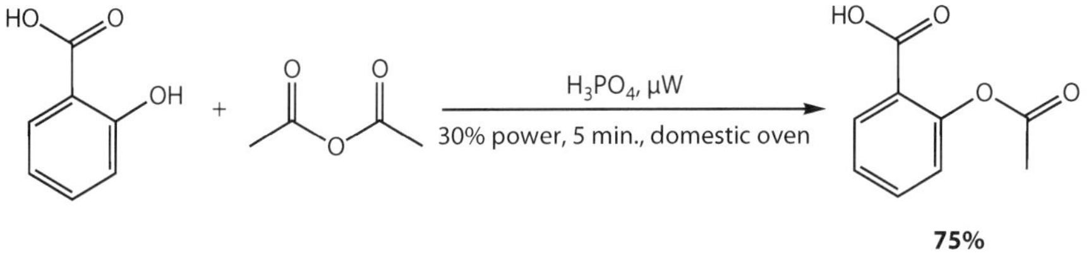

SCHEME 8.24 Aspirin synthesis via esterification of salicylic acid.

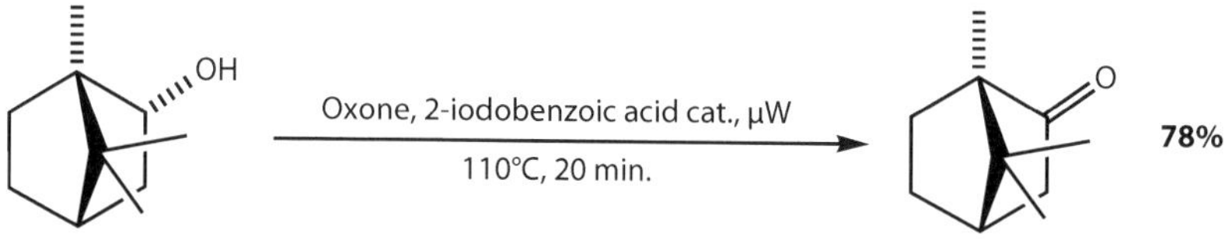

SCHEME 8.25 Oxidation of borneol to camphor.

A greener, microwave-promoted transesterification reaction is the preparation of biodiesel from vegetable oil.[52] This experiment is discussed in Section 7.7.4, along with the manufacture of other biofuels.

8.8.7 Oxidations

Oxidation reactions are very important transformations for the synthetic chemist. Murphree and Kappe have quite recently outlined an undergraduate oxidation of borneol (a chiral secondary alcohol) to camphor, using Oxone® in a monomode oven[53] (Scheme 8.25). This reaction occurs under catalytic conditions in a short timeframe. The advantages and disadvantages of using Oxone as an oxidant from a green perspective are covered in Sections 4.3.3 and 6.2.1.2.

There are many examples illustrating the ease of benzylic and allylic alcohol oxidation with the aid of microwave heating on alumina, clays, and silica gel. Varma et al. have used both MnO_2 "doped" silica gel[54] and Clayfen (montmorillonite K 10 clay-supported iron(III) nitrate)[55] to produce high yields of aldehydes and ketones on brief heating in a commercial oven (Scheme 8.26).

Benzoin oxidation to benzil products (1,2-diketones) is another transformation that has caught the attention of chemical educators (Section 6.2.8.1). The lengthy reflux times and toxic oxidizing agents that are historically required to effect these reactions can be replaced with mild, inexpensive oxidants (e.g., Oxone and $CuSO_4$ on solid supports). High yields and short reaction times characterize these microwave-accelerated oxidations. A specific example involves Oxone supported on wet neutral alumina with irradiation in a commercial oven[56] (Scheme 8.27).

8.8.8 Reductions

White and Kittredge have published the student reduction of cyclohexanone to cyclohexanol by irradiating the ketone and reductant ($NaBH_4$) in a solvent-free environment.[57] Sodium borohydride is first ground with silica gel before combination

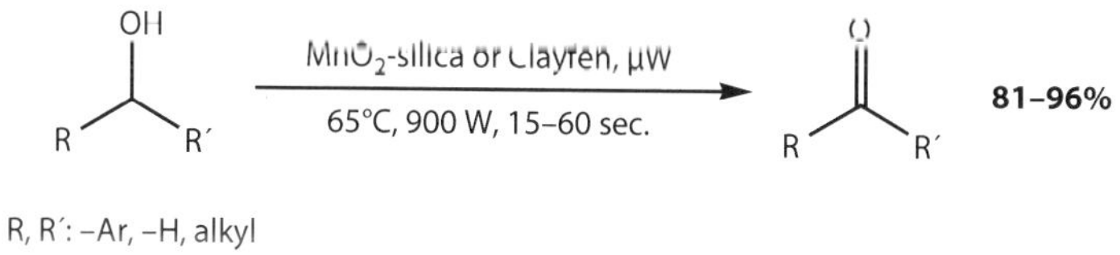

SCHEME 8.26 Oxidation of alcohols to aldehydes and ketones on solid supports.

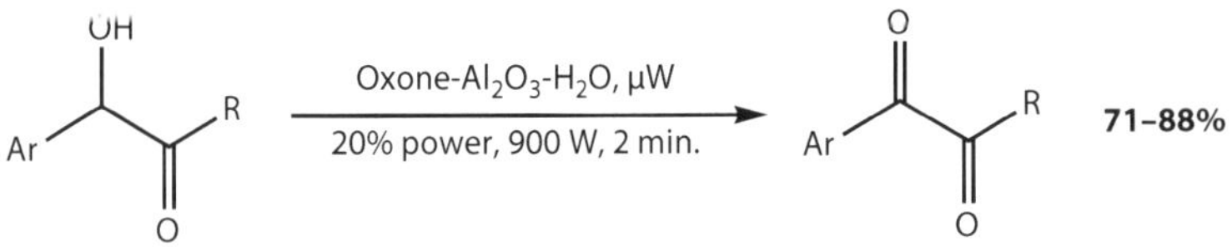

SCHEME 8.27 Solvent-free oxidation of benzoin using wet alumina-Oxone.

SCHEME 8.28 Reduction of cyclohexanone to cyclohexanol using $NaBH_4$/silica gel.

with cyclohexanone. The sample is heated in a monomode reactor, and affords the secondary alcohol product in excellent yield (Scheme 8.28). Related high-yielding reductions of aromatic ketones and aldehydes using solid supports have also been reported[58] (Scheme 8.29). This work has been extended to the reductive animation of carbonyls, which is a primary method for synthesizing amines.[59] In this approach, aromatic ketones are mixed with amines on montmorillonite K 10 clay, and heated to produce Schiff bases. These are then reduced by adding $NaBH_4$ supported on wet montmorillonite K 10 clay (Scheme 8.30). The two-step process occurs in less than five minutes in a domestic microwave oven.

Wolff-Kishner reductions are notorious for requiring high temperatures and long refluxes, hence their general lack of appearance in undergraduate laboratory sequences. However, Parquet and Lin have described a facile two-step microwave-accelerated reduction of isatin using hydrazine in ethylene glycol, followed by KOH in a domestic oven[8] (Scheme 8.31).

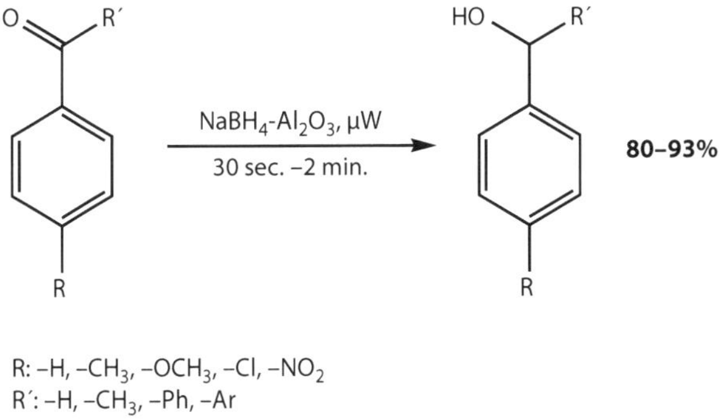

SCHEME 8.29 Reduction of aromatic aldehydes and ketones using a solid support.

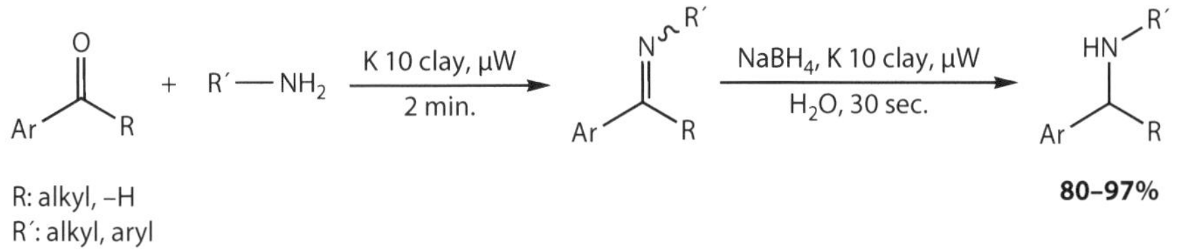

SCHEME 8.30 Reductive amination of aromatic ketones using $NaBH_4$-wet clay.

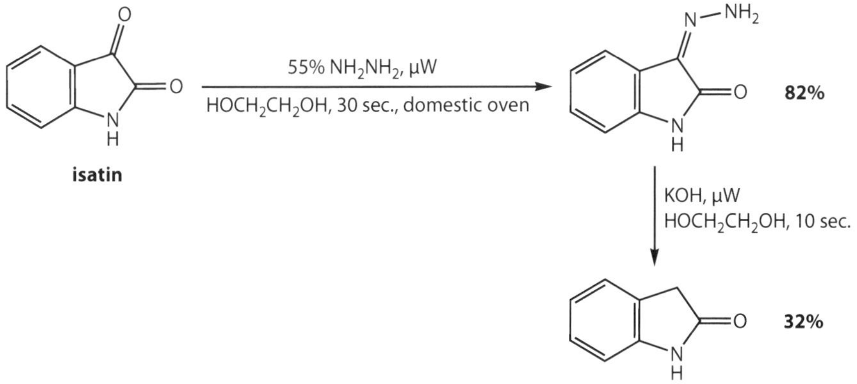

SCHEME 8.31 Two-step Wolff-Kishner reduction of isatin.

8.8.9 Hydrolyses

Carbohydrate chemistry often involves protection of hydroxy groups by acetylation, forming esters. These groups must eventually be removed during deprotection, usually by heating in an aqueous base for several hours (saponification). However, impregnating alumina with KOH as dry basic media can produce quantitative yields of the free carbohydrate on microwave irradiation[60] (Scheme 8.32).

As well as designing a microwave-enhanced borneol oxidation (Scheme 8.25), Murphree and Kappe have developed an amide hydrolysis reaction.[53] Here, benzamide is hydrolyzed to benzoic acid with aqueous sulfuric acid under microwave irradiation for seven minutes (Scheme 8.33). This reaction truly reinforces the power of microwave heating, as students are routinely taught that amide hydrolyses require extensive refluxing under concentrated acidic or basic conditions.

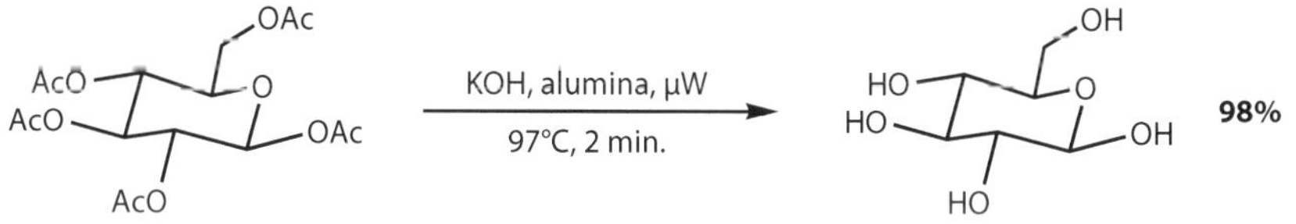

SCHEME 8.32 Hydrolyses of acetate-protecting groups by preabsorbed KOH on alumina.

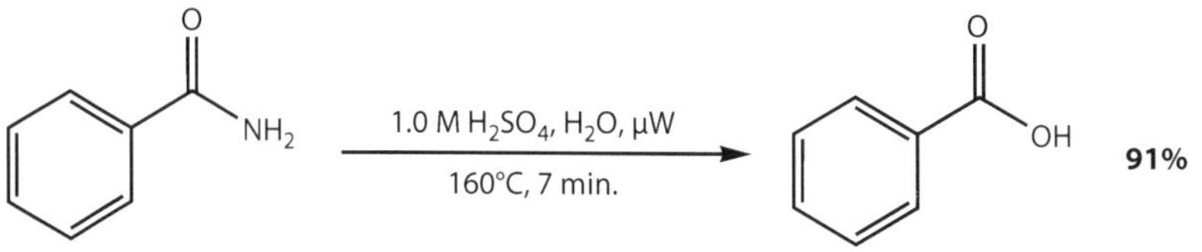

SCHEME 8.33 Hydrolysis of benzamide to benzoic acid.

8.9 CONCLUSION

I started experimenting with microwave heating eight years ago, and have not looked back since. Indeed, I now have to make a conscious effort to include at least one experiment that demonstrates a classical reflux in my laboratory curriculum. The ease and safety of microwave heating, its efficiency and associated educational benefits, the reduced environmental footprint, and the extensive array of chemistry that undergoes rate acceleration all advocate for its adoption. Transitioning to microwave heating has never been easier, with the ever-increasing number of published microwave-enhanced experiments performed in laboratory grade ovens. Required experimental conditions are routinely reported in the research literature, making it possible to translate a reaction to the teaching laboratory. The guidelines described here will hopefully encourage further incorporation of microwave heating into undergraduate curricula, and perhaps lead to the development of novel, greener organic reactivity.

REFERENCES

1. Decareau, R. V., Peterson, R. A. *Microwave Processing and Engineering.* Wiley-VCH: Weinheim, Germany, 1986, 141.
2. Gedye, R. N., Smith, F. E., Westaway, K. C., Ali, H., Baldisera, L., Laberge, L., Rousell, J. *Tetrahedron Lett.* 1986, 27, 279–282.
3. Giguere, R. J., Bray, T. L., Duncan, S. M., Majetich, G. *Tetrahedron Lett.* 1986, 27, 4945–4948.
4. Bari, S. S., Bose, A. K., Chaudhary, A. G., Manhas, M. S., Raju, V. S., Robb, E. W. *J. Chem. Educ.* 1992, 69, 938–939.
5. Elder, J. W. *J. Chem. Educ.* 1994, 71, A142, A144.
6. Elder, J. W., Holtz, K. M. *J. Chem. Educ.* 1996, 73, A104–A105.
7. Trehan, I. R., Brar, J. S., Arora, A. K., Kad, G. L. *J. Chem. Educ.* 1997, 74, 324.
8. Parquet, E., Lin, Q. *J. Chem. Educ.* 1997, 74, 1225.
9. Mirafzal, G. A., Summer, J. M. *J. Chem. Educ.* 2000, 77, 356–357.
10. Baldwin, B. W., Wilhite, D. M. *J. Chem. Educ.* 2002, 79, 1344.
11. Friebe, T. L. *Chem. Educator* 2003, 8, 33–36.
12. Shaw, R., Severin, A., Balfour, M., Nettles, C. *J. Chem. Educ.* 2005, 82, 625–629.
13. Montes, I., Sanabria, D., García, M., Castro, J., Fajardo, J. *J. Chem. Educ.* 2006, 83, 628–631.
14. Huang, W., Richert, R. *J. Phys. Chem. B* 2008, 112, 9909–9913.
15. Loupy, A., Ed. *Microwaves in Organic Synthesis.* Wiley-VCH, Weinheim, Germany, 2002.
16. Hayes, B. *Microwave Synthesis—Chemistry at the Speed of Light.* CEM Publishing, Matthews, NC, 2002.

17. Kingston, H. M., Haswell, S. J., Eds. *Microwave-Enhanced Chemistry—Fundamentals, Sample Preparation, and Applications*. American Chemical Society, Washington, DC, 1997.
18. Kappe, C. O., Stadler, A. *Microwaves in Organic and Medicinal Chemistry*. Wiley-VCH: Weinheim, Germany, 2005.
19. Loupy, A., Perreux, L., Liagre, M., Burle, K., Moneuse, M. *Pure Appl. Chem.* 2001, 73, 161–166.
20. Dallinger, D., Kappe, C. O. *Chem. Rev.* 2007, 107, 2563–2591.
21. Leadbeater, N., Torenius, H. M. *J. Org. Chem.* 2002, 67, 3145–3148.
22. Anton Paar home page. www.anton-parr.com (accessed December 23, 2010).
23. Biotage home page. http://biotage.com (accessed December 23, 2010).
24. CEM Corporation home page. www.cem.com (accessed December 23, 2010).
25. Milestone, Inc. home page. www.milestonesci.com (accessed December 23, 2010).
26. Zovinka, E. P., Stock, A. E. *J. Chem. Educ.* 2010, 87, 350–352.
27. McGowan, C., Leadbeater, N. *Clean, Fast Organic Chemistry: Microwave-Assisted Laboratory Experiments*. CEM Publishing, Matthews, NC, 2006.
28. Richter, R. *Clean Chemistry: Techniques for the Modern Laboratory*. Milestone Press: Monroe, CT, 2003.
29. University of Oregon Greener Education Materials for Chemists. http://greenchem.uoregon.edu/gems.html (accessed December 23, 2010).
30. Baar, M. R., Wustholz, K. *J. Chem. Educ.* 2005, 82, 1393–1394.
31. Pavia, D. L., Lampman, G. M., Kriz, G. S., Engel, R. G. Experiment 41: 1,4-Diphenyl-1,3-butadiene. In *Introduction to Organic Laboratory Techniques: A Small Scale Approach*, 2nd ed. Brooks/Cole: Pacific Grove, CA, 2005, 341–347.
32. Katritzky, A. R., Cai, C., Collins, M. D., Scriven, E. F. V., Singh, S. K., Barnhardt, E. K. *J. Chem. Educ.* 2006, 83, 634–636.
33. Loupy, A., Petit, A., Hamelin, J., Texier-Boullet, F., Jacquault, P., Mathe, D. *Synthesis* 1998, 1213–1234.
34. Varma, R. S. *Tetrahedron* 2002, 58, 1235–1255.
35. Lidström, P., Tierney, J., Wathey, B., Westman, J. *Tetrahedron* 2001, 57, 9225–9283.
36. Baar, M. R., Falcone, D., Gordon, C. *J. Chem. Educ.* 2010, 87, 84–86.
37. Martin, E., Kellen-Yuen, C. *J. Chem. Educ.* 2007, 84, 2004–2006.
38. Ju, Y., Varma, R. S. *Green Chem.* 2004, 6, 219–221.
39. Cherng, Y.-J. *Tetrahedron* 2000, 56, 8287–8289.
40. Williamson, K. L., Minard, R. D., Masters, K. M. Malonic Ester of a Barbiturate. In *Macroscale and Microscale Organic Experiments*, 5th ed. Houghton Mifflin, Boston, 2007, 575–585.
41. Majetich, G., Hicks, R. *Res. Chem. Intermed.* 1994, 20, 61–77.
42. Coursindel, T., Martinez, J., Parrot, I. *J. Chem. Educ.* 2010, 87, 640–642.
43. Crouch, R. D., Howard, J. L., Zile, J. L., Barker, K. H. *J. Chem. Educ.* 2006, 83, 1658–1660.
44. Musiol, R., Tyman-Szram, B., Polanksi, J. *J. Chem. Educ.* 2006, 83, 632–633.
45. Abenhaim, D., Ngoc Son, C. P., Loupy, A., Ba Hiep, N. *Synth. Commun.* 1994, 24, 1199–1205.
46. Rao, H. S. P., Jothilingam, S. *J. Chem. Sci.* 2005, 117, 323–328.
47. Wali, A., Muthukumaru Pillai, M., Satish, S. *React. Kinet. Catal. Lett.* 1997, 60, 189–194.
48. Villemin, D., Caillot, F. *Tetrahedron Lett.* 2001, 42, 639–642.
49. Dintzner, M. R., Wucka, P. R., Lyons, T. W. *J. Chem. Educ.* 2006, 83, 270–272.
50. Majetich, G., Hicks, R. *Radiat. Phys. Chem.* 1995, 45, 567–579.
51. Raner, K. D., Strauss, C. R., Trainor, R. W., Thorn, J. S. *J. Org. Chem.* 1995, 60, 2456–2460.

52. Miller, T. A., Leadbeater, N. E. *Chem. Educator* 2009, 14, 98–104.
53. Murphree, S. S., Kappe, C. O. *J. Chem. Educ.* 2009, 86, 227–229.
54. Varma, R. S., Saini, R. K., Dahiya, R. *Tetrahedron Lett.* 1997, 38, 7823–7824.
55. Varma, R. S., Dahiya, R. *Tetrahedron Lett.* 1997, 38, 2043–2044.
56. Varma, R. S., Dahiya, R., Kumar, D. *Molecules Online* 1998, 2, 82–85.
57. White, L., Kittredge, K. *J. Chem. Educ.* 2005, 82, 1055–1056.
58. Varma, R. S., Saini, R. K. *Tetrahedron Lett.* 1997, 38, 4337–4338.
59. Varma, R. S., Dahiya, R. *Tetrahedron* 1998, 54, 6293–6298.
60. Limousin, C., Cléophax, J., Petit, A., Loupy, A., Lukacs, G. *J. Carbohydr. Chem.* 1997, 16, 327–342.

Appendix: The Greener Organic Chemistry Reaction Index. Operative Reaction Mechanisms, Experimental Techniques, and Greener Principles for Each Undergraduate Reaction

The experimental references included in this appendix are primarily arranged alphabetically according to mechanistic type, followed by the nature of the reaction performed. For example, if one is interested in learning about greener ketone reductions, he or she should first consult "nucleophilic addition" as the mechanistic type, and then locate entries 75–83 inclusive. For each of the entries the following is listed: the primary literature reference, experimental techniques employed, and greener principles highlighted. One will see that this simple functional group transformation can be effected solvent-free, under aqueous conditions, with microwave heating, etc.

A total of 178 conversions are profiled in the Appendix, covering literature examples from 1978 until May 2011. All the reactions are student tested and taken from peer-reviewed publications, with an overwhelming majority from pedagogical journals. The list is not meant to be all-inclusive—rather, references have been selected to illustrate the range of greener organic chemistry possible in both introductory and advanced level laboratories. Some excellent greener reactions can also be found in laboratory textbooks, which have been highlighted throughout this book. Three that should be consulted are listed below. The University of Oregon Greener Education Materials for Chemists database (http://greenchem.uoregon.edu/gems.html) is very informative, and includes other resources from instructors that have not been subject to peer review.

1. Doxsee, K., Hutchinson, J. *Green Organic Chemistry: Strategies, Tools, and Laboratory Experiments*. Brooks/Cole, Pacific Grove, CA, 2004.
2. Kirchhoff, M., Ryan, M. A., Eds. *Greener Approaches to Undergraduate Chemistry Experiments*. American Chemical Society, Washington, DC, 2002.
3. Roesky, H., Kennepohl, D., Eds. *Experiments in Green and Sustainable Chemistry*. Wiley-VCH, Weinheim, 2009.

As a final note, it is important to appreciate that the reactions in this appendix all have green elements to them—*they are not perfectly green processes.* Indeed, some have features that would benefit from "greening." This is always a critical point to discuss with students, and will hopefully provide an impetus for even greener experiments to be developed in years to come.

Entry	Reaction Mechanism	Transformation	Journal	Techniques	Greener Principles
1.	Electrophilic addition	Alkene bromination	JCE 2005, 82, 306–310	MS, H, VF	Safer reagent, alternative reaction solvent, product recycling
2.	Electrophilic addition	Alkene bromination	JCE 2005, 82, 306–310	MS, H, VF, R	Greener reagents, alternative reaction solvent, atom efficiency, product recycling
3.	Electrophilic addition	Alkene bromination	GCLR 2010, 3, 39–47	MS, VF, H, LE, LD, RE	Catalytic, greener reagent, aqueous solvent, atom efficiency
4.	Electrophilic addition	Alkene chlorination	JCE 2003, 80, 1319–1321	MS, LD, GF, RE	Greener reagent
5.	Electrophilic addition	Alkene epoxidation	JCE 2004, 81, 1187–1190	MS, LE, LD, GF	Catalytic, greener reagent, alternative reaction solvent, atom efficiency
6.	Electrophilic addition	Alkene epoxidation	JCE 2004, 81, 1018–1019	MS, LE, LD, GF, RE	Aqueous solvent
7.	Electrophilic addition	Alkene iodochlorination	JCE 2008, 85, 962–964	H, LE, LD, GF, RE, VF	Alternative reaction solvent
8.	Electrophilic addition	Alkene oxidation/cleavage	JCE 2000, 77, 1627–1629	MS, H, VF, R	Catalytic, greener reagent, aqueous solvent
9.	Electrophilic addition	Halohydrin formation	JCE 1985, 62, 638	H, VF, R	Aqueous solvent
10.	Electrophilic addition/ nucleophilic addition	Iodolactonization	JCE 2006, 83, 921–922	MS, LE, LD, GF, RE	Aqueous solvent
11.	Electrophilic addition/ pericyclic	Hetero Diels-Alder reaction	JCE 2006, 83, 270–272	MI, VF, LE, GF, LD, D/ RE	Catalytic, greener reagent, solvent-free, atom efficiency, microwave heating
12.	Electrophilic aromatic substitution	Acylation of ferrocene	JCE 2008, 85, 261–262	MI, MS, LE, LD, GF, RE, CC	Catalytic, greener reagent, microwave heating
13.	Electrophilic aromatic substitution	Alkylation	JCE 2007, 84, 692–693	H, VF, RE	Greener, recyclable reagent

(*continued*)

Entry	Reaction Mechanism	Transformation	Journal	Techniques	Greener Principles
14.	Electrophilic aromatic substitution	Bromination	TCE 1998, 3, 1–6	MS, GF, LD, VF, LE, RE, D	Greener reagent
15.	Electrophilic aromatic substitution	Dipyrromethane formation	JCE 2006, 83, 1665–1666	H, VF	Catalytic, greener reagent, aqueous solvent, atom efficiency
16.	Electrophilic aromatic substitution	Friedel-Crafts acylation	TCE 2001, 6, 25–27	MS, LE, GF, SE	Solvent-free
17.	Electrophilic aromatic substitution	Fries rearrangement	JCE 1997, 74, 324	MI, LE, LD, GF, RE	Recyclable reagent, microwave heating
18.	Electrophilic aromatic substitution	Iodination	JCE 2008, 85, 1426–1428	ES, VF, R	Greener reagent
19.	Electrophilic aromatic substitution	Nitration	JCE 2005, 82, 616–617	MS, H, VF, R	Aqueous solvent, renewable feedstock
20.	Electrophilic aromatic substitution	Porphyrin formation	GC 2001, 3, 267–270	MI, LE, GF, RE, CC	Solvent-free, microwave heating
21.	Electrophilic aromatic substitution	Porphyrin formation	GC 2001, 3, 267–270	H, CC	Solvent-free
22.	Electrophilic substitution	α-Fluorination	JCE 2008, 85, 834–835	MS, H, VF, RE	Greener reagent
23.	Electrophilic substitution/ nucleophilic acyl substitution	Haloform reaction	JCE 2010, 87, 190–193	MS, H, VF	Aqueous solvent
24.	Electrophilic substitution/ rearrangement/nucleophilic addition/decarboxylation	Hofmann rearrangement	JCE 1999, 76, 1717	MS, H, VF, R	Greener reagent
25.	Elimination	Alcohol dehydration	JCE 1993, 70, 493–495	H, LE, LD, GF	Catalytic, greener reagent
26.	Fermentation	Alcohol production	JCE 2010, 87, 708–710	H, D	Biocatalytic, greener reagent, renewable feedstock
27.	Ionic liquid formation	Deep eutectic solvent synthesis	EC 2005, 42 (1), 12–15	ES, H	Greener product

28.	Metathesis	Cross-metathesis	JCE 2006, 83, 283–284	MS, RE, CC, VF, R	Catalytic, greener reagent
29.	Metathesis	Ring-closing metathesis	JCE 2007, 84, 1995–1997	MS, H, VF, ST, IA, CC, RE	Catalytic, greener reagent
30.	Metathesis	Ring-closing metathesis	JCE 2010, 87, 721–723	MS, IA, GF, RE	Catalytic, greener reagent
31.	Metathesis	Ring-closing metathesis	JCE 2007, 84, 1998–2000	MS, H, RE, VF, IA	Catalytic, greener reagent
32.	Metathesis	Ring-opening metathesis polymerization	JCE 1993, 70, 165–167	H, VF	Catalytic, greener reagent, aqueous solvent, renewable feedstock, product recycling
33.	Metathesis	Ring-opening metathesis polymerization	JCE 1999, 76, 661–665	MS, IA, VF, RE, VF	Catalytic, greener reagent
34.	Nucleophilic acyl substitution	Amide formation	JCE 2003, 80, 1444–1445	VF	Greener reagent
35.	Nucleophilic acyl substitution	Amide formation	JCE 2002, 79, 1344	MI, VF, R	Microwave heating
36.	Nucleophilic acyl substitution	Amide hydrolysis	JCE 2009, 86, 227–229	MI, MS, VF	Microwave heating
37.	Nucleophilic acyl substitution	Amine formylation	JCE 2009, 86, 227–229	MI, MS, VF	Microwave heating
38.	Nucleophilic acyl substitution	Aspirin synthesis	JCE 2006, 83, 628–631	MI, ES, VF, R	Catalytic, microwave heating
39.	Nucleophilic acyl substitution	Aspirin, acetanilide, phenacetin, acetaminophen synthesis	JCE 2000, 77, 356–357	MI, VF, R	Catalytic, aqueous solvent, microwave heating
40.	Nucleophilic acyl substitution	Claisen condensation	JCE 2003, 80, 1446–1447	H, LE, GF, LD, RE, T, R	Solvent-free
41.	Nucleophilic acyl substitution	Coffee ground degradation	JCE 2001, 78, 1669–1671	H, SX, RE	Consumer product recycling, renewable feedstock
42.	Nucleophilic acyl substitution	Diimide formation	JCE 2008, 85, 1649–1651	SG, H, R	Solvent-free
43.	Nucleophilic acyl substitution	Ester hydrolysis	JCE 2006, 83, 634–636	MI, MS, LE, LD, GF, RE, R	Aqueous solvent, microwave heating
44.	Nucleophilic acyl substitution	Esterification	JCE 1993, 70, 493–495	H, LE, LD, GF, D	Catalytic, greener reagent
45.	Nucleophilic acyl substitution	Esterification	JCE 2009, 86, 227–229	MI, MS, RE, LE, LD, GF	Catalytic, microwave heating
46.	Nucleophilic acyl substitution	Phthalimide synthesis	JCE 1992, 69, 938–939	SG, MI, VF, R	Microwave heating

(continued)

Entry	Reaction Mechanism	Transformation	Journal	Techniques	Greener Principles
47.	Nucleophilic acyl substitution	Polylactic acid degradation	JME 2008, 30, 257–280	MS, H, RE	Consumer product recycling, renewable feedstock
48.	Nucleophilic acyl substitution	Polyethylene terephthalate degradation	JCE 2003, 80, 79–82	MS, H, VF, GF	Consumer product recycling
49.	Nucleophilic acyl substitution	Polyethylene terephthalate degradation	JCE 1999, 76, 1525–1526	MS, H, LE, VF	Consumer product recycling
50.	Nucleophilic acyl substitution	Polyethylene terephthalate degradation	JCE 2010, 87, 519–521	MS, H, VF	Consumer product recycling
51.	Nucleophilic acyl substitution	Transesterification	JCE 2010, 87, 423–425	MS, VF, RE, D	Biocatalytic, greener reagent, low-solvent usage
52.	Nucleophilic acyl substitution	Transesterification (biodiesel synthesis)	TCE 2005, 10, 130–132	MS, H, LE, VF	Consumer product recycling, renewable feedstock, greener product
53.	Nucleophilic acyl substitution	Transesterification (biodiesel synthesis)	JCE 2007, 84, 296–298	MS, H, C	Consumer product recycling, renewable feedstock, greener product
54.	Nucleophilic acyl substitution	Transesterification (biodiesel synthesis)	JCE 2006, 83, 257–259	MS, H, LE, LD	Consumer product recycling, renewable feedstock, greener product
55.	Nucleophilic acyl substitution	Transesterification (biodiesel synthesis)	JCE 2006, 83, 260–262	MS, H, LE	Consumer product recycling, renewable feedstock, greener product
56.	Nucleophilic acyl substitution	Transesterification (biodiesel synthesis)	TCE 2009, 14, 98–104	ES, MI	Consumer product recycling, renewable feedstock, microwave heating, greener product
57.	Nucleophilic acyl substitution	Transesterification (biodiesel synthesis)	JCE 2011, 88, 197–200	MS, H	Consumer product recycling,renewable feedstock, greener product

58.	Nucleophilic acyl substitution	Transesterification (biodiesel synthesis)	JCE 2011, 88, 201–203	MS, H, LD, GF	Consumer product recycling, renewable feedstock, greener product
59.	Nucleophilic acyl substitution/ dehydration	Cu(II) phthalocyanine complex formation	JCE 2011, 88, 86–87	SG, MI	Solvent-free, catalytic, greener reagent, microwave heating
60.	Nucleophilic acyl substitution/ electrophilic aromatic substitution/dehydration	Pechmann condensation	JCE 2011, 88, 319–321	ES, H, VF, R	Solvent-free, catalytic,greener reagent, recyclable reagent, atom efficiency
61.	Nucleophilic addition	Acetal formation	JCE 2001, 78, 70–72	MS, H, VF, R	Catalytic, greener reagent, aqueous solvent
62.	Nucleophilic addition	Aldehyde reduction	JCE 2006, 83, 285–286	MS, VF	Aqueous solvent
63.	Nucleophilic addition	Aldehyde reduction	JCE 2005, 82, 1674–1675	MS, LE, LD, RE	Greener reagent
64.	Nucleophilic addition	Aldehyde reduction	JCE 2011, 88, 322–324	MS, H, LE, LD, RE	Catalytic, greener reagent, atom efficiency
65.	Nucleophilic addition	Aldol condensation	JCE 2007, 84, 475–476	MS, VF	Catalytic, greener reagent, aqueous solvent, atom efficiency
66.	Nucleophilic addition	Aldol condensation	JCE 2006, 83, 1871–1872	MS, LE, LD, GF, RE	Catalytic, greener reagent, renewable feedstock, atom efficiency
67.	Nucleophilic addition	Barbier (Grignard type) reaction	JCE 1998, 75, 85	MS, GF, LE, LD, RE	Aqueous solvent
68.	Nucleophilic addition	Cannizzaro reaction	JCE 2009, 86, 85–86	SG, VF	Solvent-free
69.	Nucleophilic addition	Cannizzaro reaction	JCE 2004, 81, 1794–1795	SG, H, LE, VF, GF, LD, RE	Solvent-free
70.	Nucleophilic addition	Cellulose degradation	JCE 2008, 85, 546–548	H, VF, D	Consumer product recycling, renewable feedstock
71.	Nucleophilic addition	Cellulose degradation	TCE 2000, 5, 315–316	H, VF	Consumer product recycling, renewable feedstock

(continued)

Entry	Reaction Mechanism	Transformation	Journal	Techniques	Greener Principles
72.	Nucleophilic addition	Creatine synthesis	JCE 2006, 83, 1654–1657	MS, VF, R	Catalytic, greener reagent, aqueous solvent, atom efficiency
73.	Nucleophilic addition	Grignard reaction	JCE 2009, 86, 227–229	MI, MS, LE, LD, GF, RE, R	Microwave heating
74.	Nucleophilic addition	Horner–Wadsworth–Emmons reaction	TCE 2005, 10, 300–302	MS, H, VF, R	Aqueous solvent
75.	Nucleophilic addition	Ketone reduction	JCE 2005, 82, 1055–1056	SG, MI, LE, LD, GF, RE	Solvent-free, microwave heating
76.	Nucleophilic addition	Ketone reduction	JCE 1986, 63, 909	MS, LE, LD, GF, RE/D	Aqueous solvent
77.	Nucleophilic addition	Ketone reduction	JCE 1998, 75, 630–631	MS, GF, LD, RE, D	Biocatalytic, greener reagent
78.	Nucleophilic addition	Ketone reduction	JCE 2005, 82, 1049–1050	MS, H, VF, LE, LD, RE, CC	Biocatalytic, greener reagent
79.	Nucleophilic addition	Ketone reduction	JCE 2006, 83, 1049–1051	MS, GF, LE, LD, RE, CC	Biocatalytic, greener reagent
80.	Nucleophilic addition	Ketone reduction	JCE 2002, 79, 727–728	H, VF, LE, LD, GF, RE	Biocatalytic, greener reagent, aqueous solvent
81.	Nucleophilic addition	Ketone reduction	TCE 2008, 13, 344–347	MS, H	Greener reagent
82.	Nucleophilic addition	Ketone reduction	GCLR 2008, 1, 149–154	MS, H, D, GF, LD, RE	Catalytic, greener reagent
83.	Nucleophilic addition	Ketone reduction	JCE 1996, 73, A104-A105	MI, VF, R	Microwave heating
84.	Nucleophilic addition	Mannich reaction	JCE 2006, 83, 943–946	MS, H, VF, R, LE, RE	Alternative and recyclable reaction solvent, atom efficiency
85.	Nucleophilic addition	Michael reaction	JCE 2010, 87, 194–195	CC, RE, H, VF, R	Catalytic, greener reagent, atom efficiency
86.	Nucleophilic addition	Michael reaction/alkene reduction	JCE 2011, 88, 322–324	LE, LD, RE	Catalytic, greener reagent

87.	Nucleophilic addition	Oxazolidinone formation	JCE 2005, 82, 1229–1230	MS, LE, LD, GF, RE	Aqueous solvent, renewable feedstock
88.	Nucleophilic addition	Wittig reaction	JCE 2004, 81, 1492–1493	SG, VF, R	Solvent-free
89.	Nucleophilic addition	Wittig reaction	JCE 2007, 84, 119–121	MS, GF, RE	Solvent-free
90.	Nucleophilic addition	Wittig reaction	JCE 2007, 84, 2004–2006	SG, MI, LE, RE, CC	Solvent-free, microwave heating
91.	Nucleophilic addition	Wittig reaction	JCE 2007, 84, 119–121	H, MS, GF, RE, R	Solvent-free
92.	Nucleophilic addition	Wittig reaction	JCE 1978, 55, 813	H, MS, GF, VF, R	Aqueous solvent
93.	Nucleophilic addition(s)/ dehydration(s)/decarboxylation	Knoevenagel condensation/ Michael addition/annulation	JCE 2007, 84, 1477–1479	MI, MS, LE, LD, GF, RE	Aqueous/greener solvent, microwave heating
94.	Nucleophilic addition(s)/ nucleophilic acyl substitution	Passerini reaction	JCE 2009, 86, 1077–1079	MS, VF, R	Aqueous solvent, atom efficiency
95.	Nucleophilic addition/ dehydration	Aldol condensation	JCE 2009, 86, 488–493	SG, T, VF, R	Catalytic, greener reagent, solvent-free, atom efficiency
96.	Nucleophilic addition/ dehydration	Aldol condensation	JCE 1994, 71, A142, A144	MI, VF	Microwave heating, atom efficiency
97.	Nucleophilic addition/ dehydration	Aldol condensation	JCE 2007, 84, 475–476	MS, H, VF	Catalytic, greener reagent, aqueous solvent, atom efficiency
98.	Nucleophilic addition/ dehydration	Aldol condensation	JCE 2004, 81, 1345–1347	SG, VF, R	Catalytic, greener reagent, solvent-free, atom efficiency
99.	Nucleophilic addition/ dehydration	Aldol condensation	TCE 2011, 16, 23–25	ES, VF, R	Solvent-free, atom efficiency
100.	Nucleophilic addition/ dehydration	Biginelli reaction	JCE 2009, 86, 730–732	MS, H, SG, VF	Catalytic, greener reagent, solvent-free, improved energy efficiency, atom efficiency
101.	Nucleophilic addition/ dehydration	Coumarin formation	PAC 2001, 73, 1257–1260; GC 2000, 2, 245–247	SG, VF	Catalytic, greener reagent, solvent-free
102.	Nucleophilic addition/ dehydration	Coumarin formation	PAC 2001, 73, 1257–1260; GC 2000, 2, 245–247	MS, VF	Catalytic, greener reagent, aqueous solvent

(continued)

Entry	Reaction Mechanism	Transformation	Journal	Techniques	Greener Principles
103.	Nucleophilic addition/ dehydration	Dioxolanone synthesis	TCE 2003, 8, 33–36	MI, CC, RE	Microwave heating
104.	Nucleophilic addition/ dehydration	Hantzsch reaction	JCE 2010, 87, 628–630	MS, H, SG, VF, R	Low-solvent usage, improved energy efficiency, atom efficiency
105.	Nucleophilic addition/ dehydration	Imidazole synthesis	JCE 2006, 83, 1658–1660	MI, VF, R	Microwave heating
106.	Nucleophilic addition/ dehydration	Imine formation	JCE 2006, 83, 929–930	SG, R	Solvent-free, atom efficiency
107.	Nucleophilic addition/ dehydration	Imine formation	JCE 2006, 83, 1221–1224	VF, MS, H, VD	Alternative reaction solvent, atom efficiency
108.	Nucleophilic addition/ dehydration	Knoevenagel condensation	TCE 2007, 12, 324–326	MS, VF	Catalytic, greener reagent, aqueous solvent, atom efficiency
109.	Nucleophilic addition/ dehydration	Knoevenagel condensation	JCE 2009, 86, 227–229	MI, MS, VF	Alternative reaction solvent, atom efficiency, microwave heating
110.	Nucleophilic addition/ dehydration	Semicarbazone formation	JCE 2004, 81, 108	SG, VF, R	Solvent-free
111.	Nucleophilic addition/ dehydration or nucleophilic acyl substitution/dehydration	Heterocyclic syntheses (e.g., benzimidazole, phthalimide)	JCE 2006, 83, 632–633	SG, MI, VF, R	Solvent-free, atom efficiency, microwave heating
112.	Nucleophilic addition/ dehydration or nucleophilic acyl substitution/dehydration	Heterocyclic syntheses (e.g., pyrrole, oxazoline, thiazoline, indole)	JCE 2006, 83, 634–636	MI, MS, LE, LD, GF, RE, R, CC	Solvent-free, microwave heating
113.	Nucleophilic addition/ dehydration/nucleophilic addition	Aldol condensation/Michael reaction	JCE 2005, 82, 468–469	SG, MS, H, VF	Solvent-free, renewable feedstock, atom efficiency

114.	Nucleophilic addition/dehydration/nucleophilic addition/hydrolysis	Coumarin formation	JCE 2004, 81, 874–876	MS, H, VF, R	Catalytic, greener reagent, aqueous solvent, atom efficiency
115.	Nucleophilic addition/electrophilic addition	Alkene epoxidation	JCE 2011, 88, 322–324	MS, LE, LD, GF, RE	Catalytic, greener reagent, atom efficiency
116.	Nucleophilic addition/elimination	Alcohol oxidation	JCE 2003, 80, 907–908	SG, H, LE, VF, LD, GF, RE	Solvent-free
117.	Nucleophilic addition/elimination	Alcohol oxidation	TCE 2004, 9, 30–31	SG, H, LE, LD, GF, RE	Solvent-free
118.	Nucleophilic addition/elimination	Alcohol oxidation	JCE 2001, 78, 66–67	MS, H, LE, LD, RE	Catalytic, greener reagent, aqueous solvent
119.	Nucleophilic addition/elimination	Alcohol oxidation	TCE 2010, 15, 115–116	MS, LE, LD, GF, R	Product recycling
120.	Nucleophilic addition/elimination	Alcohol oxidation	JCE 1981, 58, 824	MS, H, D, LE, LD	Greener reagent
121.	Nucleophilic addition/elimination	Alcohol oxidation	JCE 1982, 59, 862	D	Greener reagent
122.	Nucleophilic addition/elimination	Alcohol oxidation	JCE 1982, 59, 981	MS, H, D, LE, LD	Greener reagent
123.	Nucleophilic addition/elimination	Alcohol oxidation	JCE 1985, 62, 519–521	D, LE, LD	Greener reagent
124.	Nucleophilic addition/elimination	Alcohol oxidation	QN 2009, 32, 1667–1669	MS, H, LE, LD, RE	Greener reagent
125.	Nucleophilic addition/elimination	Alcohol oxidation	JCE 1991, 68, 1048–1049	MS, LE, LD, RE, R	Catalytic, greener reagent
126.	Nucleophilic addition/elimination	Alcohol oxidation	JCE 1980, 57, 438	MS, VF, H, GF, RE, R	Greener reagent, recyclable reagent
127.	Nucleophilic addition/elimination	Alcohol oxidation	JCE 1987, 64, 371–372	MS, VF, H, GF, RE, R	Greener reagent, recyclable reagent

(continued)

Entry	Reaction Mechanism	Transformation	Journal	Techniques	Greener Principles
128.	Nucleophilic addition/ elimination	Alcohol oxidation	JCE 2006, 83, 268–269	MS, H, VF, RE, R	Greener reagent, recyclable reagent
129.	Nucleophilic addition/ elimination	Alcohol oxidation	JCE 2001, 78, 951–952	MS, VF, RE, CC	Greener reagent
130.	Nucleophilic addition/ elimination	Alcohol oxidation	JCE 2010, 87, 981–984	MS, H, LE, LD, RE, CC, IA, VF	Greener reagent, recyclable reagent
131.	Nucleophilic addition/ elimination	Alcohol oxidation	JCE 2009, 86, 227–229	MI, MS, GF, LE, LD, RE	Catalytic, greener reagent, microwave heating
132.	Nucleophilic addition/ elimination	Alcohol oxidation	JCE 2011, 88, 652–656	MS, LE, LD, GF, RE, SB	Catalytic, greener reagents, alternative reaction solvent, renewable feedstock
133.	Nucleophilic addition/ elimination	Aldehyde oxidation	JCE 2007, 84, 852–854	MS, H, VF, R	Aqueous solvent
134.	Nucleophilic addition/ elimination	Benzylic oxidation	JCE 2004, 81, 388–390	MS, VF, CC	Catalytic, greener reagent
135.	Nucleophilic addition/ elimination	Ketone oxidation	TCE 2004, 9, 370–373	MS, H, VF, R	Greener reagent
136.	Nucleophilic addition/ elimination	Ketone oxidation	JCE 2010, 87, 190–193	MS, H, VF	Greener reagent, aqueous solvent
137.	Nucleophilic addition/ elimination	Wolff-Kischner reaction	JCE 1997, 74, 1225	MI, VF, LE, LD, GF, RE, R	Microwave heating, alternative reaction solvent
138.	Nucleophilic addition/ nucleophilic acyl substitution	Hydantoin synthesis	JCE 2010, 87, 640–642	MI, GF, RE	Microwave heating
139.	Nucleophilic addition/ rearrangement	Baeyer-Villiger reaction	JCE 2005, 82, 1837–1838	LE, LD, GF, RE, VF, R	Solvent-free
140.	Nucleophilic addition/ rearrangement	Epoxide rearrangement	JCE 2008, 85, 1274–1275	MS	Catalytic, greener reagent, atom efficiency

141.	Nucleophilic substitution	Ether synthesis	JCE 2010, 87, 1233–1235	MS, RE, H, GP	Greener reagent
142.	Nucleophilic substitution	Ether synthesis	JCE 2006, 83, 285–286	MS, H, GF, RE	Catalytic, greener reagent
143.	Nucleophilic substitution	Ionic liquid synthesis	JCE 2009, 86, 856–858	H, LE, LD, GF, D	Aqueous solvent
144.	Nucleophilic substitution	Ionic liquid synthesis	JCE 2010, 87, 196–201	MS, MI, LE, RE	Greener products, microwave heating
145.	Nucleophilic substitution	Ionic liquid synthesis	JCE 2006, 83, 943–946	MS, H, LE, LD, GF, RE	Greener products
146.	Nucleophilic substitution	Williamson ether synthesis	JCE 2005, 82, 1839–1840	SG, H, VF, SB	Solvent-free
147.	Nucleophilic substitution	Williamson ether synthesis	JCE 1980, 57, 822	MS, H, LE, LD, D/RE, D	Catalytic, greener reagent, aqueous solvent
148.	Nucleophilic substitution	Williamson ether synthesis	JCE 2009, 86, 850–852	H, VF	Alternative reaction solvent
149.	Nucleophilic substitution	Williamson ether synthesis	JCE 2006, 83, 634–636	MI, MS, LE, LD, GF, RE	Microwave heating
150.	Nucleophilic substitution	Williamson ether synthesis	JCE 2010, 87, 84–86	MI, MS, ES, VF, R	Microwave heating
151.	Nucleophilic substitution	Wittig salt synthesis	JCE 2010, 87, 84–86	MI, MS, VF	Microwave heating
152.	Nucleophilic substitution(s)	Diazotization, alcohol formation	JCE 2010, 87, 623–624	MS, VF, R	Renewable feedstock, aqueous solvent
153.	Oxidative addition/reductive elimination	Ullmann coupling	JCE 2011,88, 331–333	SG, VF, CC, RE	Solvent-free, catalytic, greener reagent
154.	Oxidative addition/syn addition/β-hydride elimination/reductive elimination	Heck reaction	TCE 2007, 12, 77–79	MS, H, GF, VF, R	Catalytic, greener reagent, aqueous solvent
155.	Oxidative addition/ transmetallation/reductive elimination	Sonogashira reaction	JCE 1999, 76, 74–75	MS, H, SE, CC, RE, R	Catalytic, greener reagent, alternative reaction solvent
156.	Oxidative addition/ transmetallation/reductive elimination	Suzuki reaction	JCE 2008, 85, 555–557	MS, H, VF, GF, R	Catalytic, greener reagent, aqueous solvent

(*continued*)

Entry	Reaction Mechanism	Transformation	Journal	Techniques	Greener Principles
157.	Oxidative addition/ transmetallation/reductive elimination	Suzuki reaction	TCE 2007, 12, 414–418	MI, VF, LD, GF, RE, CC	Catalytic, greener reagent, aqueous solvent, microwave heating
158.	Oxidative addition/ transmetallation/reductive elimination	Suzuki reaction	TCE 2009, 14, 258–260	MS, H, LE, LD, GF, RE, CC, R	Catalytic, greener reagent, aqueous solvent
159.	Pericyclic	1,3-Dipolar cycloaddition	JCE 2005, 82, 1833–1836	MS, H, VF, LE, LD, GF, RE	Catalytic, greener reagent, atom efficiency
160.	Pericyclic	Diels-Alder reaction	JCE 2009, 86, 488–493	MS, MI, VF, R	Alternative reaction solvent, atom efficiency, microwave heating
161.	Pericyclic	Diels-Alder reaction	JCE 2009, 86, 488–493	MS, H, VF	Aqueous solvent, atom efficiency
162.	Pericyclic	Diels-Alder reaction	JCE 2006, 83, 634–636	MI, MS, RE, R	Microwave heating, atom efficiency
163.	Pericyclic	Diels-Alder reaction	JCE 2005, 82, 625–629	MI, VF, R	Microwave heating, atom efficiency
164.	Pericyclic	Diels-Alder reaction	JCE 1992, 69, 938–939	SG, MI, VF	Microwave heating, atom efficiency
165.	Pericyclic	Diels-Alder reaction	JCE 2010, 87, 84–86	MI, MS, VF	Alternative reaction solvent, microwave heating, atom efficiency
166.	Pericyclic	Diels-Alder reaction	TCE 2010, 15, 28–31	MI, VF, R	Microwave heating
167.	Pericyclic	Hetero Diels-Alder reaction	JCE 1998, 75, 1285–1287	D, MS, H, LE, LD, GF, RE, VF	Aqueous solvent, atom efficiency
168.	Pericyclic	Hetero Diels-Alder reaction	JCE 2008, 85, 1538–1540	MS, H, LE, LD, GF, RE	Aqueous solvent, atom efficiency

169.	Pericyclic	Photochemical 2 + 2 cycloaddition	JCE 2005, 82, 1679–1681	H, GF, VF, SG, VI, LE, LD, RE	Solvent-free, sunlight irradiation, atom efficiency
170.	Pericyclic/elimination	Diels-Alder reaction	JCE 1994, 71, A142–144	MI, VF, R	Microwave heating, atom efficiency
171.	Pericyclic/nucleophilic acyl substitution	Diels-Alder reaction	JCE 2009, 86, 488–493	ES	Solvent-free, atom efficiency
172.	Polymerization	Adhesive synthesis	JCE 2008, 85, 972–975	MS, H, VF	Renewable feedstock, greener product
173.	Polymerization	Polylactic acid formation	JCE 2008, 85, 258–260	MS, H, RE	Greener product
174.	Polymerization	Polysuccinimide formation	JCE 2005, 82, 1380–1381	H, VF	Solvent-free, renewable feedstock, greener product
175.	Radical	Biaryl formation	JCE 2008, 85, 411–412	MS, H, VF	Aqueous solvent
176.	Radical	Biaryl formation	JCE 2004, 81, 1636–1640	MS, H, VF, LE, LD, GF, RE, R	Aqueous solvent
177.	Radical	Biaryl formation	JCE 2010, 87, 526–527	MS, H, VF	Catalytic, aqueous solvent, greener reagent, atom efficiency
178.	Radical	Bromination	JCE 2005, 82, 120–121	H, VI	Alternative reaction solvent

Journal abbreviations: EC = *Education in Chemistry*, GC = *Green Chemistry*, GCLR = *Green Chemistry Letters and Reviews*, JCE = *Journal of Chemical Education*, JME = *Journal of Materials Education*, PAC = *Pure and Applied Chemistry*, QN = *Quimica Nova*, TCE = *The Chemical Educator*.

Experimental technique abbreviations: C = centrifugation, CC = column chromatography, D = distillation, ES = experimenter stirring, GF = gravity filtration, GP = gas phase reactivity, H = heating, IA = inert atmosphere, LD = liquid drying, LE = liquid extraction, MI = microwave irradiation, MS = magnetic stirring, R = recrystallization, RE = rotary evaporation, SB = sublimation, SE = steam bath evaporation, SG = solid grinding, ST = Schlenk techniques, SX = Soxhlet extraction, T = trituration, VD = vacuum distillation, VF = vacuum filtration, VI = visible light irradiation.

Index

B

C

D

H

I

M

N

O

P

Q

R

S

T

U

V

Printed and bound by CPI Group (UK) Ltd, Croydon, CR0 4YY
21/10/2024
01777083-0010